CARPENTRY AND JOINERY
BOOK 1

Carpentry and Joinery
Book 1

David R. Bates
Former Head of Construction Practice Studies
Nene College, Northampton

Routledge
Taylor & Francis Group
LONDON AND NEW YORK

First published 1982 by Pearson Education Limited

2 Park Square, Milton Park, Abingdon, Oxon OX14 4RN
711 Third Avenue, New York, NY 10017, USA

Routledge is an imprint of the Taylor & Francis Group, an informa business

First issued in hardback 2016

ISBN: 978-0-582-99479-9 (pbk)
ISBN: 978-1-138-83544-3 (hbk)

Foreword

It is said that every man is a builder by instinct and, while this may be true to a degree, the best way to become a craftsman is first to be an apprentice.

To be an apprentice means to learn a trade, to learn skills and have knowledge which will both be a means of earning a living and provide an invaluable expertise for life. But whereas this used to be the main purpose and advantage of serving an apprenticeship, nowadays many people in high positions in industry have made their way to the top after beginning with an apprenticeship.

The once leisurely pace of learning a trade has now been replaced by a much more concentrated period of learning because of the shortened length of apprenticeship. In addition, there is an ever-increasing number of new materials and techniques being introduced which have to be understood and assimilated into the craftsman's daily workload. There is therefore a large and growing body of knowledge which will always be essential to the craftsman, and the aim of this craft series of books is to provide this fundamental knowledge in a manner which is simple, direct and easy to understand.

As most apprentices nowadays have the advantage of attending a college of further education to help them learn their craft, the publishers of these books have chosen their authors from experienced craftsmen who are also experienced teachers and who understand the needs of craft apprentices. The learning objectives and self-testing questions associated with each chapter will be most useful to students and also to college lecturers who may well wish to integrate the books into their teaching programme.

The needs have also to be kept in view of the increasing numbers of late entrants to the crafts, people who for various reasons did not serve an apprenticeship and are entering a trade as adults, probably under a government sponsored or other similar scheme. Such students will find the practical, down-to-earth style of these books to be an enormous help to them in reaching craftsman status.

Preface

This book is an attempt to lay before the apprentice carpenter and joiner a simple, straightforward and reasonably readable outline of some of the skills, tools, materials and methods he is likely to use or encounter during his apprenticeship.

It is hoped that the book will prove to be of particular use to those students of woodwork, both young and adult, who are studying to attain a Craft Certificate in Carpentry and Joinery, and in this direction, this work goes approximately half way, the remainder being covered in a second volume.

In writing this book, I have been mindful of the changes in methods and materials which have come about during the past few years, and have tried to set down those aspects of the craft which are still very much in evidence at the present moment, and which are likely to remain with us for many years to come.

We live in a world which changes rapidly with the progress of technology and no-one can say – with certainty – what the ultimate effect of these changes may be upon the craftsman. But one point at least stands out clearly amongst a welter of ever moving ideals, fashions and ways of life: the quality of our buildings and their value to society, both real and aesthetic, depend very largely upon the quality of the workmen engaged in building them, and therefore we should never underestimate the need for good craftsmen, now or in the future.

As a lecturer, I am very much aware of the wide areas of work undertaken by the employees of the many companies, both large and small, which can be grouped broadly under the heading of carpentry and joinery. Also, I am aware of the trend towards specialisation – the carpenter, the joiner, the machinist, and the formworker. All these craftsmen have their place in the construction industry, and if their technical expertise is analysed carefully, it is surprising how much "common ground" becomes apparent.

It is hoped that this volume will go some way towards encouraging the would-be craftsman in his endeavour to acquire the skills and understanding of the craft which he needs to achieve the utmost job satisfaction from his chosen career.

I would like to express my gratitude to the following organisations and companies who have been of material assistance in the preparation of the manuscript:

The Controller of Her Majesty's Stationery Office for permission to use extracts from The Building Regulations 1985, The Woodworking Machines Regulations 1974 and The Health and Safety at Work etc. Act 1974;

The Director General, British Standards Institution, for permission to use material from BS 1192 – Building Drawing Practice;

Messrs British Industrial Fastenings Ltd., for their help in providing details of their nailing machines;

Messrs Rawlplug Co. Ltd. for their help in providing me with illustrations of their fixing devices and tools;

Messrs Record, Ridgeway Tools Ltd. for their help in providing me with illustrations of woodworking tools;

Messrs Spear and Jackson (Tools) Ltd. for their help in providing material and illustrations on saws;

Messrs Wadkin Ltd. for their help with the chapter dealing with woodworking machines.

Finally, I should like to express my sincere personal thanks to Mr L. Jacques for his help in reading and correcting the manuscript, to Mr B.J. Bates of Tresham College, Kettering, for his considerable assistance with the illustrations, and lastly, my special thanks to my wife, Mrs P.M. Bates for her never-ending patience in typing the manuscript.

1982 DRB

Contents

1. Hand Tools and Equipment

After completing this chapter the student should be able to:

1. Name, describe and state the use of the woodworker's marking, measuring and testing tools.
2. Name, describe and state the function of the various types of saw used by the woodworker.
3. Name, describe and state the function of the various types of chisel used by the woodworker.
4. Name, describe and state the function of the various types of plane used by the woodworker.
5. Describe the sharpening procedures for planes, chisels, scrapers and gouges.
6. Select the most suitable type of "brace bit" for a particular job.
7. Sketch and describe the use of the bench hook, mitre box, shooting board, straight edge, winding strips, sawing stool, G cramps and sash cramps.

THE WOODWORKER'S TOOL KIT

There can be little doubt that a great deal of skill and expertise is required by any person who wishes to produce good quality items of woodwork. This is true, whether the job in hand be a piece of furniture, an article of joinery, the cutting and erection of a roof or the construction of formwork for a large concrete structure. Whatever type of work is involved, the quality of the finished product depends ultimately upon the ability of the craftsman to cut and shape his material quickly and accurately, either by hand or machine. It is also true to say that the ability of the skilled craftsman to do this depends to a very great extent upon the efficiency of the tools he uses. Even a highly skilled craftsman would have difficulty in producing good work using inferior or badly maintained tools; to the less skilled – the beginner, the apprentice or the trainee – such tools would present wellnigh insuperable difficulties. In truth, the woodworker's job is demanding and difficult enough without the handicap of poor tools and equipment.

In this respect, the student craftsman is urged most strongly to purchase only the finest quality tools available, to keep them sharp and in good order, and generally to look after them and see that they come to no harm in the often arduous conditions under which they may be used.

In short, the craftsman's tool kit is his means of earning a living, and, perhaps equally important, his prime means of obtaining job satisfaction in his chosen career, and it should therefore be treated accordingly.

TOOLS FOR MARKING, MEASURING AND TESTING

Rules

These are available in various forms such as bar rules, folding rules and flexible tapes. The type most convenient in use depends to a large extent upon the sort of work on which the craftsman is mainly engaged.

Generally speaking, the folding rule or tape will be found most useful for work on site where the rule is carried around in the pocket, and the bar rule most useful in the workshop where extra length and rigidity are no encumbrance.

Bar and folding rules

These are made of steel, boxwood and plastic. The steel rule lends itself better to fine, accurate work, and serves also as a straight edge.

Fig. 1.1. *Types of rule.*

(a) Folding rule.

(b) Bar rule.

(c) Flexible tape rule.

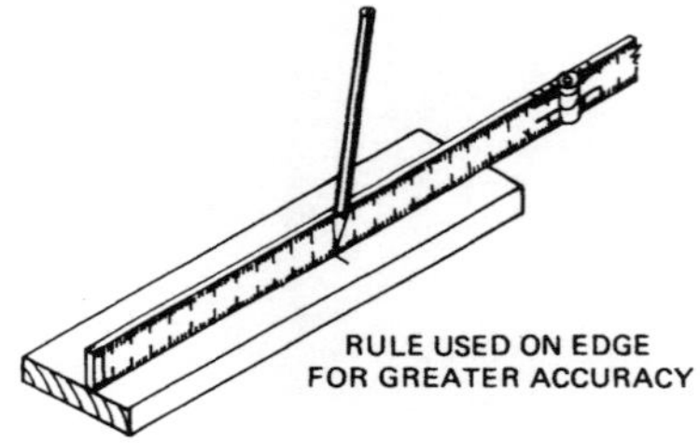

Fig. 1.2. *Avoiding parallax.*

Flexible tapes

These are available up to 30 m in length and are extremely useful for large scale setting out of rooms, buildings, partitions, etc., and for measuring long lengths of timber. Flexible pocket rules up to 3 m in length are always useful for board and sheet measure both on site and in the workshop.

The various types of rule described are shown in Fig. 1.1.

Metric and Imperial units

Rules are available graduated in either metric or Imperial units, i.e. marked in metres/millimetres or feet/inches. Some rules are marked in both, but whilst this may be useful on occasions, they suffer the severe drawback of not having an overall length representing both metric and Imperial units. Where both types of unit are in use, separate rules are generally more satisfactory.

Where choice of rule is concerned, the best advice that can be given is to ensure that the rule is well made, clearly marked, and as long as can be conveniently used in the practical situations previously mentioned. Having purchased a rule, use it exclusively, as far as possible, in order to gain familiarity and confidence. It is not good policy to borrow, or to use an unfamiliar rule, especially where the overall unit length differs from one's own, as this is undoubtedly one of the easiest ways of falling into error. Figure 1.2 shows how to use a rule on edge, in order to avoid errors due to parallax, whilst Fig. 1.3 shows how to divide a board into a number of parts by the proportionate method.

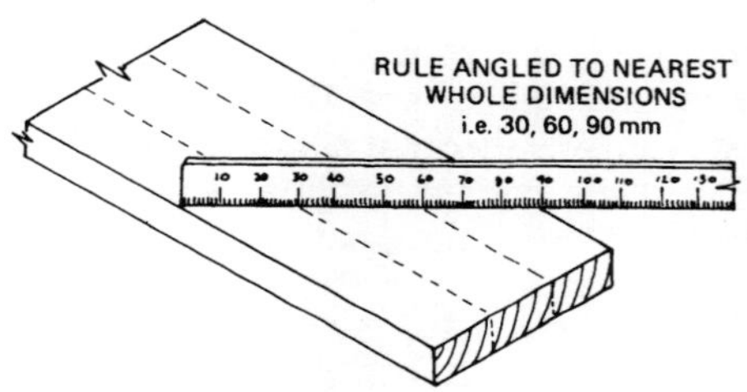

Fig. 1.3. *Dividing a board into equal parts.*

Fig. 1.4. *Types of woodworker's try square.*

(a) Wood and brass try square.

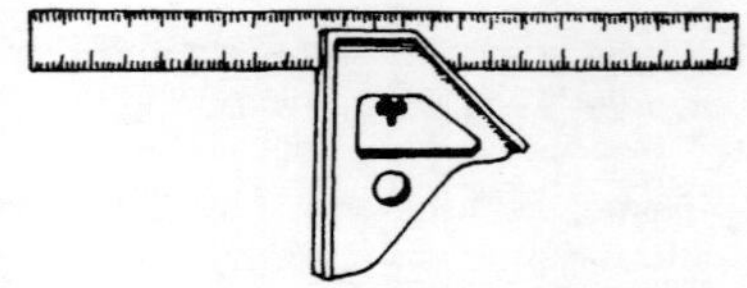

(b) Adjustable try square.

Squares and bevels

Try squares

These are used to mark and test angles of 90° (right angles) and therefore need to be both well made and accurate. They should be well looked after and handled with care in order to retain their accuracy.

Try squares may be obtained made entirely of steel or with a steel blade riveted into a beech or rosewood stock, with brass facings to reduce wear. There is little to choose between the two types, the choice being mainly a matter of personal preference. The size of a try square is denoted by the length of the blade, i.e. the maximum length of a line which can be marked with it, this varying from 100 mm to 300 mm. Most craftsmen possess at least two try squares – one small or medium sized and one large.

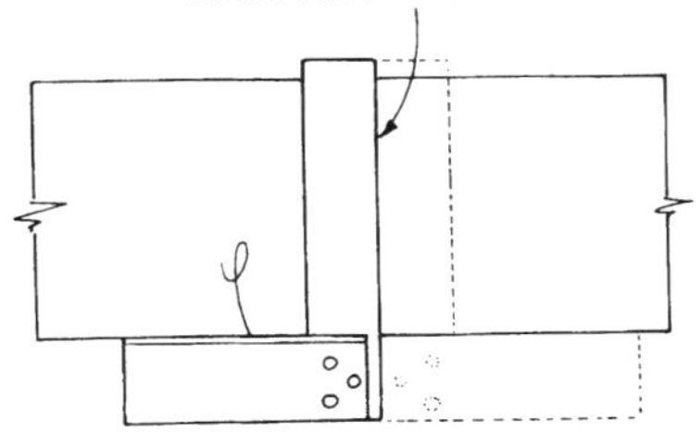

Fig. 1.5. *Testing a try square for accuracy.*

Figure 1.4 shows two types of woodworker's try square, the adjustable variety finding most favour amongst site carpenters.

Figure 1.5 shows a method of testing a try square for accuracy, a procedure which should always be carried out after purchasing a new tool, and thereafter at frequent intervals throughout its life. The method adopted is to square a line across the face of a piece of wood which has a true, accurate edge. The square is then turned over and the line re-marked. The error, if any, must be corrected by filing, and is one half of that shown by the test.

The mitre square

(*See* Fig. 1.6.) This is basically similar to the try square but used to mark and test angles of 45° (mitres).

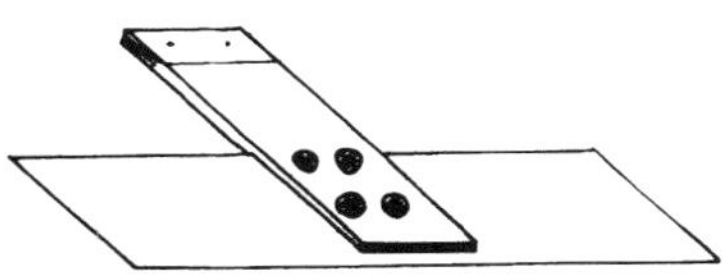

Fig. 1.6. *Mitre square.*

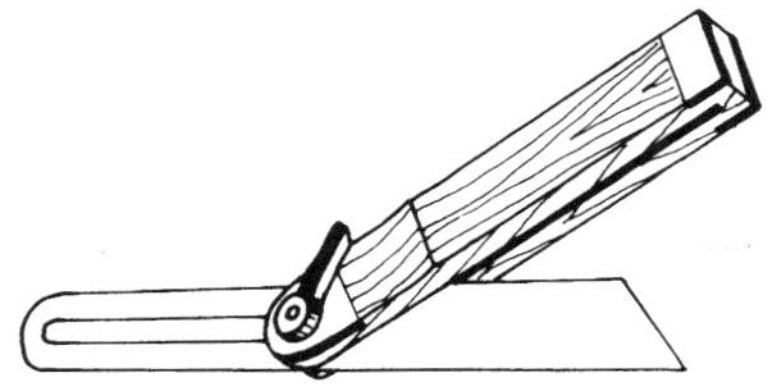

Fig. 1.7. *Sliding bevel.*

Sliding bevel

This tool, shown in Fig. 1.7, is a kind of adjustable try square which is used for marking and testing angles other than 90°, though on occasions it may serve the purpose of a try square or mitre square. In use the blade is set to the required angle and locked by means of a thumb screw or set screw in the stock.

Gauges

Marking gauge

This simple tool, shown in Fig. 1.8*(a)*, is made of beechwood and is used for gauging parallel lines, as when marking the width or thickness of a piece of wood, or when marking recesses for hinges. The gauge marks by means of a sharp point (some woodworkers prefer a cutting edge) being drawn along the workpiece, and is kept parallel to the face or face edge by means of the stock which is pressed tightly against the edge of the wood. Most joiners possess at least two of these useful tools. Figure 1.9 shows the marking gauge in use.

Cutting gauge

This tool is shown in Fig. 1.8*(b)* and, as can be seen, is very similar to the marking gauge.

Fig. 1.8. *Types of gauge.*

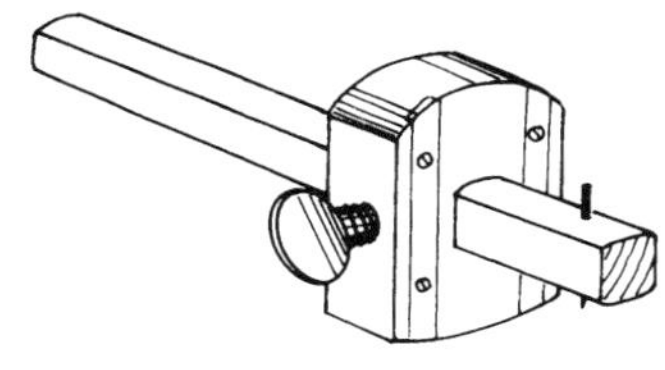

(a) Marking gauge.

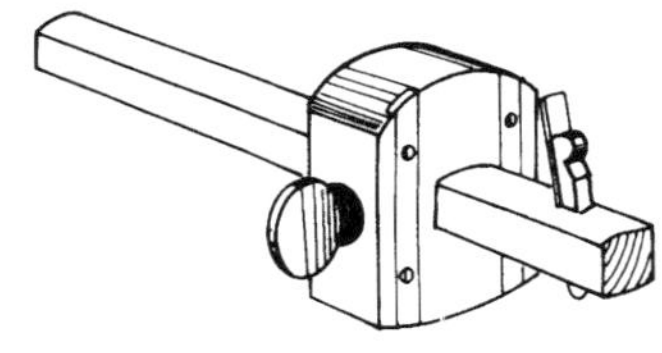

(b) Cutting gauge.

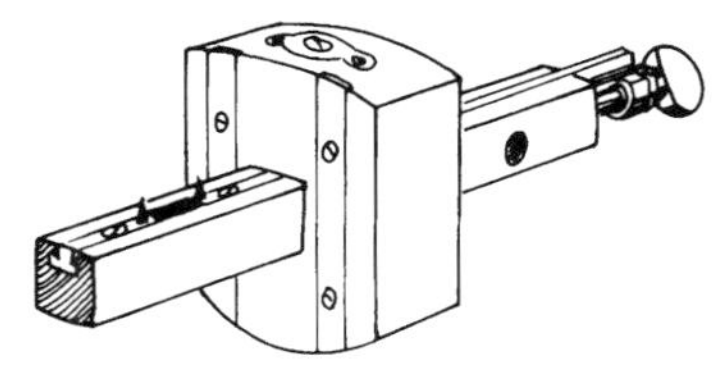

(c) Mortice gauge.

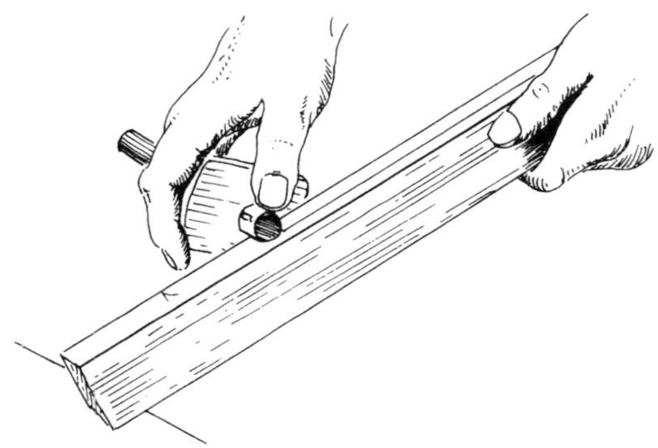

Fig. 1.9. *Using the marking gauge.*

The cutting gauge, however, has a blade in place of the point found in the marking gauge. It is used to cut relatively deep lines in timber. It is a particularly useful tool for marking lines across the grain, where the blade gives a clean, precise and deep cut such as is required when marking dovetail joints and cross grain rebates, tongues, etc.

Mortice gauge

This tool is used for marking the double lines required when setting out mortice and tenon joints, hence its name. This gauge, shown in Fig. 1.8*(c)*, is usually made of rosewood with brass inlays to prevent excessive wear.

The mortice gauge has a sliding stock, similar to the two previously mentioned types, a fixed point and an adjustable sliding point which can be set to suit the particular mortice chisel in use. Figure 1.10 shows how the gauge is set to the width of the chisel blade.

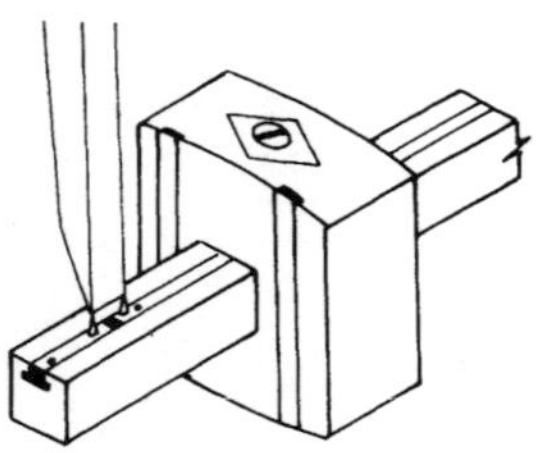

Fig. 1.10. *Setting a mortice gauge to chisel blade.*

Marking knife

This tool, shown in Fig. 1.11*(a)*, is used for marking out, where a "cut" line is preferable to a pencil line, as is the case when marking the shoulders of a tenon or a groove

Fig. 1.11. *Marking knife.*

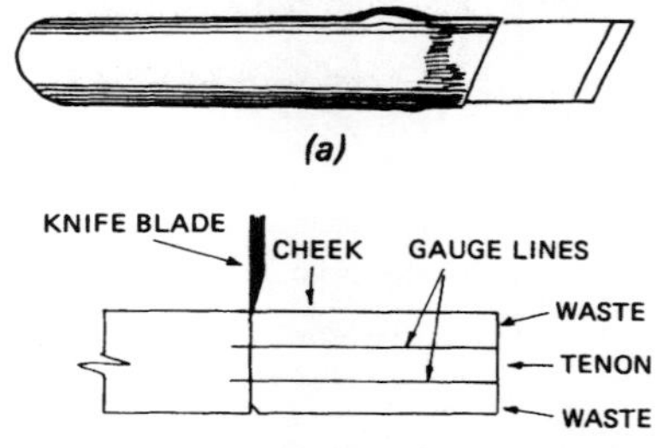

(b) Use of marking knife for tenon shoulders.

across the grain. The marking knife gives a clean, accurate cut, thus making the follow up with saw or chisel much more precise. Figure 1.11*(b)* shows how the "cut" produced by the marking knife is applied in relation to the "waste" side of the line.

Compasses or dividers

These are always useful, both in the tool bag and on the bench, and serve a variety of purposes including:

Fig. 1.12. *Uses of dividers.*

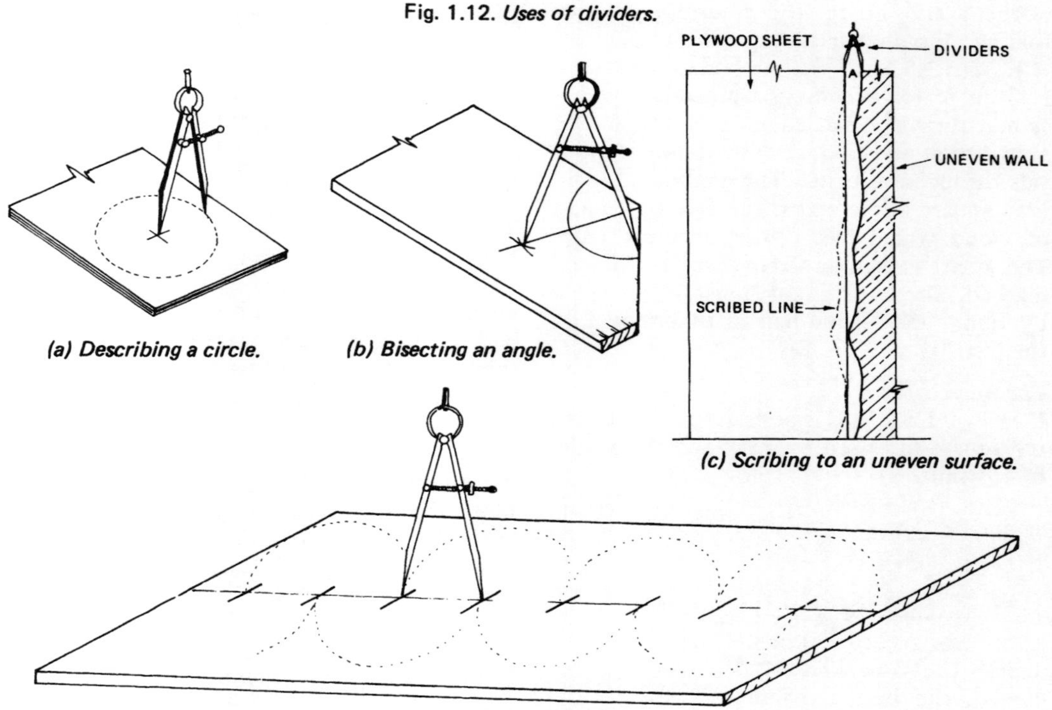

(a) Describing a circle.

(b) Bisecting an angle.

(c) Scribing to an uneven surface.

(d) Stepping off.

(*a*) describing arcs and circles,
(*b*) bisecting angles,
(*c*) scribing to an irregular surface,
(*d*) "stepping off" equal divisions,

as shown in Fig. 1.12.

Trammel

This tool, shown in Fig. 1.13, comprises of two trammel heads on a wooden bar. The heads can be moved along the bar to any desired position and there locked. In use, this tool serves the purpose of a large pair of compasses, i.e. for describing large arcs and circles, and for stepping off. If three heads are available the trammel may be used to set out elliptical curves as explained in Chapter 4.

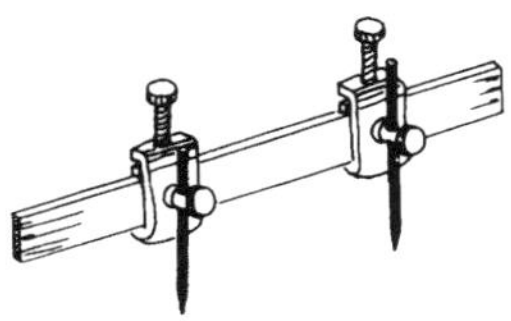

Fig. 1.13. *Trammel heads.*

Plumb rules and levels

Spirit levels

These tools which vary in length from 200 mm to 1 m or more, are used for marking and testing level surfaces, or, in some instances, for marking and testing vertical surfaces (plumbing). They work on the principle that a bubble, enclosed in a glass tube containing spirit, will rise to the highest possible point within the tube. The tube, being slightly curved, can be set within the body of the spirit level so that when placed on a horizontal or vertical surface, the bubble will come to rest between two calibrations, thus indicating the accuracy or otherwise of the surface. Good quality levels have adjustable bubble tubes, which can be reset from time to time should an inaccuracy develop, or which can be replaced if broken. For site work the spirit level is essential, most carpenters possessing two such tools as shown in Fig. 1.14.

Fig. 1.14. *Spirit levels.*

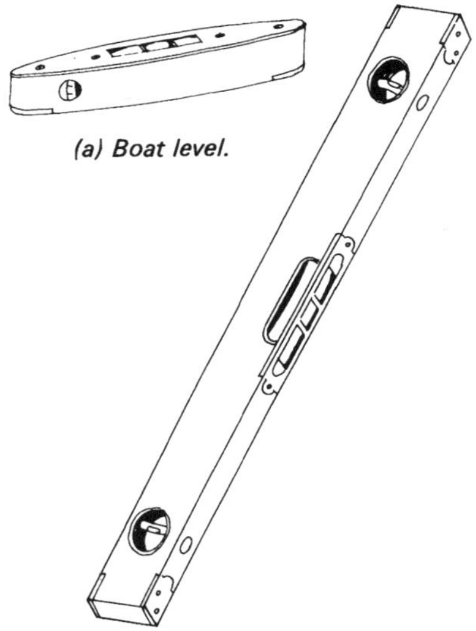

(a) Boat level.

(b) Long spirit level.

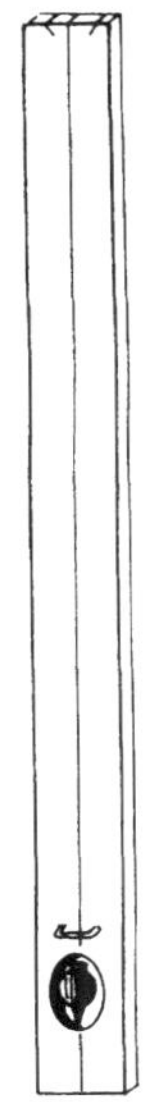

Fig. 1.15. *Plumb rule and bob.*

Plumb bob and rule

This is an item of equipment which the carpenter can make for himself, and is invaluable on the site when fixing door frames and linings. The rule – generally about 1.8 m long – serves not only as an extremely accurate tool for marking and testing vertical lines and surfaces, but also as a straight edge. With practice, the plumb bob can be used almost as quickly as the shorter "spirit level" type of plumb rule and is, of course, very accurate. The experienced craftsman has no need to wait for the bob to stop swinging before knowing whether or not the rule is "plumb" – he can tell by the degree of swing to either side of the gauged centre line. A plumb bob and rule is shown in Fig. 1.15.

CUTTING TOOLS

Saws

There are several types of saw used by the woodworker, each of which is designed to suit a particular aspect of his work.

Rip saw

This tool, shown in Fig. 1.16(*a*), is 600 mm to 700 mm in length, with either a "straight" or "skew" back, this latter feature being really a matter of personal preference. The

Fig. 1.16. *Rip saw.*

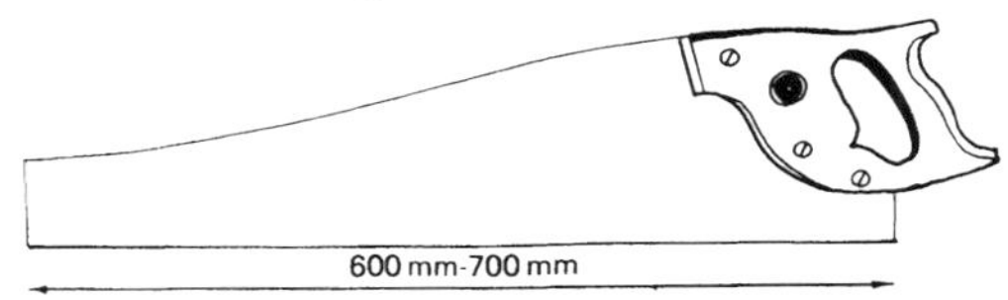

(a)

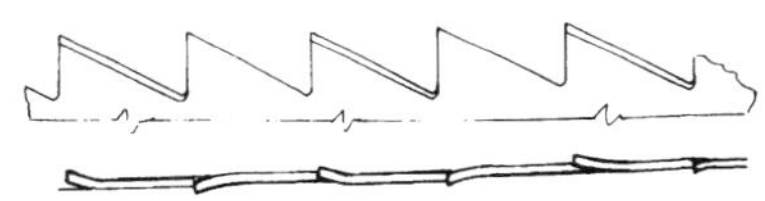

(b) Rip saw teeth.

main features of a rip saw are the size and shape of the teeth, these being spaced at about 4 per 25 mm and quite fierce as shown in Fig. 1.16*(b)*. Rip saw teeth are designed for cutting down the grain – i.e. for "ripping" along the length of the wood fibres. The teeth are sharpened to chisel edges rather than knife points and require only a minimum of set in order to cut cleanly. Properly set and sharpened, a rip saw makes light work of sawing down the grain, a job which it does speedily and efficiently.

Cross cut or hand saw

(*See* Fig. 1.17*(a)*.) This is a general purpose type of saw which can be used for both ripping and cross cutting. It is, however, designed mainly for cutting across the grain, its teeth, 6 or 7 per 25 mm, being sharpened to knife points for severing the wood fibres.

Fig. 1.17. *Cross cut saw.*

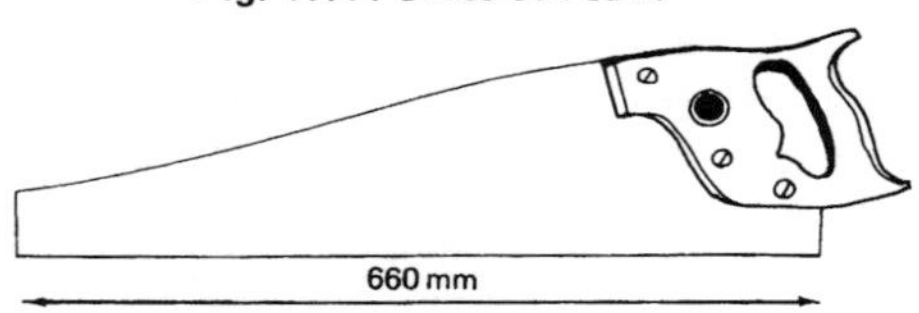

(a)

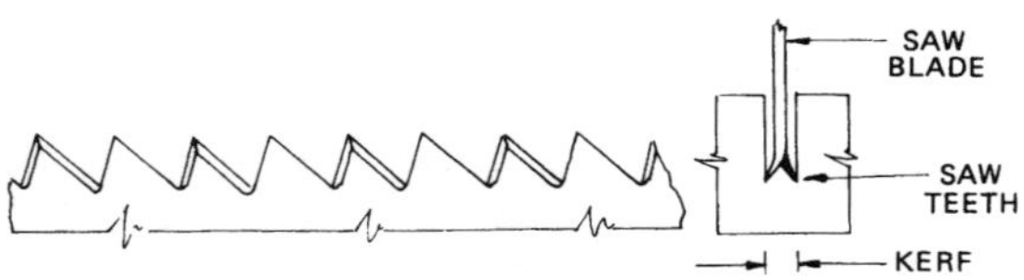

(b) Cross cut teeth.

The cross cut saw is about 660 mm long, either straight or skew backed and with teeth which are less fierce than those of the rip saw. Cross cut saws require rather more "set" than rip saws, in order to give sufficient clearance in the "kerf" (the actual slot formed by the cutting action of the saw teeth) to prevent binding. The shape of cross cut teeth and their cutting action are shown in Fig. 1.17*(b)*.

Panel saw

(*See* Fig. 18*(a)*.) This is a smaller, slightly stiffer bladed version of the cross cut saw, and is generally about 500 mm long with approximately 10 teeth per 25 mm. The teeth of the panel saw are designed primarily for cross cutting and are similar in shape to those of the cross cut saw, although somewhat smaller.

The panel saw is undoubtedly one of the handiest and most useful saws which the carpenter can carry in his tool bag. It is large enough for quite heavy cuts, with or across the grain, and yet is sufficiently fine for cutting small sections, mouldings, architraves, skirting boards, and thin panels. From this latter purpose the saw derives its name.

Tenon saw

This is a smallish saw about 300 mm to 350 mm long with a thin parallel blade and fine cross cut type teeth, about 16 per 25 mm. The thin blade is stiffened by means of a brass or steel strip along the back edge, brass backed saws generally having a better balance and feel than the steel backed types. Brass backed saws are also rather more expensive, but are well worth the extra cost which is practically negligible when set against a lifetime's use.

The tenon saw, shown in Fig. 1.18*(b)*, is essential to the bench worker, the joiner or cabinet maker, and is used for fine, accurate

Fig. 1.18. *Other types of saw.*

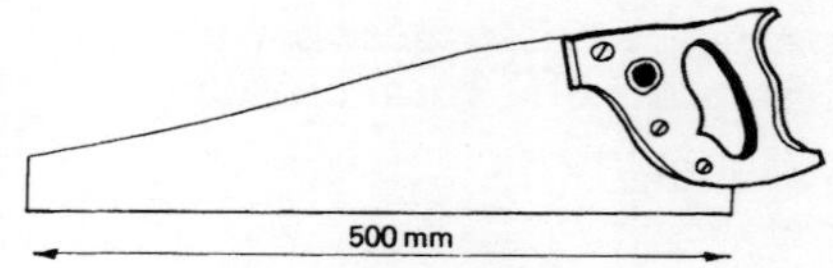

(a) Panel saw.

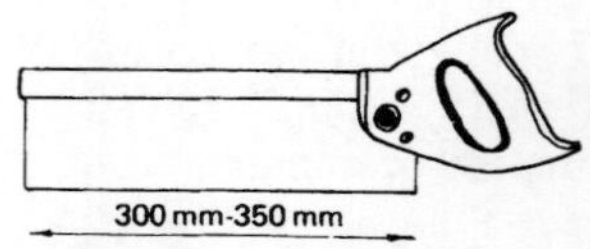

(b) Tenon saw.

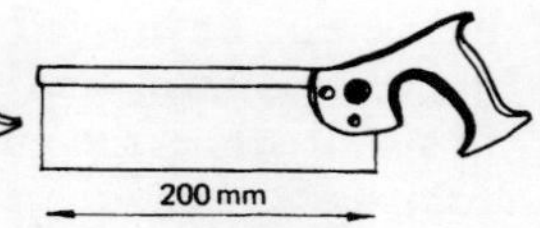

(c) Dovetail saw.

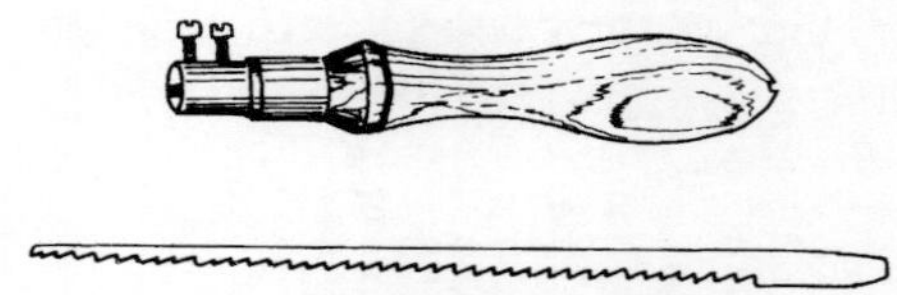

(d) Pad saw and blade.

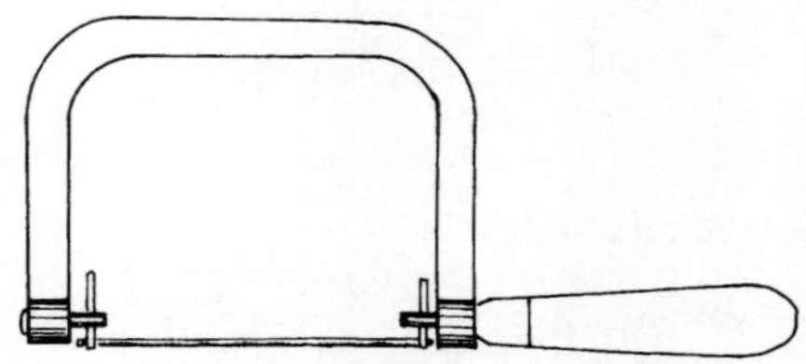

(e) Coping saw.

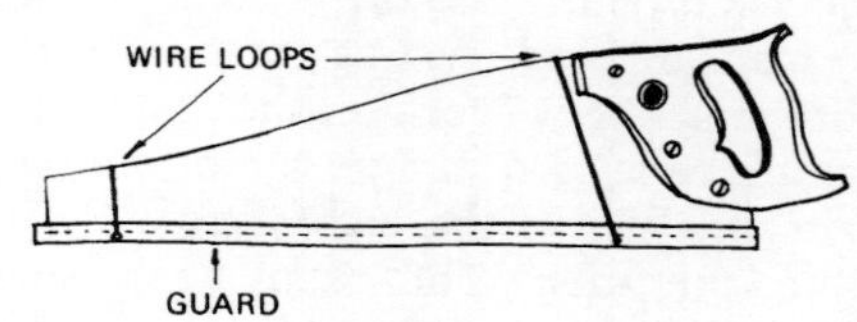

(f) Saw guard.

work such as tenons, shoulders, halvings, housings – anywhere in fact where a fine, accurate saw cut is required. When used in conjunction with a marking knife, a clean, sharp edge to the kerf is obtainable. As with any of the hand saws, the least possible set which will prevent the saw binding is the ideal to aim for, as the greater the set to the teeth, the slower and harder the work.

Dovetail saw
The dovetail saw, shown in Fig. 1.18*(c)*, is a smaller version of the tenon saw and is generally about 200 mm long with up to 18 teeth per 25 mm. Again, the brass backed saws are generally the better quality and have a better feel. The saw is not normally carried by the "site" carpenter, and is used for small fine work such as cutting small mouldings and dovetail joints. It is from this use the saw derives its name.

Pad saw or keyhole saw
This tool, shown in Fig. 1.18*(d)*, has a thin, narrow, replaceable blade locked into its handle by means of a pair of set screws. Although these are not the easiest nor the most efficient of the craftsman's saws to use, they are nevertheless indispensable where a "closed" cut has to be made as shown in Fig. 1.19. As the name suggests, they are also useful in cutting or cleaning out a keyhole. The saw blade should be retracted into the hollow handle when not in use.

Coping saw
This is a small, inexpensive but extremely useful little saw which is used for cutting shapes and contours, scribing and turning sharp corners. It is the ideal tool for removing waste material when cutting dovetails or tenons with a central haunch (*see* Chapter 3). The coping saw, shown in Fig. 1.18*(e)*, uses thin replaceable blades which are held in tension by the metal frame. The blade can be turned through 360° to facilitate cutting in any direction. The throat of the saw allows for making cuts which may be as far as 140 mm away from the edge of a board. The saw is most effective when used on relatively thin timbers, up to 20 mm or so in thickness, but with care, can be used on material up to 50 mm thick.

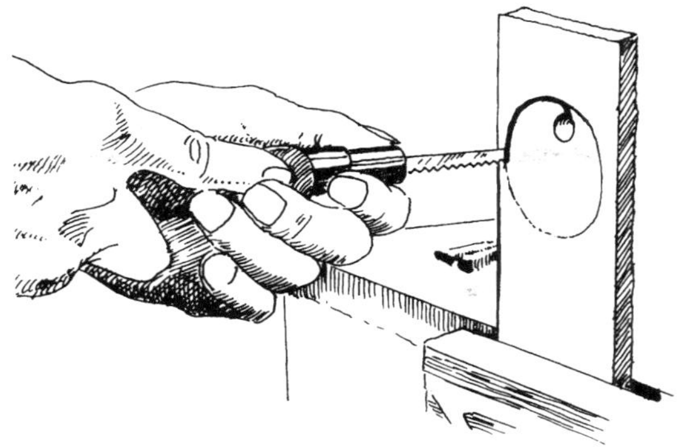

Fig. 1.19. *Using a pad saw for a "closed" cut.*

Care of saws
In order to get the best use from a saw, it must be well looked after. The blade should be kept clean and bright with no trace of rust or tarnish to create friction. It should be oiled or waxed frequently to keep it cutting freely and enable it to resist the occasional shower. When carried in the tool bag, the teeth of the saw should be protected with a guard strip as shown in Fig. 1.18*(f)*.

The saw handle should be kept clean, rubbed occasionally with linseed oil and above all, kept smooth to avoid blistering the hand. Saws should be kept sharp – a well maintained saw being a delight to use, a poorly maintained one quite the opposite. Extra care should be taken when cutting old, nailed or gritty timber and cutting close to brickwork (plugs, etc.) to avoid damaging the teeth.

Good quality saws have beech, walnut or rosewood handles, into which the blade is secured by removable saw screws. Cheaper, poorer quality saws are often held into the handle by rivets, which cannot be removed, a feature which may make sharpening difficult when the saw has seen a good deal of use.

Chisels
Like saws, chisels are available in a variety of shapes and sizes as illustrated in Fig. 1.20, each designed to do a particular type of work.

Firmer chisel, flat chisel
This is a general purpose type of chisel with a strong flat blade which is equally suitable for paring (pushing with the hand) or driving with a mallet. Firmer chisels are obtainable in widths from 3 mm to 35 mm (the size of a chisel being denoted by the width of the blade).

Figure 1.20*(a)* shows the firmer chisel with its various parts named.

As with any wood cutting tool, the steel blade must be of the finest quality in order to obtain and hold a razor sharp edge in use. The tang, on the other hand, needs a much softer temper so that it will not break under

Fig. 1.20. *Types of chisel.*

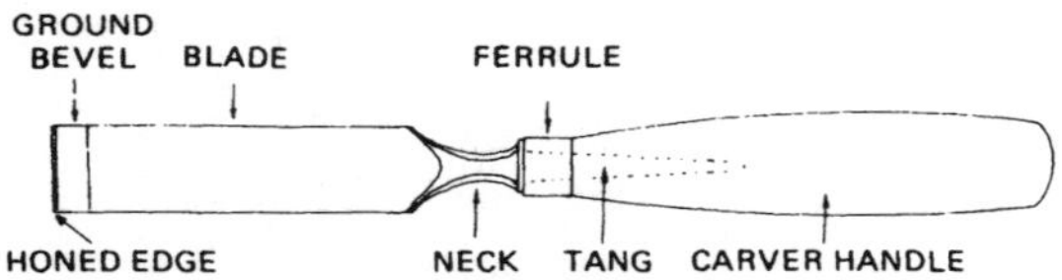

(a) Firmer/flat chisel.

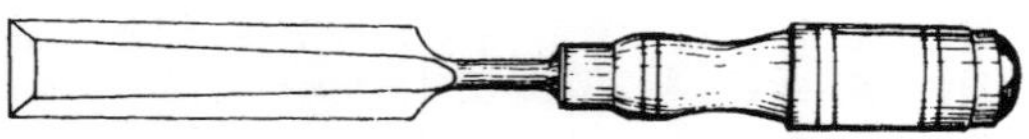

(b) Bevel edge chisel.

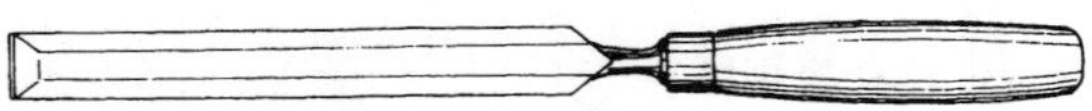

(c) Paring chisel.

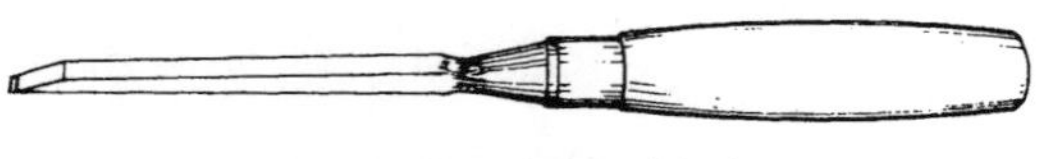

(d) Sash mortice chisel.

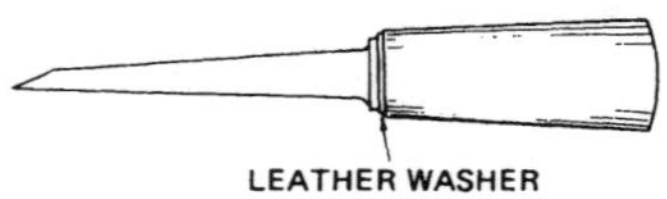

(e) Carpenter's mortice chisel.

(f) Mortice lock chisel.

stress. The handle, which is nowadays generally a "carver" type, should be reinforced around the tang with a good ferrule – either steel or brass – and may be made of wood or plastic. Wooden handles are best made of polished boxwood. The main criteria, as far as the chisel handle is concerned, are that it should be tough, comfortable to the hand and smooth so as not to cause blisters with prolonged use. Chisels should never be driven with a hammer, as this practice will only lead to damage or roughness – and leading ultimately to poorer grip and thus deterioration in the quality of the work.

Figure 1.21 shows the firmer chisel *(a)* being used with a mallet, and *(b)* for paring respectively.

Bevelled edge chisel

These are very useful chisels, basically similar to the firmer type, but with bevelled edges (*see* Fig. 1.20*(b)*). This feature enables the chisel to work closely into acute angles. The reduction in cross sectional area of the blade also makes for easier working. Bevelled edge chisels are undoubtedly the most popular type amongst craftsmen, most of whom possess a range of them in sizes from 3 mm to 35 mm. It should be noted that bevelled edge chisels, especially in the smaller sizes, are less robust than firmer chisels, and should be treated accordingly.

Paring chisel

This tool (*see* Fig. 1.20*(c)*) is available with a flat or bevelled blade in widths from 12 mm to 44 mm. The paring chisel has a long, finely tapered blade and is used solely for paring as its name suggests. The long blade enables the woodworker to work in deep recesses and across long housings where the shorter blade of the previously mentioned types could not reach. It is also possible to obtain greater control over the angle of cutting with the paring chisel. If used in conjunction with a mallet however, the fine, highly tempered blade may snap.

Fig. 1.21. *Uses of the firmer chisel.*

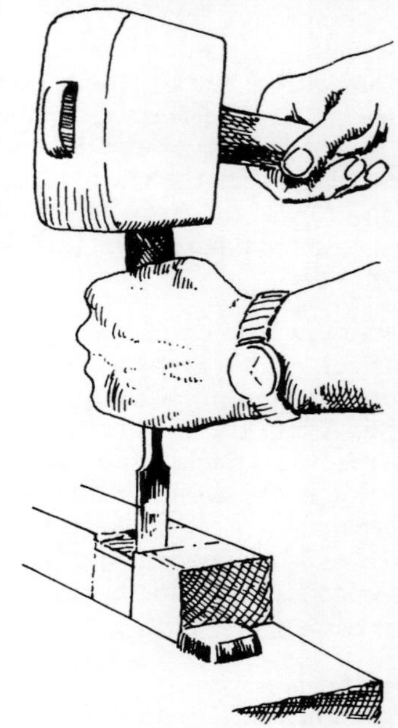

(a) Using firmer chisel with a mallet.

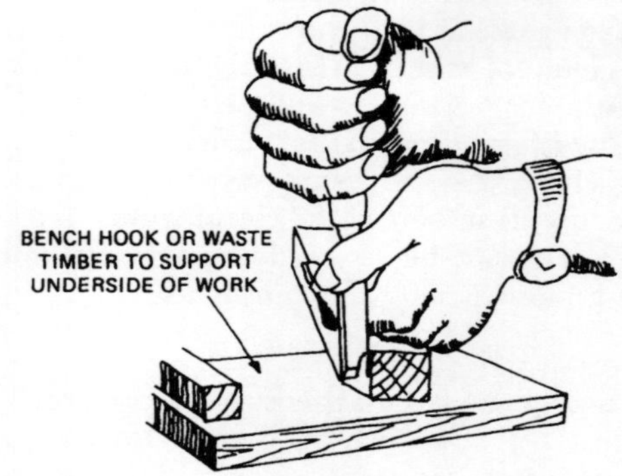

(b) Paring with a firmer chisel.

Fig. 1.22. *Morticing a door stile by hand.*

Mortice chisel

This tool has a long thick blade to enable it to withstand the heavy work to which it is put. Mortice chisels are used mainly for "chopping" mortices, and are driven with heavy blows of the mallet. The best mortice chisels have a tough leather washer between the shoulder of the tang and the handle to help absorb the shock of repeated blows.

Figure 1.20*(d)* and *(e)* shows the two principal types of mortice chisel, the "sash mortice" chisel and the more old fashioned "carpenter's mortice chisel" respectively. Both types are highly efficient, the thickness of the blade helping to keep the chisel cutting in a straight line. Figure 1.22 shows a sash mortice chisel being used to mortice a door stile.

Mortice lock chisel

This tool (*see* Fig. 1.20*(f)*), is a specialised tool which is used to cut deep mortices for long, horizontal mortice locks. The mortice required for this type of lock often penetrates into the end grain of a lock rail. The mortice lock chisel is designed to facilitate this, the tool being used with a levering action. Sometimes this tool is referred to as a "swan neck chisel".

Gouges

Gouges are a form of chisel with a blade which is curved in cross section, facilitating the cutting or paring of concave shapes. Basically, gouges take two separate forms: inside ground or scribing gouges as shown in Fig. 1.23*(a)*, and outside ground types as shown in Fig. 1.23*(b)*.

Fig. 1.23. *Gouges.*

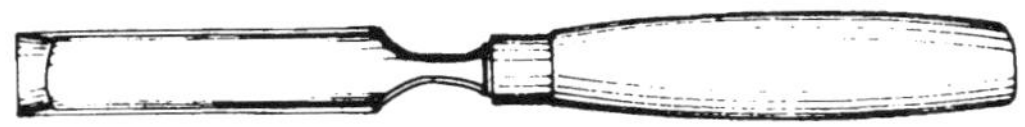

(a) Inside ground gouge.

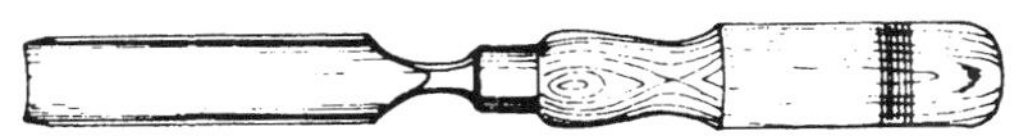

(b) Outside ground gouge.

Gouges, of either type are available in a wide variety of widths and radii to suit the curvature of any particular piece of concave work.

Figure 1.24 shows typical uses of *(a)* the inside, and *(b)* the outside ground gouges. Scribing gouges with extra long blades are

Fig. 1.24. *Using gouges.*

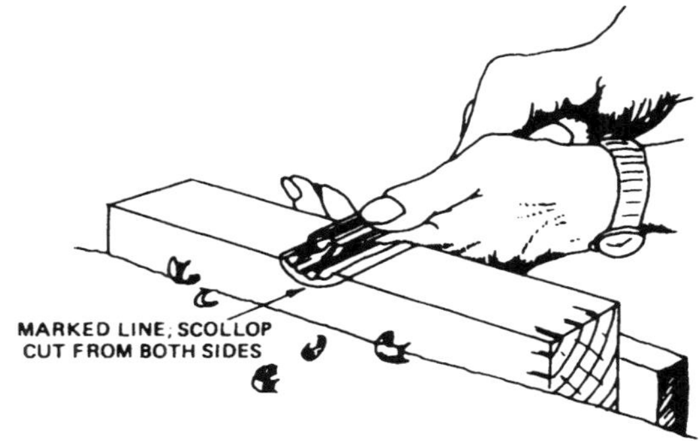

(a) Use of an inside ground gouge.

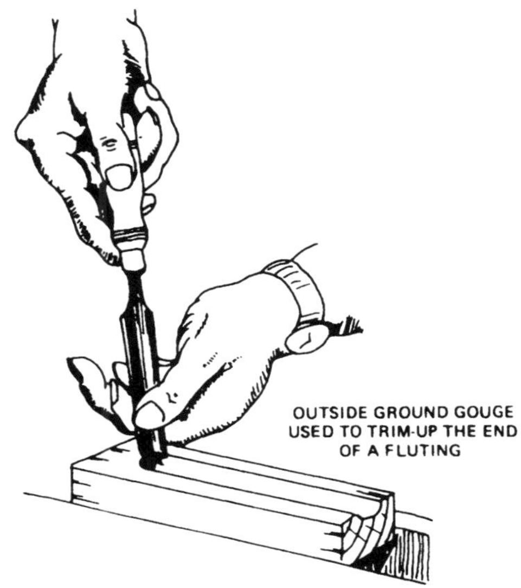

(b) Using an outside ground gouge.

available, these generally being known as "paring gouges".

Planes

The plane, in one or other of its many forms (*see* Figs. 1.25 and 1.28) is one of the most essential and useful of the woodworker's tools. It almost certainly originated several thousands of years ago when some enterprising craftsman thought of fixing a chisel into a block of wood in order to clean up a flat surface. By thus accurately controlling the depth of the cut made by the chisel, a true shaving was produced.

The modern plane is truly a masterpiece of scientific design and fine engineering, but still requires the understanding, hand control and "feel" for the material that our forebears needed and had to acquire by experience in order to produce good quality work. Indeed, lest we become too complacent, it would be well to realise that many of the woodworker's tools used by the Romans two thousand years ago – planes included – would not look unduly out of place on a modern joiner's bench.

Bench planes

By reason of the frequency of their use, three of the planes most commonly used in everyday work have become known as the "bench planes". These are the smoothing plane, the jack plane and the trying plane, shown in Fig. 1.25*(a)*, *(b)* and *(c)* respectively. It will be seen that these three planes differ considerably in length, but in fact they have identical cutting actions, each being fitted with double irons – a cutting iron and a back iron.

Fig. 1.25. *The bench planes.*

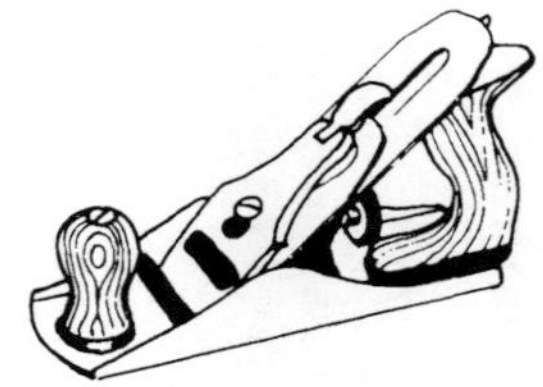

(a) Smoothing plane.

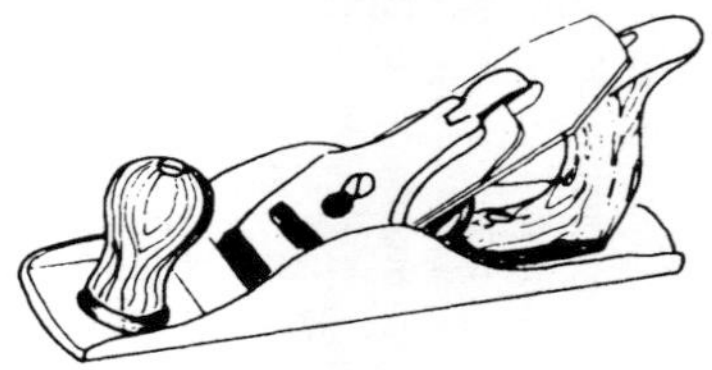

(b) Jack plane.

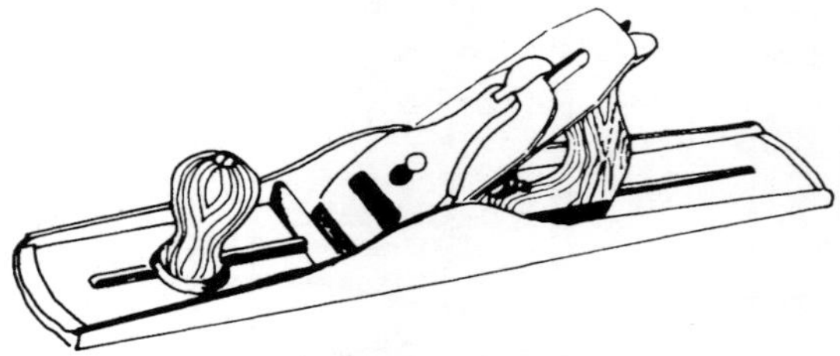

(c) Trying plane.

Cutting action of the plane

Figure 1.26 shows in diagrammatic form a longitudinal section through the mouth of a typical bench plane. It will be seen that the thickness of the shaving is controlled by the amount of projection of the cutting iron below the sole, whilst the back iron pushes the shaving forward as it is cut, breaking it up, and preventing it from developing any strength and thus tearing up in advance of the cut. The back iron can be adjusted at

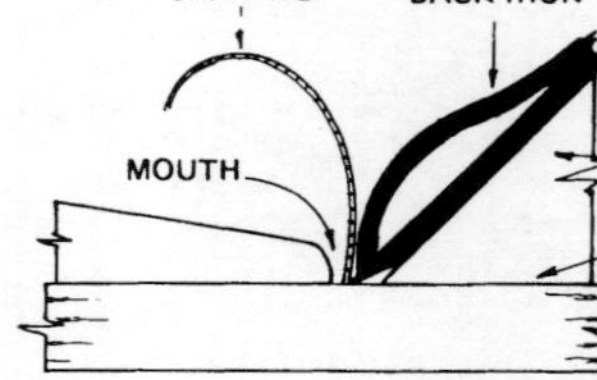

Fig. 1.26. *Cutting action of a jack plane.*

will, the closer it is set to the cutting edge, the less likelihood there is of the grain tearing. Generally, a fine shaving requires a fine "set" of the back iron and vice versa.

The mouth of the plane is also adjustable. This is done by moving the "frog" – the part of the plane which holds the plane irons – either backwards or forwards, thus increasing or decreasing the width of the mouth. The sole of the plane immediately in front of the mouth presses on the wood and helps to prevent the shaving lifting in advance of the cut. With certain timbers, or types of grain, the shaving has a tendency to "split off", rather than being sliced off – a fine mouth prevents this. A fine mouth in conjunction with a finely set back iron and a razor sharp blade normally makes it possible to clean up all but the most difficult timbers. On "easier" timbers, a wider mouth in conjunction with a more coarsely set back iron greatly reduces the physical effort required to remove a shaving. The blade or cutting iron should always be sharp.

Smoothing plane

This is the smallest of the bench planes and is about 250 mm long with a cutting iron from 50 mm to 60 mm wide. The plane is used mainly for "cleaning up", its sole being short enough to enable the craftsman to "follow the grain", or to use the plane with one hand as is often required when working away from the bench.

Jack plane
This plane is about 400 mm long, with a cutter 50 mm-60 mm wide and has been rightly described as the "maid of all work", such is its usefulness and versatility. Its main use is the rapid and accurate removal of waste material, such as when facing, edging and bringing timbers to size by hand, and when "shooting" doors for hanging. The plane is sufficiently long to enable it to produce a reasonably accurate surface on long timbers, whilst being short enough for use as a smoothing plane when necessary. No carpenter's tool kit is complete without this tool.

Trying plane
Generally about 550 mm long with a cutter up to 66 mm wide, the trying plane, or "jointer" as it is often called, is used for "shooting" long straight edges and surfaces. Its length will not allow it to work into a hollow, and thus it can produce (when properly handled), extremely accurate surfaces, such as required when shooting long boards for edge jointing (*see* Chapter 3).

Wooden planes
The bench planes previously described are the steel planes and are those most commonly used by the modern craftsman. However, it is only fair to remind the reader that each has its counterpart in wood, generally well selected beech. Wooden planes, whilst less easy to set and adjust, have certain advantages over the steel planes, namely:

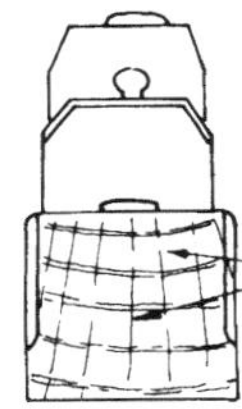

Fig. 1.27. *End view of a wooden plane.*

(*a*) they are less easily damaged by accidental misuse;
(*b*) they are generally lighter to carry around;
(*c*) they tend to work with less effort as there is less friction between the sole and the work than in the case of metal planes.

Wooden planes, when new, should be well soaked in raw linseed oil to preserve and protect the wood and to help their working properties, although some manufacturers presoak their planes, so relieving the craftsman of this task. In any case, a wooden plane should be rubbed over with raw linseed oil fairly frequently for the same reasons. A well-made wooden plane should always be so fashioned that its end section shows the medullary rays (*see* Chapter 2) radiating towards the sole, as shown in Fig. 1.27. This is essential if the sole of the plane is to remain true.

It would be true to say that whilst the wooden trying plane has become largely obsolete by reason of its bulk and cumbersome handling, there are many site carpenters who prefer the wooden jack plane, and quite a few joiners and cabinet makers possess a wooden smoothing plane in addition to the steel, which they use for the "special" job or the difficult timber.

Block plane
This extremely useful little plane (*see* Fig. 1.28(*a*)) is about 160 mm long with a blade in the region of 44 mm wide. The block plane has a single iron (no back iron) which fits into the body with the ground edge uppermost. The main feature of the block plane is the low angle of the blade, pitched at about 20°, which thus gives a very fine shearing cut. Most modern block planes are fully adjustable for depth of cut and alignment of the blade, and usually have an adjustable mouth for the reasons already stated.

Block planes are used mainly for trimming end grain, shooting mitres, etc., and in fact for any small fine work. Their small size enables them to be used comfortably in one hand.

Bench rebate plane
Sometimes referred to as a badger plane, this tool is about the same length as a jack plane, but somewhat narrower, usually about 54 mm in width. The bench rebate plane (*see* Fig. 1.28(*b*) has double irons and a similar cutting angle to the jack or smoothing plane, its main feature being that the blade extends the full width of the sole (in fact it should project slightly on either side) enabling it to work right into the corner of a rebate. The main function of the bench rebate plane is the cutting or cleaning up of rebates for door and window frames etc., for which purpose it is almost indispensable to the craftsman. The cutting edge of the blade must be dead straight and square in order to cut efficiently

Fig. 1.28. *Types of plane.*

(a) Block plane.

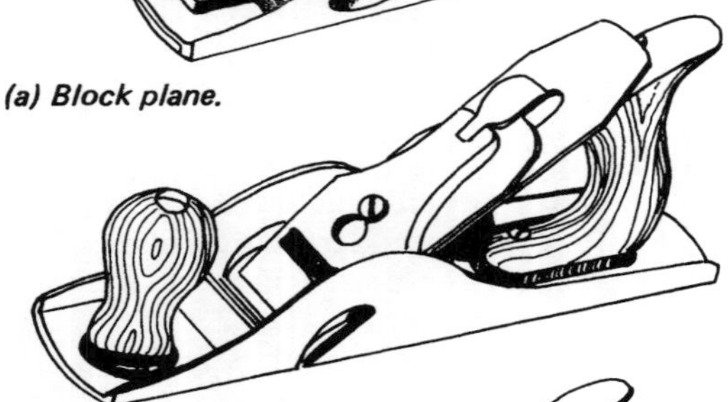

(b) Bench rebate plane.

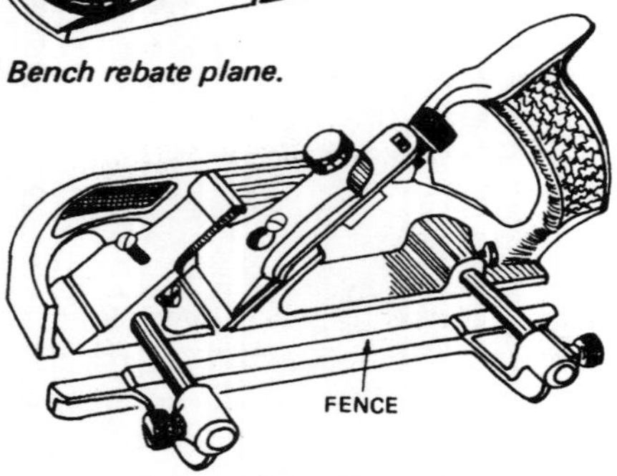

(c) Side fillister or rebate plane.

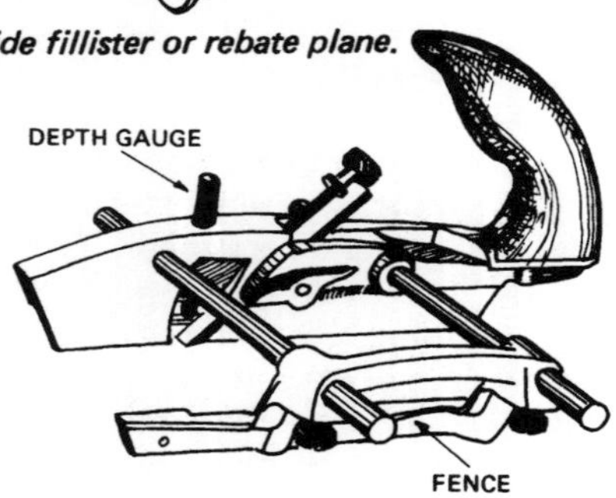

(d) Plough.

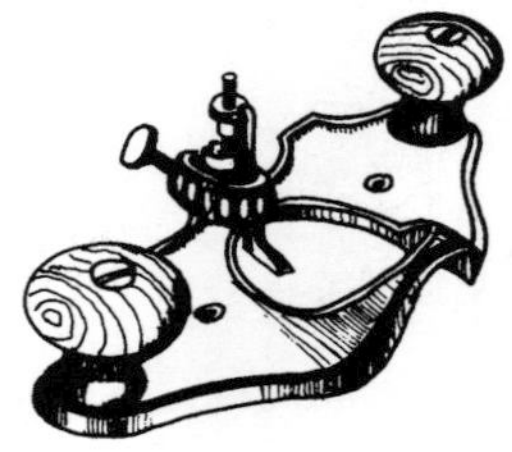

(e) Router.

(f) Spoke shave.

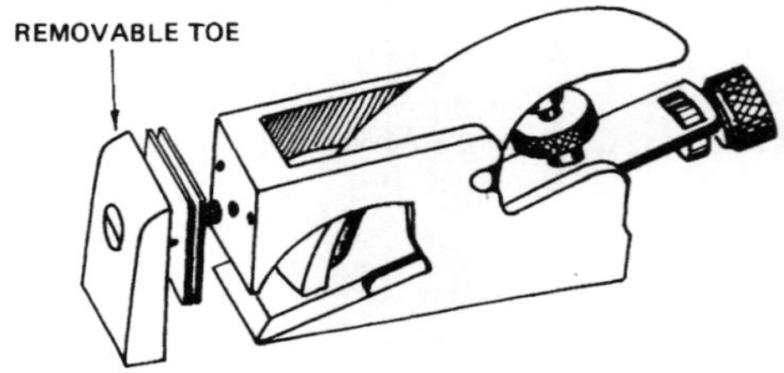

(g) Bull nose plane.

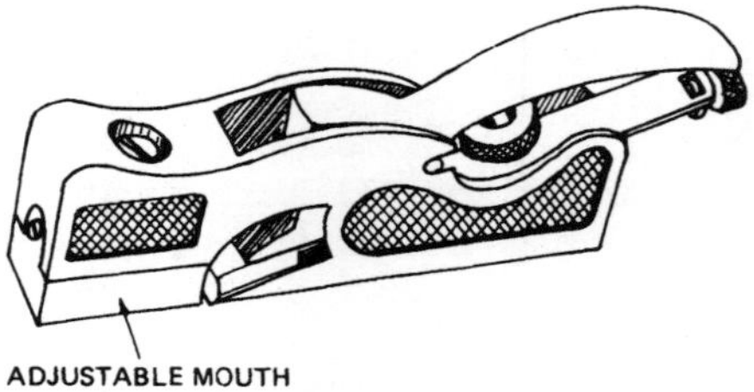

(h) Shoulder plane.

on both sides (left or right hand) without constant adjustment.

Side fillister or rebate plane

This plane (*see* Fig. 1.28(*c*)) is used almost exclusively for cutting rebates, and has a fence attached to the body by a pair of slide bars which enables it to be set to cut a parallel rebate up to about 45 mm in width.

A depth gauge is provided so as to prevent the plane cutting below the desired depth. Figure 1.29(*a*) shows a rebate in the process of being cut with the side fillister and indicates the functions of the fence and depth gauge.

Fig. 1.29. *Use of side fillister.*

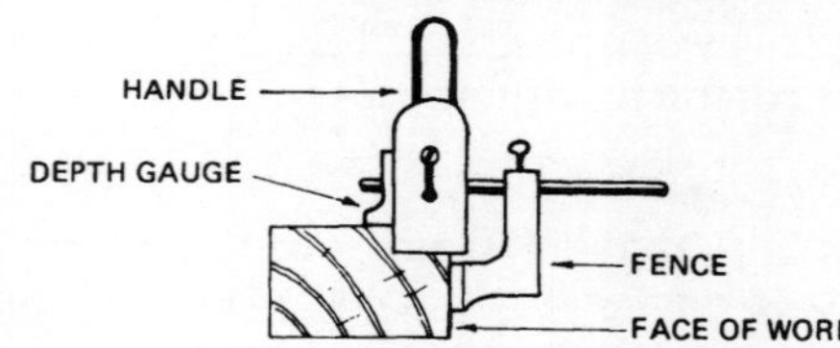

(a) Working on face.

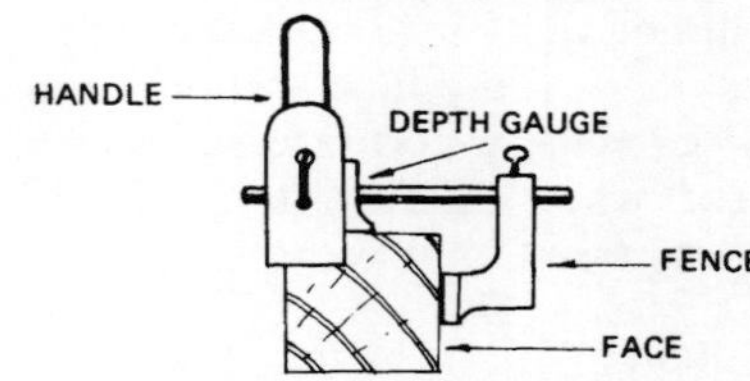

(b) Working away from face.

Side fillisters have a single iron set in the body at a normal cutting angle (45°) with the ground edge downwards. It may also be fitted with "spurs" – small vertical cutters level with the edge of the blade – enabling it to work across the grain without tearing.

To cut an accurate rebate with the side fillister, it is absolutely essential to hold the plane level – that is, with the sole of the plane parallel to the bottom of the rebate – a technique which is not particularly easy. For this reason, gauge lines should be applied to mark the extent of the rebate. Most craftsmen would use the side fillister to remove the greater part of the waste material from the rebate and then finish accurately to the gauge lines with the bench rebate plane which, having greater length, finer adjustment and the advantage of a back iron, tends to cut more cleanly.

A further feature of the side fillister is the provision to cut a rebate on the side of the wood remote from the face side. This is the proper procedure when preparing moulded and rebated sash material by hand (*see* Fig. 1.29*(b)*).

Plough

This plane (*see* Fig. 1.28*(d)*) is used for cutting grooves. The plane is supplied with a set of blades varying in size, enabling grooves to be cut from 3 mm to 16 mm or so in width (wider than this if the operation is repeated) and generally to a depth of about 18 mm.

A fence is provided to ensure that the groove is cut parallel to the edge of the wood and there is also a depth gauge which can be set to limit the depth of the groove. As with the side fillister, the plough must be held upright if it is to perform properly.

The router

This tool (*see* Fig. 1.28*(e)*) is essential to the joiner and cabinet maker, though perhaps less so to the carpenter working on site. In essence, the router is a very narrow bladed plane which is used to give a clean, level finish to the bottom of grooves, housings and sinkings, etc.

The router blade can be set to project substantially below the sole in order to reach the surface on which the levelling or smoothing is required.

To avoid time-consuming use of the router, it is normal practice first to remove as much waste as possible with the chisel, and then use the router for finishing off. For the cleanest results, the router should be set so as to cut nearly down to the required depth and, having levelled off to this setting, the cutter should be advanced a further half millimetre or so to finish off with a fine cut.

There are still many wooden routers in use which, although less efficient than their steel counterparts, nevertheless perform satisfactorily. The wooden router is often known as an "old woman's tooth".

Spokeshave

Originally a wheelwright's tool for shaving wooden spokes, the spokeshave (*see* Fig. 1.28*(f)*) is really a plane with a short sole which enables it to work on concave and convex surfaces. Spokeshaves are available with a rounded face, making the tool useful on internal curves of quite small radii.

The blade of a spokeshave should be inserted into the body with the ground edge downwards, and in the modern metal tool is fully adjustable both laterally and for depth of cut.

Considerable skill is required to use the spokeshave to good effect, the shortness of the sole and the position of the handles having a tendency to cause the tool to roll and chatter, especially in inexperienced hands. In use, the spokeshave should be pushed rather than pulled, and it will generally be found advantageous to make the cut at a slight angle; this gives a shearing cut and helps to stabilise the sole.

Bull nose plane

This is a very small type of rebate plane (*see* Fig. 1.28*(g)*). Bull nose planes have a single cutting iron held in the body at a low angle and with the ground edge upwards as in the block plane. The main feature of the bull nose plane is the shortness of the sole in front of the cutter, enabling the plane to work within a few millimetres of a corner. As in any form of rebate plane the blade extends the full width of the sole, protruding slightly on either side of the body. Altogether a very useful little plane for use on the bench or on site.

Shoulder plane

This is also a form of rebate plane, generally about 230 mm long with a blade up to 28 mm wide (*see* Fig. 1.28*(h)*). The blade is set in the body ground edge upwards and at an extremely low angle so as to make clean work of end grain cuts.

The shoulder plane is heavy, accurate and expensive, and is used mainly, as its name implies, for trimming the shoulders of tenons and similar joints. Whilst this can be regarded as a specialist item for the joiner or cabinet maker, it is nevertheless a very useful tool.

The scraper

The scraper is a tool which is used for cleaning up timber which has very difficult grain where even the most finely set plane tends

Fig. 1.30. *Scraper.*

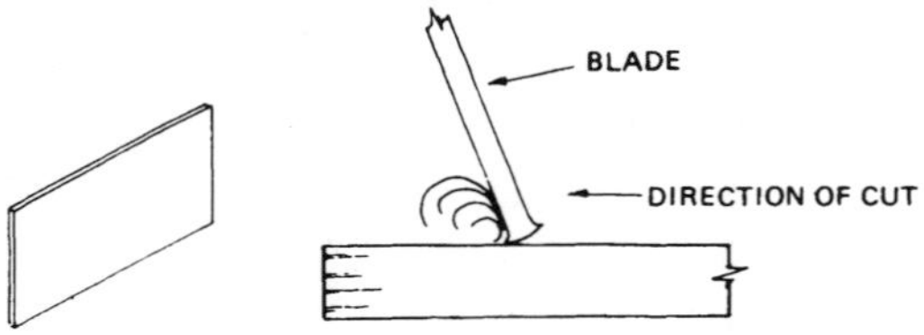

(a) Flat scraper. *(b) Cutting action of scraper.*

to tear the surface, and is thus indispensable to the worker in hardwoods. It should not be used where a plane could perform satisfactorily. In its simplest form, the scraper is a flat piece of steel about 125 mm long by 60 mm wide and 1 mm thick (*see* Fig. 1.30*(a)*) which cuts by reason of a sharp burr formed on the cutting edge. The cutting action of the scraper is shown in Fig. 1.30*(b)*, and when sharpened and used correctly, proper shavings – not scrapings – are removed from the surface of the work.

Sharpening the scraper

The two long edges are first filed or rubbed straight on the edge of the oilstone, and the sides then flattened off on the oilstone face. The scraper is then laid flat on the bench and a burr is drawn out by rubbing the edge, at a very slight angle, with a piece of hard, smooth steel – a nail punch is ideal. The burr is then turned over towards the side by drawing the sharpening tool several times across the edge of the scraper. This procedure is repeated on all four cutting edges, and the tool is ready for use. The four stages in sharpening the scraper are shown in Fig. 1.31*(a)-(d)*.

Fig. 1.31. *Sharpening the scraper.*

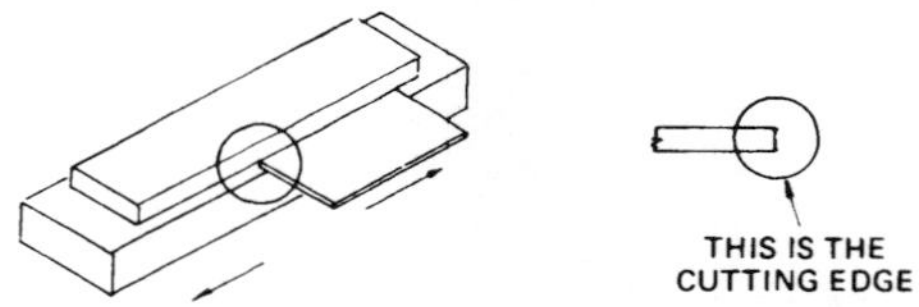

(a) Straightening the scraper edge.

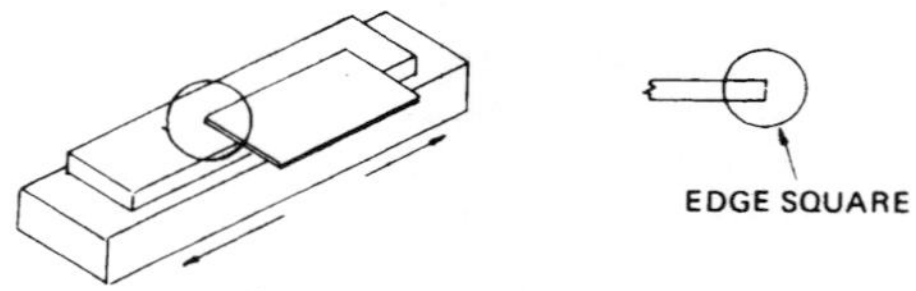

(b) Removing the old burr.

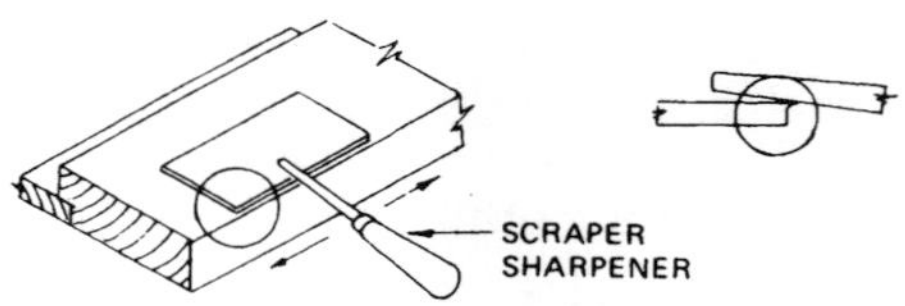

(c) Drawing out the burr.

(d) Turning the burr.

Using the scraper

The scraper is held with both hands, the two thumbs pressed into the back and the fingers curled round the front to bend the steel blade slightly. The burred edge is then pushed along the wood, the front of the tool being tilted slightly forward to push the shaving away from the cutting edge, as shown in Fig. 1.32.

Fig. 1.32. *Using a scraper.*

In use the scraper soon gets very hot due to friction, and "handled" scrapers are available which cut in exactly the same way but eliminate the discomfort to the hands of holding a hot steel blade. These "handled" scrapers have a short sole, 75 mm or so long, which prevent "digging in", and generally help in the production of a true, flat surface.

Sharpening planes and chisels

All good craftsmen have learned, by example and experience, that a fine sharp edge to a cutting tool is the prime factor in producing good, accurate work. Indeed, so important is this aspect of the woodworker's "edge" tools that one would be hard put to achieve worthwhile, economical results without keeping such tools in razor sharp condition.

Oilstones

These are the means whereby keen cutting edges can be obtained, and are available in

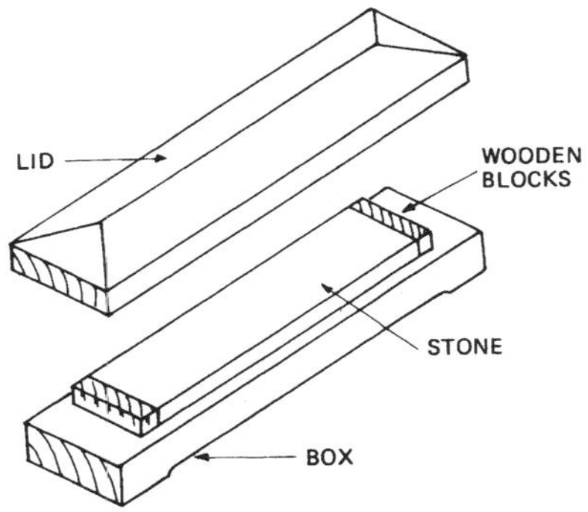

Fig. 1.33. *Oilstone fitted in wooden box.*

the form of man-made abrasive stones such as silicon carbide and aluminium oxide, or as natural stones such as Arkanas and Washita. The man-made stones are the most popular type for general bench and site work, in either fine or medium grit, whilst natural stones are generally finer, more expensive and have a slower cut. They do, however, produce very fine, sharp edges which are perhaps ideally suited for cabinet making in difficult hardwood.

Of the man-made stones, silicon carbide gives a very fast cutting action, the stone being rather softer than the somewhat slower cutting aluminium oxide variety.

Whichever type of oilstone is used, it must be large enough for the user to obtain a good long stroke with the plane iron or chisel, 200 mm long by 50 mm wide being the usual size. The stone should be housed in a wooden box as shown in Fig. 1.33. Note the wooden blocks at either end of the stone to allow the use of the full length of the stone without risk of accidently "jagging" the blade against the end. A pair of panel pins driven into the underside of the box and snipped off to leave a millimetre or so protruding will enable it to grip the bench firmly. The oilstone should be kept clean and used with a thin clear oil which floats the minute particles of steel and prevents clogging.

Grinding and honing angles

Reference to Fig. 1.34 will show clearly the two angles formed by grinding and honing a plane iron.

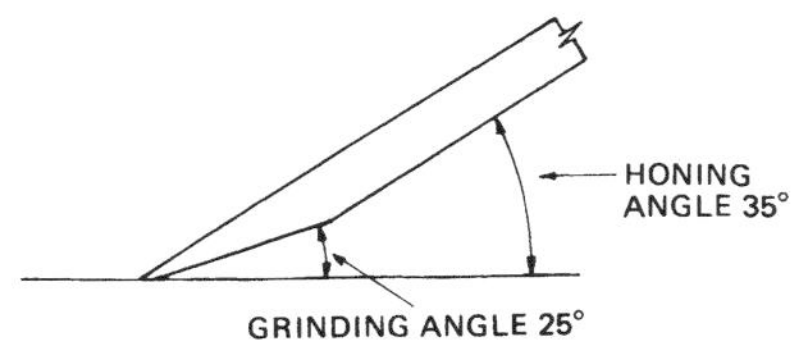

Fig. 1.34. *Honing and grinding angles.*

The grinding angle

About 25°, this is the angle produced when the blade is "ground", as is necessary from time to time when the edge becomes "thick" or has been accidentally "snicked".

The honing angle

About 35°, this is the angle at which the blade is held in the hands when "honing" (rubbing on the oilstone). After repeated sharpening, the honed edge becomes wide or thick and the blade then requires regrinding.

Using the oilstone

Place the oilstone on the bench, apply a few drops of thin oil, and place the blade in contact with the stone, honed edge downwards, holding the blade as shown in Fig. 1.35*(a)*. Exerting a light pressure on the stone, commence to move the blade backwards and forwards, using the full length of the stone, (keep wrists as stiff as possible during honing). Since most plane irons are wider than the oilstone, the blade must be moved from side to side during the honing process to cover its full width.

Fig. 1.35. *Using the oilstone.*

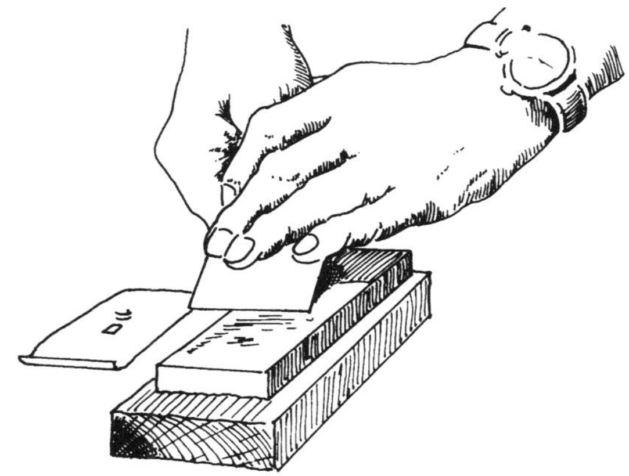

(a) Holding a plane iron for sharpening (honing).

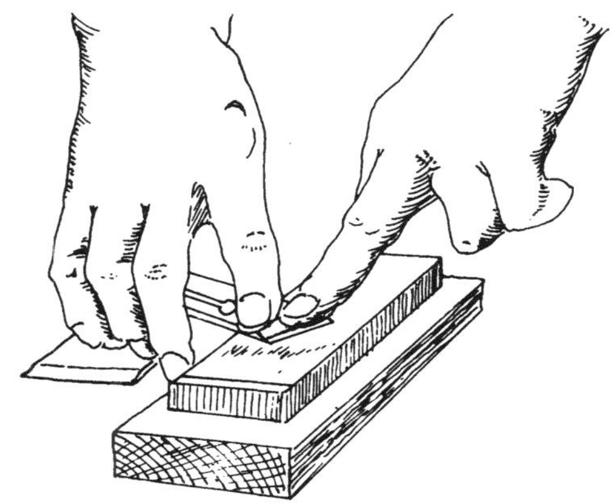

(b) Removing the wire edge from a plane iron.

After a few moments rubbing on the stone it should be possible to feel a slight burr or wire edge if the flat side of the blade edge is tested with the thumb. This wire edge must now be removed by "flatting" the blade as shown in Fig. 1.35*(b)*. During the flatting process it is absolutely essential to hold the blade flat – in close contact with the stone. Any tendency to lift the back of the blade will result in a "face bevel" which will invariably cause a plane to "choke" (shavings to clog in the mouth of the plane) or a chisel to wander from the line when paring. Having flatted the blade wipe it clean on a rag and examine the edge under a good light. If the edge is sharp, it will be quite invisible; if not quite sharp, it will show up as a thin line of reflected light and the honing process must be repeated until a satisfactory edge is obtained.

Many experienced craftsmen further improve the keenness of the edge by gently "stropping" the iron on the palm of the hand or on a barber's leather strop. This gives an edge *par excellence* and can make all the difference where tough or difficult wood is involved.

NOTE: Hand stropping of chisels and plane irons is a practice which, although common, can be extremely dangerous if carried out by an inexperienced person. The novice is strongly advised to use a leather strop if an exceptionally keen edge is required.

Shape of honed edge
Where chisels and rebate planes are concerned, the honed edge should be perfectly straight and square, only the most minute deviation being acceptable, but for the

Fig. 1.36. *Profile shape of plane irons.*

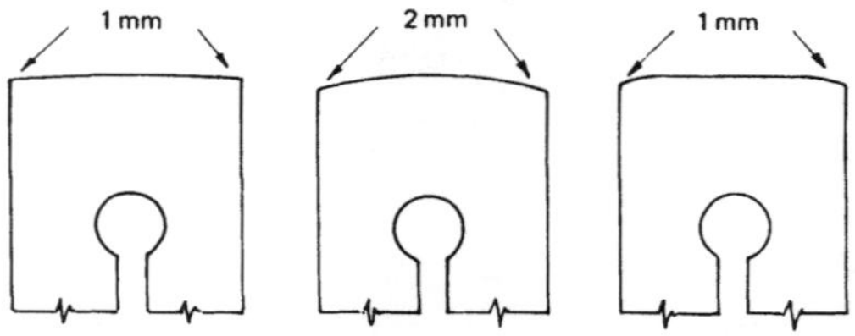

(a) Smoothing plane. *(b) Jack plane.* *(c) Trying plane.*

bench planes – smoothing, jack and trying – the edge produced by honing must be slightly convex, as shown in Fig. 1.36, in order to prevent "cornering" (ridges produced on the wood surface by the edge of the blade) and to give a degree of bevel control when shooting edges.

NOTE: When sharpening chisels, especially narrow ones, use should be made of the edges of the oilstone. Otherwise the oilstone would soon wear hollow, making wide, straight, honed edges difficult to obtain. Properly used, an oilstone will give several years use before requiring truing up by rubbing its face on a stone slab sprinkled with sand and water.

Slip stones
These are used to sharpen the edges of gouges and other tools with concave profiles, and are available in a variety of shapes and sizes to suit the radii of the cutters concerned. Generally speaking, slip stones are worked backwards and forwards over a stationary tool – the reverse of normal oilstone use, but requiring the same care in preserving the correct honing angle as with planes and chisels.

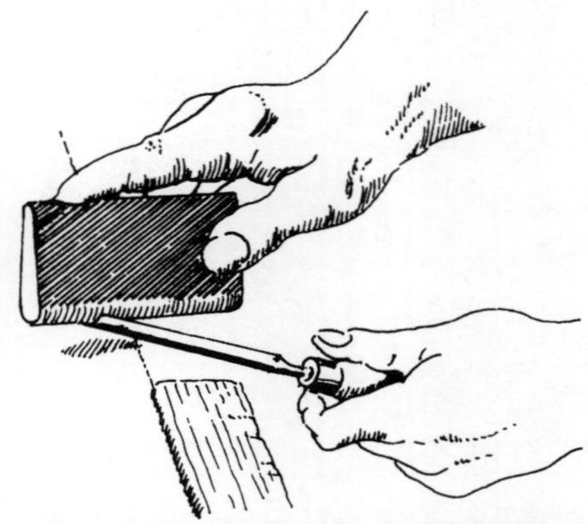

Fig. 1.37. *Using a slip stone to sharpen a gouge.*

The use of the slip stone to sharpen an inside ground gouge is shown in Fig. 1.37.

BORING TOOLS

The chief boring tools used by the woodworker are the bradawl, the hand drill and the brace and bit, illustrated in Fig. 1.38.

The bradawl
This is a simple but effective tool (*see* Fig. 1.38*(a)*), used for boring small, shallow holes in wood such as are required for starting small to medium sized screws. Every woodworker should possess at least one good quality bradawl–cheap bradawls are worthless as the tang of the blade is sure to part company from the handle after a minimum of use.

Drills

Hand drill
This tool (*see* Fig. 1.38*(b)*) sometimes

Fig. 1.38. *Boring tools.*

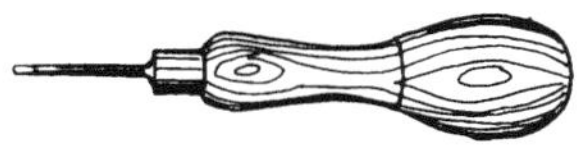

(a) Bradawl.

(b) Hand drill.

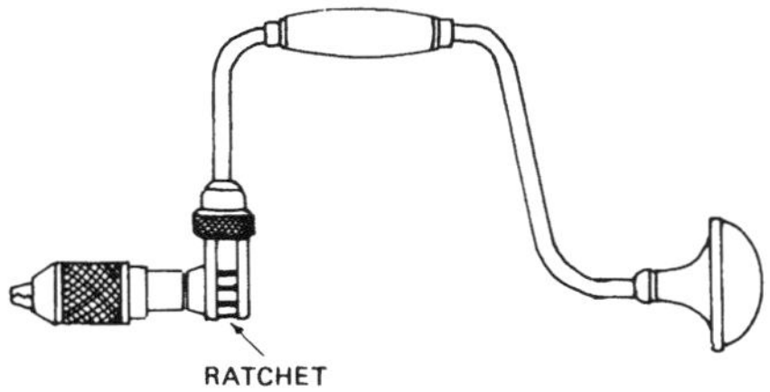

(c) Ratchet brace.

known as a wheel brace, is an essential part of every craftsman's kit. It is used for boring screw holes up to 6 mm in diameter, for which purpose a selection of twist drills is required.

Ratchet brace

This tool (*see* Fig. 1.38*(c)*) is another essential item of equipment. Basically it is a handle for the variety of "bits" with which it is used. Most modern braces have a ratchet action enabling them to be used close to an obstruction, as when boring a hole through a floor close to a wall. The brace is also useful for inserting or removing large screws, in conjunction with a screwdriver bit. For this job, the "swing" of the brace handle gives tremendous leverage which is very handy when dealing with stubborn screws.

Bits

Jennings pattern twist bit

(*See* Fig. 1.39*(a)*.) An excellent bit for boring clean, deep, accurate holes from 4 mm diameter upwards. The bit has a double spiral, two cutting edges and two spurs which should be kept sharp with a file.

Irwin pattern twist bit

(*See* Fig. 1.39*(b)*.) Similar to the Jennings pattern bit but with a single spiral, the Irwin twist bit is very free cutting, especially if the wood is at all wet. In practice there is little to choose between the two types, this being mainly a matter of personal preference.

Centre bit

(*See* Fig. 1.39*(c)*.) A short, fast cutting bit with a single spur, the centre bit is used to bore accurate but relatively shallow holes, as when cutting stopped housings and recesses, etc. The absence of a spiral renders them inaccurate for deep holes. Available in diameters from 6 mm to 38 mm.

Expansion bit

(*See* Fig. 1.39*(d)*.) Basically similar to the centre bit, the expansion bit has an adjustable cutter and spur, enabling holes up to 75 mm diameter to be cut. Although useful on occasions this tool, however, is not the easiest to use.

Forstner bit

(*See* Fig. 1.39*(e)*.) This is a very specialised type of centre bit which is guided by its sharpened rim. Having no central screw, it can be used to bore holes part way through thin timber without risk of breaking through.

Fig. 1.39. *Bits and countersinks.*

(a) Jennings pattern twist bit.

(b) Irwin pattern twist bit.

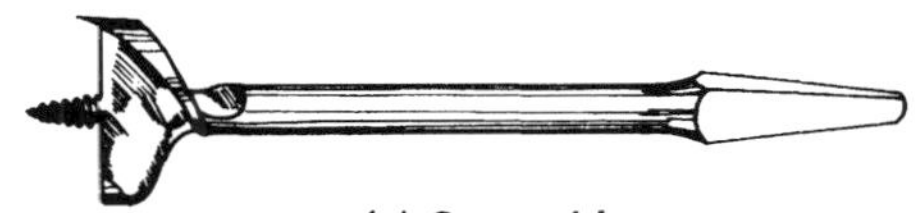

(c) Centre bit.

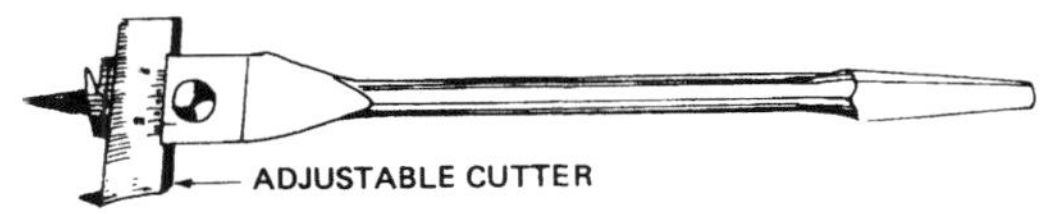

(d) Expansion bit.

(e) Forstner bit.

(f) Snail countersink.

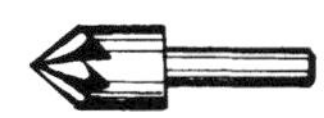

(g) Rose countersink.

(h) Screwdriver bit.

Countersinks

These are used to form recesses for screw heads. A snail type countersink, for use in a brace, is shown in Fig. 1.39*(f)* whilst Fig. 1.39*(g)* shows a rose type bit for use in a hand drill or electric drill.

Screwdriver bit

This is shown in Fig. 1.39*(h)*. Its use has already been described. The use of the ratchet to permit a partial swing of the brace handle makes the insertion or removal of screws easier.

Care of twist bits

The woodworker requires a comprehensive set of twist bits of various types and sizes, and as these are expensive items they should be well looked after.

Twist bits should be kept clean and bright, and sharpened when necessary with a fine saw file. Care should be taken to avoid damaging the spurs and cutting edges, as is certain to happen if they inadvertently contact nails, screws, brickwork, etc., in use. To avoid damage in the tool bag, tool box or on the bench, twist bits should be kept in a specially made wallet known as a "bit roll".

PERCUSSION TOOLS

Hammers

Hammers are available in various types and sizes, and are obviously essential to the craftsman. They are illustrated in Fig. 1.40.

Warrington hammer

This type of hammer (*see* Fig. 1.40*(a)*) is a favourite amongst bench joiners and cabinet makers, and is classified by the weight of the head, about 600 g (18-24 oz) being average. Good quality Warrington hammers have a slightly rounded face to avoid leaving hammer marks (half crowns) on the surface of the wood, and have a smooth, well-shaped handle of ash or hickory.

Fig. 1.40. *Hammers.*

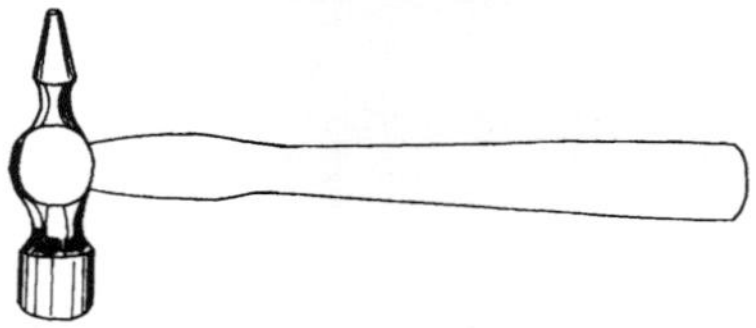

(a) Warrington hammer.

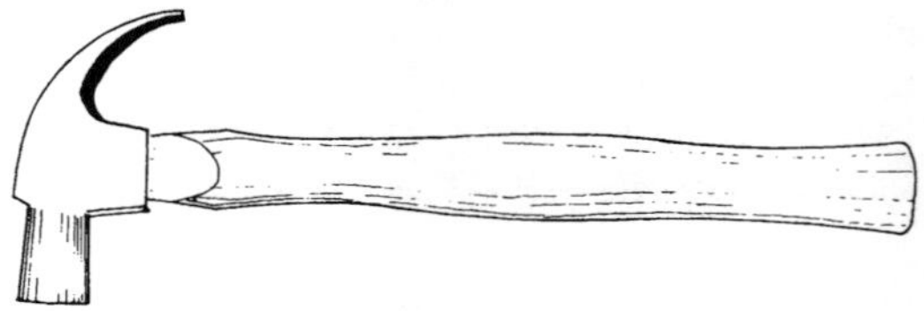

(b) Claw hammer.

(c) Nail punch.

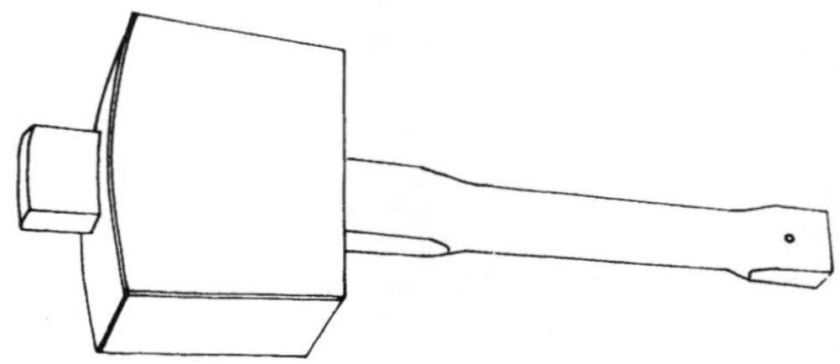

(d) Carpenter's mallet.

When using a hammer, the handle or shaft should be grasped near the end well away from the head, and swung from the wrist rather than the forearm. The face of the head should be kept clean to avoid skidding and bending the nail.

Pin hammer

This is a smaller version of the Warrington hammer, the head being considerably smaller and lighter. The pin hammer is used for light work, such as driving panel pins, for which purpose the "pein" – the thin end of the head – is useful for getting small pins started.

Claw hammer

This type of hammer (*see* Fig. 1.40*(b)*) is favoured by site carpenters since the generally heavier head is more suited for driving long nails, while the claw is also useful for withdrawing bent or unwanted nails. The temper of a claw hammer is less simple than that of a Warrington hammer, the face needing to be very hard, whilst the claw must be very tough and springy. For this reason it is important to buy the best one can afford – cheap ones soon give trouble. The shaft may be of ash or hickory, or sometimes of steel with a rubber grip. The choice is largely a matter of personal preference, the main consideration being that the tool has a good balance and feels right in the hand.

Nail punches

Also called nail sets (*see* Fig. 1.40*(c)*) these are small but essential tools used to "punch" nail heads below the surface of the wood. They are available in various sizes to suit the diameter of the nail head, and have a hollow tip to prevent them slipping off the nail.

Mallet
This is used for a variety of "persuading" purposes such as assembling joinery and driving chisels. The main function of the mallet (*see* Fig. 1.40*(d)*) is to deliver a heavy blow without causing damage to the work piece or chisel handle. Generally made with a beech head, they are also available with metal shafts and a head made of hard rubber.

SCREWDRIVERS AND PINCERS

Screwdrivers

Screwdrivers are available in a variety of types and sizes (*see* Fig. 1.41), all of which serve a similar purpose – inserting and withdrawing screws. It is important to ensure that the tip of the screwdriver blade is properly ground, with clean sharp edges, and also that the blade is the correct size for the screw being used, otherwise the tool is bound to slip, damaging the screw head.

Fig. 1.41. *Screwdrivers.*

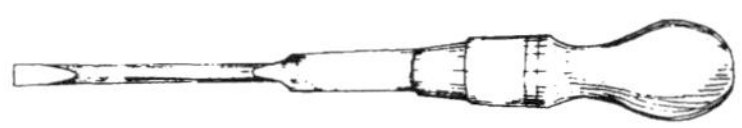

(a) Cabinet screwdriver.

(b) Ratchet screwdriver.

(c) Spiral ratchet screwdriver.

Cabinet screwdriver
(*See* Fig. 1.41*(a)*.) This type of screwdriver has a polished beech or box handle and is perhaps the most usual screwdriver used by the woodworker, being both powerful and positive in action. Most craftsmen have three – small, medium and large.

Ratchet screwdriver
(*See* Fig. 1.41*(b)*.) This type is very useful for driving small to medium sized screws and is a worthwhile addition to the tool kit. It is, however, generally less positive in action than the ordinary screwdriver.

Spiral ratchet screwdriver
(*See* Fig. 1.41*(c)*.) This type is very popular and takes a great deal of the effort out of continuous screwdriving. The tool works with a "pump" action and is supplied with two or three different sized blades. This type of screwdriver is best suited to small and medium size screws and its use is best avoided on any job where a "slip" might cause severe damage to the workpiece, i.e. on polished work or plated metal parts.

NOTE: Screwdrivers can be obtained with blades shaped to suit either the normal "slotted" or the "Phillips" type head.

Pincers

This is another essential woodworker's tool used mainly for pulling out nails, though there are many other uses too numerous to describe. "Tower" pattern pincers about 200 mm long will generally be found most serviceable.

WORKSHOP EQUIPMENT

Several specialised items of equipment are required by the woodworker, both on the bench and on site. Some of these are provided by the employer; others craftsmen make for themselves.

Mitre blocks and boxes

Shown in Fig. 1.42*(a)* and *(b)*, these are used to guide the saw when cutting angles and mitres. They are made as required to suit the job in hand or simply as replacements when badly worn.

Fig. 1.42. *Saw guides.*

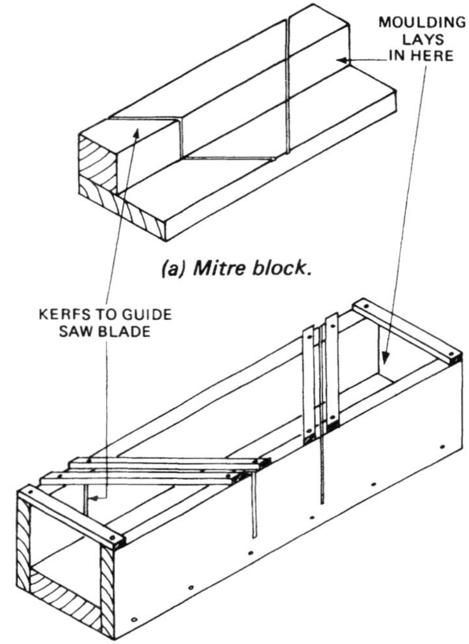

(a) Mitre block.

(b) Mitre box.

Bench hook

Figure 1.43 shows a simple but essential item of equipment made by the craftsman himself to hold the wood firmly on the bench when sawing. It is made either left or right handed to suit the individual.

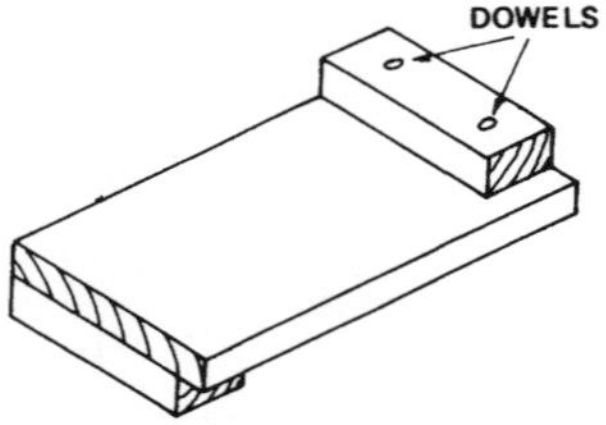

Fig. 1.43. *Bench hook.*

Shooting boards

These are used in conjunction with a jack or trying plane as shown in Fig. 1.44 for the accurate trimming of squared ends, mitres, etc. Made by the craftsman, they require truing up from time to time, and may be adapted for special angles.

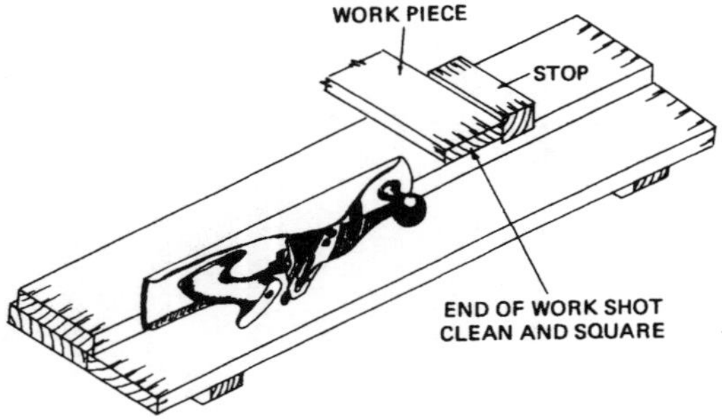

Fig. 1.44. *Use of shooting board.*

Straight edges

These are used mainly for marking and testing straight lines and surfaces. A straight edge must be both straight and parallel to be of any value and should be hung up by the end when not in use to prevent distortion. In the workshop, two or three straight edges, up to 2 m in length will be found sufficient for normal use, but on the building site longer versions up to 4 m or more, and proportionately heavy, are commonly required. They are made by the craftsman from good quality redwood as required.

Winding strips

These are short wooden straight edges used in pairs as shown in Fig. 1.45 to test surfaces for "twist" when planing up by hand. They are normally used by the beginner or the apprentice to whom they prove most useful until such time as the eye, through experience, no longer requires such assistance.

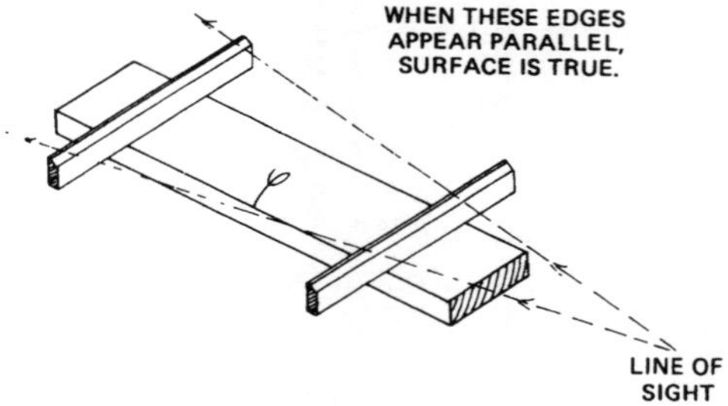

Fig. 1.45. *Use of winding strips.*

Sawing stool

This item of equipment, shown in Fig. 1.46, is quite useful in the workshop but "comes into its own" on site where it serves the purposes of a short step ladder, a bench for holding and resting materials being worked on, and as a vice for holding the edges of doors whilst "shooting them in" for hanging. The making of a good solid sawing stool by hand provides a very satisfactory test for the young craftsman.

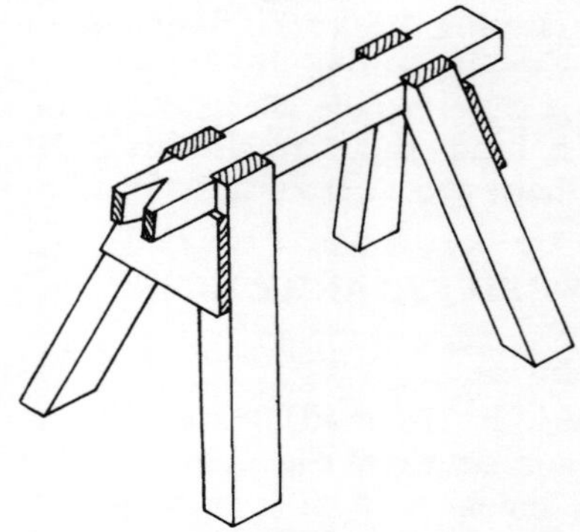

Fig. 1.46. *Sawing stool.*

Cramps

G cramps

Shown in Fig. 1.47*(a)*, these are normally provided by the employer and are available in a range of sizes to grip material up to 300 mm in thickness.

Sash cramps

These are made from steel bars, either flat or T-shaped in section (*see* Fig. 1.47*(b)* and *(c)* respectively), with a sliding shoe which can be opened or closed to any position along the length of the bar. Sash cramps vary in length from 600 mm to 1.8 m or more to suit the size of the framework being cramped up. Extension bars are available to enable extra long items to be cramped.

A good selection of sash cramps is an essential requirement in a work shop, or indeed to the home craftsman.

Fig. 1.47. *Cramps.*

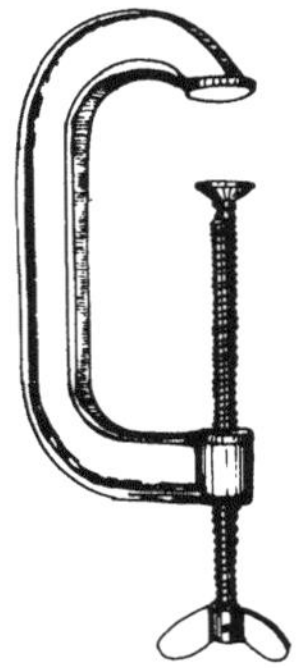

(a) G. cramp.

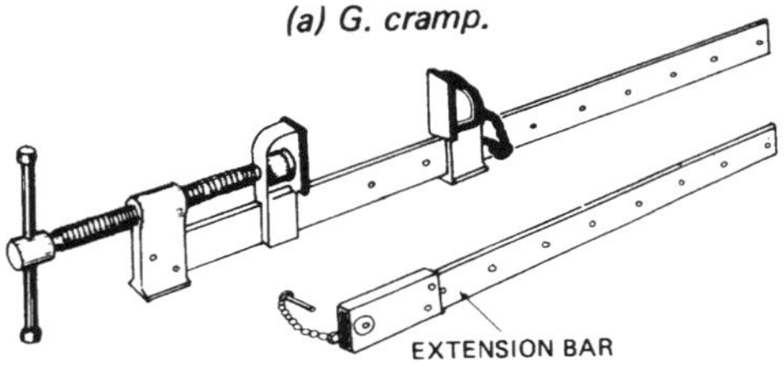

(b) Sash cramp – flat bar.

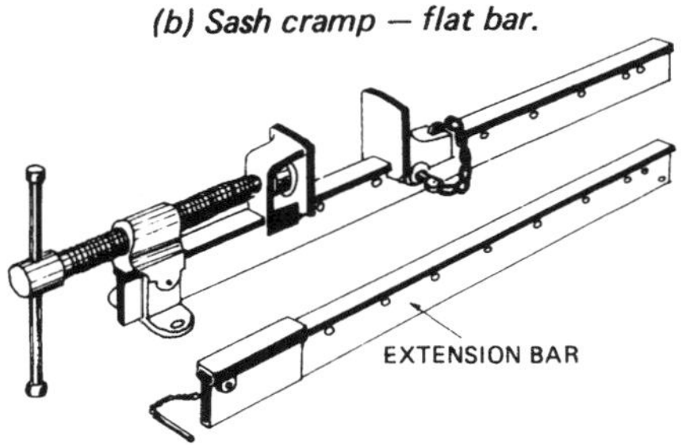

(c) Tee bar sash cramp.

FURTHER READING

Jackson, Albert and Day, David. *The Complete Book of Tools.* (Michael Joseph)

Woodworker – magazine issued monthly
Working Wood – magazine issued quarterly

SELF-TESTING QUESTIONS

All the information required to answer the following questions is contained within this chapter. Attempt each section *as fully or as briefly* as the question demands, and then check your answers against the information given in the chapter.

1. Name three types of rule, used by carpenters and joiners, stating in each case where the particular type of rule could be used to its best advantage.

2. *(a)* State the purpose of the following tools: *(i)* try square; *(ii)* mitre square; *(iii)* sliding bevel.

(b) Explain with a sketch the method of testing a try square for accuracy.

3. *(a)* Explain briefly the main advantage that a cutting gauge has over a marking gauge.

(b) State *two* specific uses for a cutting gauge.

(c) Show by means of a sketch how the points of a mortice gauge are set to suit a particular chisel.

4. *(a)* State three typical uses for a pair of compasses (or dividers).

(b) State where a "cut" line made by a marking knife is preferable to a pencil line.

5. *(a)* Sketch saw teeth designed for: *(i)* ripping; *(ii)* cross cutting.

(b) Describe briefly the cutting action of *(a)(i)* and *(ii)*.

(c) State typical uses for: *(i)* a panel saw; *(ii)* a tenon saw; *(iii)* a pad saw; *(iv)* a coping saw.

6. *(a)* State typical uses for the following: *(i)* firmer chisel; *(ii)* bevelled edge chisel; *(iii)* swan neck chisel; *(iv)* paring chisel; *(v)* inside ground gouge.

(b) Sketch a firmer chisel naming its various parts.

7. *(a)* Name the three planes which are commonly known as "the bench planes". Describe a typical use for each.

8. State typical uses for the following tools: *(a)* side fillister; *(b)* bull nose plane; *(c)* bench rebate plane; *(d)* router; *(e)* spokeshave; *(f)* scraper.

9. Define with sketches:

(a) the grinding and honing angles for a plane iron;

(b) the profile shapes of smoothing, jack and trying plane irons.

10. *(a)* Name "bits" suitable for boring:

(i) a 12.5 mm hole right through a 75 mm thick board;

(ii) a shallow recess half way through a a piece of wood 9 mm thick;

(iii) a 53 mm diameter hole through a 38 mm thick board.

(b) State one particular use for the "ratchet" on a carpenter's brace.

11. Name six items of equipment made by the carpenter or joiner to facilitate his work on site or in the workshop. State the particular use of each item.

12. Describe briefly the special precautions a carpenter takes to avoid damage to: *(a)* handsaws; *(b)* twist bits; *(c)* oilstones; *(d)* chisels; and *(e)* screw heads when inserting or removing screws.

2. Timber

After completing this chapter the student should be able to:

1. Describe the growth of a tree.
2. Sketch, name and state the function of the cells in hardwoods and softwoods.
3. Name, describe and state the distribution and uses of at least six hardwoods and four softwoods.
4. Describe the seasoning of timber by air and kiln drying methods.
5. Calculate the moisture content of a timber sample.
6. Sketch and describe the methods used in the conversion of timber from logs to boards.
7. Sketch and explain the distortion which occurs when timber loses or absorbs moisture.
8. Sketch and describe the most common timber defects.
9. Describe the symptoms, treatment and preventative measures associated with insect and fungal attack.
10. Describe the construction, uses and advantages of man-made boards.

TREES

Timber, the basic raw material of the carpenter and joiner, is the product of the woody perennial plant which we all know as a tree. Trees differ from most other types of plant, not only on account of their size and strength, but also in the way in which they grow.

Growth of a tree

A tree starts life as a seed which, given the basic necessities for life – air, water, food, sunlight and a suitable temperature, soon germinates, throws out leaves on a stem and becomes recognisable as a young tree or sapling.

This is perhaps the way in which most plants begin life, but trees, shrubs and similar woody stemmed plants differ in that they are "exogenous", which is to say that from this stage onwards further growth takes place in the form of new layers of material which are built up around the original stem. This new growth is made each year in a special part of the tree just beneath the bark known as the "cambium".

Each year's growth, therefore, is really in the form of a cone as shown in Fig. 2.1 and it will be readily understood why the trunks of most trees are tapered towards the top. In later years the tree tends to increase its rate of growth towards the top, and the trunk or bole gradually assumes a more parallel shape.

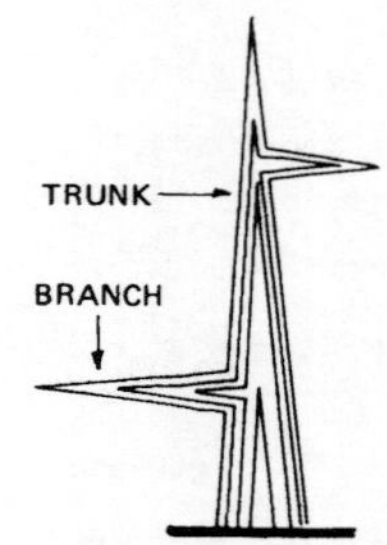

Fig. 2.1. *Conical growth of a tree.*

Parts of a tree

A tree may be regarded as having three main parts as shown in Fig. 2.2, namely the roots, the trunk and the crown. The roots serve to anchor the tree firmly into the ground, support its weight and absorb water and soluble mineral salts from the soil. The function of the trunk is to support the crown, and store and conduct food and water. The crown of the tree – the branches, leaves and twigs – manufactures the food which the tree needs for its growth.

Briefly, the tree absorbs water and mineral salts through its roots, conducts them through the outer part of the trunk (the sapwood) to the crown and thus into the leaves. Here, by means of a complicated chemical process known as "photosynthesis", the leaves absorb carbon dioxide from the air and with the aid of sunlight convert it into food (cellulose). The food is passed down the tree in the "bast", a layer between the bark and the cambium, to wherever growth is taking place. Growth occurs during certain times of the year, when the cells in the cambium layer divide to form new wood, this process continuing throughout the growing season.

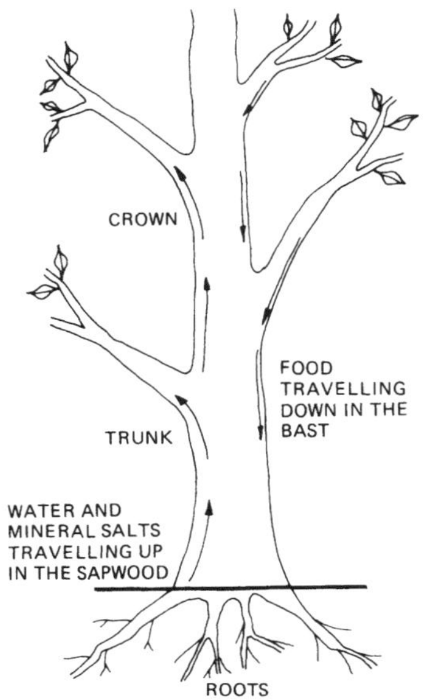

Fig. 2.2. *Parts of a tree.*

Annual rings

During the spring and early summer of each year, therefore, cells are formed in the cambium layer thus making new wood. The cells formed during the spring are relatively large and have thin walls, but as the season progresses, newly formed cells become progressively smaller with thicker walls until finally growth ceases, to start again with a surge the following spring.

Each year's growth therefore can often be readily seen on a cross-section of timber as shown in Fig. 2.3. In the case of tropical timbers where growth has been more or less continuous throughout the year, the "annual ring" may be difficult to see.

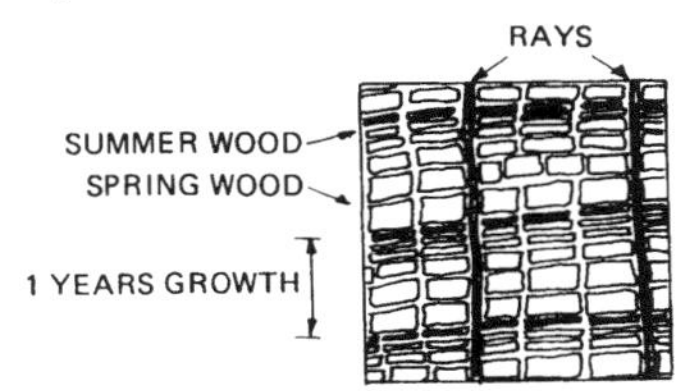

Fig. 2.3. *Yearly growth.*

Heartwood and sapwood

A cross-section through the trunk of a tree may show two distinct areas, a central core surrounded by a layer of lighter coloured wood.

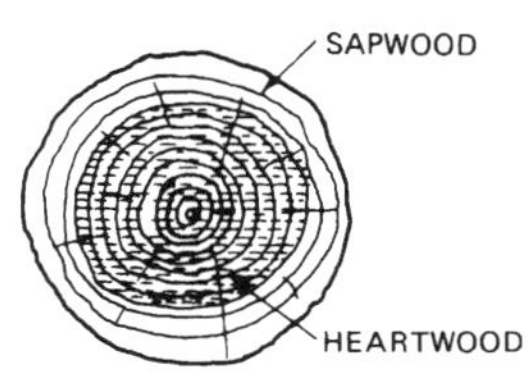

Fig. 2.4. *Sapwood and heartwood.*

The outer layer is the "sapwood" (*see* Fig. 2.4), the living part of the tree, and contains a high proportion of starch and food. The central portion, the "heartwood", is composed of cells which ceased to function and became hardened as the tree matured. It thus forms a strong skeleton enabling the tree to support a heavy crown and to stand up to high winds. Heartwood is not always easily distinguished from sapwood, but it provides the best and most durable timber.

Hardwoods and softwoods

Trees are divided botanically into two main groups, the " angiosperms" and the "gymnosperms". The angiosperms are the broadleafed trees such as oak, ash, teak and mahogany, and are commonly referred to as *hardwoods.*

The gymnosperms include the conifers whose leaves are generally in the form of needles or scales as borne by the pines, firs and spruces. These trees bear their seeds in cones and are usually known as *softwoods.*

The classification of timbers as "hardwood" or "softwood" is not particularly apt, since many hardwood species have a fairly soft texture (balsa is actually a hardwood), whilst several of the softwoods are quite hard. The nomenclature has, however, been in use for many years since its introduction in Britain to distinguish the indigenous broadleafed timbers from the generally softer imported coniferous varieties. Given a little practice and experience, the craftsman should have no difficulty in determining to which of the two groups a particular timber belongs.

Structure of softwoods

In common with all other living tissues, wood is composed of masses of tiny cells which vary in shape and size to suit their particular function.

Softwoods are made up of two kinds of cell, "tracheids" and "parenchyma", as shown in Fig. 2.5*(a)*.

Fig. 2.5. *Structure of softwoods.*

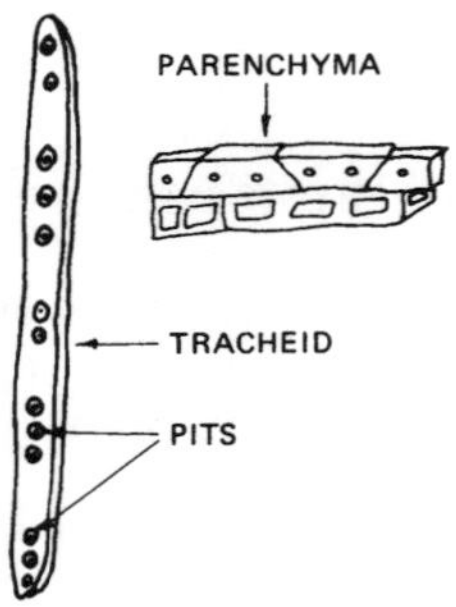

(a) Softwood cells.

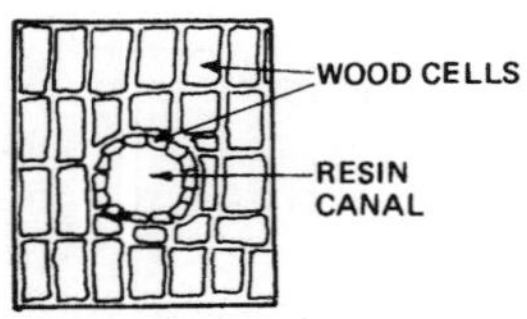

(b) Resin canal.

Tracheids
These are greatly elongated cells, all closely interlocked with one another and make up the main bulk of the timber. Their function is to give the tree its strength and to conduct the sap, for which purpose the cell is equipped with small holes or "pits" to allow the sap to pass from one tracheid to another.

Parenchyma
The other type of cell found in softwoods is relatively small, usually rectangular in shape, and its function is to store food. Parenchyma also have "pits" to allow sap to move from one cell to another and are generally arranged in radial rows known as rays. The rays in a softwood can usually be seen with a hand lens.

Resin canals
These features of softwood structure are not actually cells in themselves, but rather tubes formed by the grouping together of cells, thus surrounding and containing a tiny column of resin (*see* Fig. 2.5*(b)*).

Structure of hardwoods
The cell structure of a hardwood is not quite as simple as that of the softwood, and is composed of four types of cell instead of two (*see* Fig. 2.6).

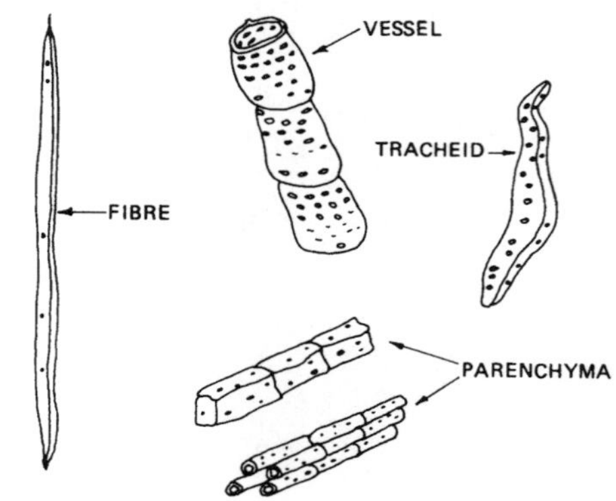

Fig. 2.6. *Hardwood cells.*

The fibre
This is a long sharply pointed cell which makes up the main bulk of the wood and serves much the same purpose as the tracheid in a softwood, i.e. to conduct sap and to give the tree its strength. Fibres generally have fewer pits than tracheids.

Parenchyma and tracheids
These types of cell are found in both hardwoods and softwoods and serve a similar function in both types.

In a hardwood, the parenchyma are frequently easily visible with the naked eye, being grouped in some quantity in the radial rows known as rays. These rays are most obvious in certain timbers such as oak or beech where they make identification of the timber easy. Hardwood tracheids are less regular in shape than their softwood counterparts.

Vessels or pores
These cells are found only in hardwoods. They are cylindrical in shape, and are joined end to end to form long tubes which run longitudinally through the timber. The function of this cell is the transportation of sap, pits again allowing sap movement through the wood.

The size of the pore or vessel varies from one species of timber to another, providing yet another means of identification.

Ring porous and diffuse porous timbers
In certain timbers the size of the pore varies between spring and summer wood, giving the appearance of pores arranged in annual rings. Such timbers are known as ring porous (*see* Fig. 2.7*(a)*).

Other types of timber have pores of similar size distributed more or less evenly throughout the wood, irrespective of spring and summer rates of growth. These are known as diffuse porous timbers (*see* Fig. 2.7*(b)*).

It should be noted that the pore is present in all hardwoods, whether it is visible with the naked eye or not. In many cases the pores may appear as no more than a multitude of fine scratches on the face of the timber, but since they are *not* found in a softwood they serve as the principal factor in determining whether the timber is a hardwood or a softwood.

Fig. 2.7. *Timber pores.*

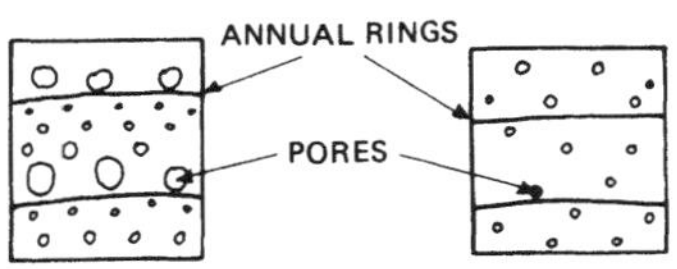

(a) Ring porous wood. *(b) Diffuse porous wood.*

COMMERCIAL TIMBERS

Of the several thousand species of timber which are known to science, only a mere handful are commercially important in any one country. On the other hand, as the world's forests become further and further depleted, lesser known species which until now have been used for little other than localised work in their country of origin are beginning to find favour as imported timbers. Many of these have been found to be excellent timbers for both general and specific purposes.

Since timber is the basic raw material of any woodworking craft, it cannot be too strongly emphasised that the ability to recognise various species, and understand something of their character and general usefulness in service, must be regarded as of prime importance.

It will be found that one of the best and easiest ways to become familiar with and to be able to identify a reasonably wide range of timber species is to form a personal collection. Small, match box size samples should be taken whenever opportunity arises and identified – fellow craftsmen and college lecturers are generally only too pleased to assist in identifying the species. Samples collected should be smoothed on three sides (face, edge and end) and labelled with their name, country of origin and recommended uses. The texture and workability will be noted whilst cleaning up the three faces.

All craftsmen – students in particular – are most strongly urged to take a keen interest in the timbers of their trade since this will not only be found to be quite fascinating and satisfying in itself, but also of great material help in the development of woodworking skills.

The following list outlines the general appearance, characteristics, distribution and uses of a representative sample of those timber species most commonly in use in today's woodworking industries.

NOTES:

(a) The abbreviation 540 kg/m^3 means that one cubic metre will weigh 540 kilograms. If the densities of various timbers are compared they help in identification; for example, beech is quite a heavy wood at about 700 kg/m^3, Douglas fir is a medium density timber at 540 kg/m^3 and obeche is comparatively light at 380 kg/m^3.

(b) The botanical names given to the various timbers (or trees) may be likened to our own surnames and first names; for example, three brothers may be listed as Smith, Harold; Smith, Edward and Smith, John. All are members of the same family (Smiths) but are individually identified positively by their own specific names.

In the same way the sugar maple, for instance, is botanically named Acer saccharum. This tells us firstly that the tree is one of the maples (Acer), and secondly which one it is (saccharum).

(c) The abbreviation spp. means "species"; thus oak, Japanese (Quercus spp.) signifies that there are several closely related species of this tree, the timber of which may be marketed under the broad or general name of "Japanese oak". It may, on occasions, be necessary to use or obtain a timber for a specific job for which the particular qualities of a definite species are required. In such an instance the use of the proper botanical name will avoid any possible error when ordering.

Softwoods

Douglas fir (Pseudotsuga taxifolia)

Other names – British Columbian pine, Oregon pine.

Distribution – Canada and USA.

Characteristics – a pinkish-brown timber with very distinct grain and annual rings and a clean sweet smell. In tangentially sawn boards (tangential and radial sawing are described later in this chapter under methods of conversion) the grain may be so pronounced as to make painting unsuccessful, such timber being more appropriately polished or varnished. Radially sawn boards are usually straight grained. Large resin canals and ducts are commonly encountered, the resin tending to bleed through paint, unless a suitable sealer coat is applied. The timber is very strong, has a good natural resistance to decay, takes nails, screws and glue very well, but small sections have a tendency to split when fixed with nails or pins. Douglas fir is fairly easily worked by hand and machine but sharp, finely set planes are needed to obtain a smooth finish on tangentially sawn boards. Available in large sizes.

Uses – good quality softwood joinery, strip flooring, stairs, heavy constructional work and plywood.

Average density about 540 kg/m^3.

Larch (Larix decidua)

Distribution – Europe, including Great Britain.

Characteristics – a strong, durable, resinous, reddish-brown timber with a pale sapwood. Generally straight grained, the tree grows to large sizes and has been found sufficiently useful as to make home grown timber of some economic importance.

Uses – gates, fences, farm buildings and rustic work; little used for joinery purposes.

Average density about 580 kg/m^3.

Parana pine (Araucaria augustifolia)

Other names – Chile pine.

Distribution – South America.

Characteristics – varies in appearance and density but is generally brownish-yellow in colour with a very even texture and an indistinct grain. Pink to purple streaks and patches are quite common, this generally being the harder, tougher timber. Parana pine tends to be rather brittle and warps badly on drying. It also has a tendency to split when nailed near the end of a board. Sharp tools, and a good deal of effort, are required to obtain a good finish, but with care the surface can be brought to a very smooth, silky lustre. The timber glues, paints and polishes extremely well, but is not durable when used outside. Available in boards up to 300 mm wide.

Uses – good class interior joinery, stairs and kitchen furniture and fitments.

Average density about 500 to 550 kg/m^3.

Pitch pine (Pinus palustris & spp.)

Other names – southern pine, longleaf pine.

Distribution – Southern USA.

Characteristics – a hard, tough, heavy timber which is extremely resinous. The grain and annual rings are very distinct, generally showing up as dark red-brown summer wood and pink to brown spring wood. The timber is easily recognised by its resinous, turpentine smell when worked. The wood is fairly easy to work by hand but the resin tends to clog tools, especially machine cutters and rollers. Pitch pine is too resinous for a painted finish, the resin tending to bleed through even several coats, and is normally varnished where a finish is required. The best figure is obtained by cutting the boards tangentially. Undoubtedly, this is one of the strongest and most durable of the softwoods.

Uses – heavy structural work, shoring, shipbuilding, polished softwood joinery and church furniture.

Average density about 660 kg/m^3.

Redwood (Pinus sylvestris)

Other names – pine, yellow pine, red deal, Scots pine.

Distribution – Russia, Scandinavia, Baltic region, United Kingdom.

Characteristics – pinkish-white timber with distinct orange-red grain, and clearly marked annual rings. The timber is resinous, has a clean, pine smell, and knots are present in all but the best selected boards. Redwood is fairly easy to work by hand and machine and can be brought to a fine finish. Nails, screws, glue and paint all take well. The timber is generally quite strong and straight grained with a moderately good natural resistance to decay.

Uses – probably the most widely used timber in the construction industry for such purposes as roofs, floors, partitions and general interior and exterior painted joinery.

Average density about 500 kg/m^3.

Sitka spruce (Picea sitchensis)

Other names – silver spruce.

Distribution – Canada, USA and some from Scotland.

Characteristics – a white, generally straight grained, elastic timber which finishes easily to a smooth silky lustre. Available in quite large sizes, the timber is quite strong but not particularly durable — less than redwood. The grain is generally similar to that of redwood, but much paler as the timber contains far less resin. It gives off a sweet, faintly peppery smell when worked, and is generally quite free from knots. Takes nails, glue, screws, paint and varnish well.

Uses — general interior joinery, kitchen fitments, ladders and sports equipment.

Average density about 450 kg/m^3.

Western hemlock (Tsuga heterophylla)

Other names — hemlock, Canadian hemlock.

Distribution — Canada and USA.

Characteristics — a pale brownish timber with grain and annual rings similar to Douglas fir but less distinct. The wood has a fine even texture but is somewhat woolly, requiring sharp tools to produce a clean cut. Hemlock has a slightly musty smell when being worked and takes nails, screws, glue and paint well. Less strong and durable than Douglas fir, the timber has good resistance to mechanical abrasion. Available in large sizes.

Uses — interior joinery, stair treads, mass produced doors and interior cladding boards.

Average density about 480 kg/m^3.

Western red cedar (Thuja plicata)

Other names — British Columbian red cedar, western cedar, Pacific cedar.

Distribution — Canada and USA.

Characteristics — a light, soft, rather spongy timber with a woolly texture. Generally straight grained with reasonably distinct markings. The colour is pink to reddish-brown, darkening to grey when exposed to the weather. Western red cedar is non-resinous, requires sharp tools to obtain a good finish, and has a fragrant and characteristic smell when being cut. Although soft and light, the timber is exceptionally durable and does not generally require a protective finish, even when used outside. Contact with iron may cause staining and corrosion. The timber is not tough or strong enough for use in heavy, structural work.

Uses — external cladding boards, shingles, greenhouses, window frames and good quality timber buildings.

Average density about 380 kg/m^3.

Whitewood (Picea abies)

Other names — European spruce, white deal.

Distribution — Northern Europe.

Characteristics — similar in appearance to redwood but paler and lacks the resinous smell. The timber is strong and light, less easily worked than redwood and suffers occasionally from large dead knots. Its natural durability is less than that of redwood and it is generally considered to be somewhat inferior.

Uses — joists, rafters, floor boards, formwork, packing cases and centring. Clear timber is used for internal joinery as it takes paint, varnish, nails, glue and screws well.

Average density about 420 kg/m^3.

Hardwoods

Abura (Mitragyna ciliata)

Other names — bahia, subaha.

Distribution — West Africa.

Characteristics — an orange-brown to pink timber with a very even texture and straight grain. Grain and annual rings are not distinct although interlocked grain may occur on quarter sawn boards. The timber has a dulling effect on tools and has a musty smell when worked, the dust occasionally giving rise to nasal irritation. Generally, abura is a mild, easily worked timber which takes nails, screws, glue and polish well. Not durable when untreated.

Uses — good class joinery, shopfitting, furniture and stock mouldings.

Average density about 560 kg/m^3.

Afrormosia (Pericopsis elata)

Other names — kokrodua.

Distribution — West Africa.

Characteristics — a yellowish-brown timber which darkens to a rich dark to golden brown colour on exposure to air. The wood which bears a superficial resemblance to teak is quite variable in texture from fine to coarse and often exhibits a distinct stripe when quarter sawn. The grain, which is fairly distinct, is frequently interlocked, making for difficult working especially on stripy boards. The timber is tough and strong but great care is needed to prevent splintering and spelching (the tendency for the grain to break away from a cut where it is not firmly supported), especially when working on end grain. Afrormosia is too hard to nail unless holes are pre-bored, but takes screws, glue and polish well, and is very durable with good resistance to fire.

Uses – furniture, high class joinery, boat building and shopfitting. Often used as an edging or backing to teak veneered boards.

Average density about 690 kg/m³.

Ash, European (Fraxinus excelsior)

Other names – English ash, French ash, etc., according to country of origin.

Distribution – Europe.

Characteristics – a creamy white timber with occasional dark streaks (black heart). The figure is quite distinct owing to the large pores in the spring wood. Generally straight grained, but rather coarse. Ash is one of the toughest and most elastic of the European timbers but it has little resistance to fungal or insect attack. The wood takes nails, glues, screws and polish well.

Uses – furniture, veneers, shopfitting, boat-building, sports equipment and tool handles.

Average density about 690 kg/m³ but variable.

Beech, European (Fagus sylvatica)

Other names – English beech, French beech, etc., according to country of origin.

Distribution – Europe.

Characteristics – a cream to pale brown timber which tends to darken a little on exposure to the air. Central European beech is often steamed during seasoning, a process which turns the wood a pinkish colour. Beech is one of the toughest and strongest of the European timbers but has little resistance to fungi or to attack by furniture and death watch beetle. There is no really distinct grain, but the large medullary rays are easily visible with the naked eye on end grain, and show up as small "flowers" on rift sawn surfaces and as short dark flecks on tangentially cut boards. Beech is moderately easy to work with hand tools, although its natural toughness and hardness make hard work of heavy cuts. It is a very even grained timber, a property which makes it ideal for turning. Beech can be glued, stained and polished to good effect but pre-boring is necessary for nailing.

Uses – furniture, tools, kitchen utensils and occasionally for high class internal joinery and wood block floors.

Average density about 700 kg/m³ but very variable.

Iroko (Chlorophora excelsa)

Other names – mvule, odum.

Distribution – West Africa.

Characteristics – a yellowish-brown timber which darkens on exposure. The vessels are distinct, being somewhat lighter in colour than the general background. The grain is irregular and generally interlocked making for rather difficult working, especially on quarter sawn surfaces. The timber has a superficial resemblance to teak but is much coarser in texture. Occasionally stone deposits (calcium carbonate) are found in the timber – sometimes in sizeable lumps, but more often as a gritty deposit in the pores – and this has a severe blunting effect on tools and machines. Iroko is tough, strong and very durable, with good resistance to acids and fire but has a tendency to splinter. It can be glued and polished satisfactorily.

Uses – high class joinery, bench tops, laboratory fittings, shop-fitting and boat building.

Average density about 640 kg/m³.

Jelutong (Dyera costulata)

Distribution – South East Asia (Malaya, Brunei, Sarawak).

Characteristics – a soft, light, pale yellow to straw coloured timber with a straight grain and fine even texture. The annual rings are very indistinct, thus there is no pronounced grain or figure on the face of a board. The timber is non-durable, very easy to work and can be brought to a fine finish, but is generally too soft and perishable for use for exterior joinery. The tree grows to great size, thus the timber is available in large dimensions. Jelutong is characterised by latex canals – small slits up to 10 mm long by 1 mm wide – which occur on all faces, sometimes in sufficient quantity to spoil the otherwise clear face. The timber takes glue, nails, screws, paint and polish well.

Uses – pattern making, drawing boards,

handicraft, carving, curved and ornamental work.

Average density about 430 kg/m^3.

Mahogany, African (Khaya ivorensis)

Other names – Lagos wood, Benin wood, ogwango, khaya.

Distribution – West Africa (mainly).

Characteristics – a reddish brown timber with a rather grey sapwood, which may well have been attacked by pinhole borers. African mahogany has an attractive grain, often stripy when cut on the quarter. Interlocked grain on rift sawn faces can make cleaning up difficult but otherwise the wood is reasonably mild and easy to work, although sometimes there is a tendency to woolliness. African mahogany takes nails, screws and glue well and when carefully cleaned up takes an excellent polish.

Uses – high class joinery, furniture, shopfitting, boat building and plywood.

Average density about 700 kg/m^3.

Mahogany, American (Swietenia spp.)

Other names – British Honduras mahogany, Central American mahogany, baywood, Cuban or Spanish mahogany.

Distribution – British Honduras, Costa Rica, Guatemala, Nicaragua, Brazil, Peru, West Indies.

Characteristics – a superb timber which is variable in colour from yellowish-brown to a deep rich red. The Cuban or Spanish mahogany is usually deep in colour with reasonably distinct annual rings and often a very attractive curl grain. The timber is generally more even in texture, less coarse and rather easier to work than the African species.

Uses – as for African mahogany but generally considered to be superior in terms of working qualities, finish obtainable and appearance.

Average density about 540 kg/m^3 (Honduras); 770 kg/m^3 (Cuban).

Maple, rock (Acer saccharum)

Other names – hard maple, sugar maple.

Distribution – Canada, North East USA.

Characteristics – a creamy white to pale brown timber with a close, even texture and a curly grain, generally resembling sycamore to which it is closely related. Maple is a strong, tough timber which takes glue and polish well and can be worked moderately easily with sharp hand tools. It has excellent resistance to abrasion and mechanical wear but little to fungal attack. Occasionally, the wood is figured with a multitude of small round burrs, such timber being highly prized for use as veneer and referred to as "bird's eye maple".

Uses – mainly for hardwood strip flooring and veneers (figured varieties).

Average density about 720 kg/m^3.

Oak, American red (Quercus spp.)

Distribution – Canada and the USA.

Characteristics – There are two main species – the Northern red oak (Quercus rubra) and the Southern red oak (Quercus falcata). The American red oaks have a reddish pink tinge to the heartwood and much less pronounced silver grain than European oaks (due to their smaller rays). The timber is generally coarser in texture than European oak, and is considered to be less strong and durable, but in fact it is quite variable in this respect, depending on its rate of growth and grain. Fast grown oak is generally stronger than slow grown.

Uses – flooring, panelling, furniture and interior joinery. Generally considered inferior to white oak.

Average density about 770 kg/m^3.

Oak, European
(Quercus robur and Q. petraea)

Other names – English oak, Polish oak, Yugoslavian oak, according to country of origin.

Distribution – Europe including Great Britain.

Characteristics – of the several species of oak indigenous to Europe, the most obvious difference as far as the woodworker is concerned is probably the texture and toughness, English oak generally being acknowledged the hardest and toughest, whilst Polish or Slavonian oak is usually the mildest and most easily worked. The timber, one of the most useful and valuable of the European hardwoods, is a pale yellowish brown with a distinct and attractive darker figure. Quarter sawn or rift sawn boards show the medullary rays to maximum advantage, this being a feature characteristic of oak and referred to as "flower" or "silver grain". The working properties of oak vary considerably from relatively easy to fairly difficult depending on whether the timber is fast or slow grown. One of the toughest and most durable European timbers, seasoned oak has good resistance to fire and fungal attack. The paler sapwood is liable to be attacked by the lyctus beetle during air seasoning but the more durable heartwood is more or less immune. Oak contains a good deal of acid, and this tends to corrode iron in contact, especially if the timber has a high moisture content (brass screws are thus a better choice than steel), and also causes it to react badly against casein glues and other alkaline materials. The wood glues easily and takes an excellent polish.

Uses — high class joinery, panelling, door and window cills, fences, gates, shopfitting, furniture and veneers.

Average density about 670-720 kg/m^3.

Oak, Japanese (Quercus spp.)

Characteristics — Very similar in appearance to European oak but generally somewhat paler in colour and milder in texture. The timber has a good figure and distinctive silver grain when cut on the quarter, but often contains a fair proportion of sapwood. It is a fairly easily worked species, but not generally as tough and durable as European varieties.

Uses — high class joinery, shopfitting, furniture and veneers.

Average density about 660 kg/m^3.

Obeche (Triplochiton scleroxylon)

Other names — wawa, arere, Nigerian whitewood.

Distribution — West Africa.

Characteristics — a light, soft, yellowish-white timber. Obeche has a distinct interlocked grain especially when quarter sawn, but a fairly even texture, making for relatively easy working and enabling it to be brought to a fine silky lustrous finish. Since the wood is soft, sharp tools are needed to prevent tearing the grain, but it can easily be sanded to a smooth finish. Obeche has a characteristic sweet smell when being worked, takes nails, screws and glue well and can be polished successfully after careful filling of the rather open grain. The wood is relatively weak and has little natural durability or resistance to wear.

Uses — due to lack of strength and durability limited to plywood manufacture and interior fitments and cabinet work. It can be stained and polished to resemble mahogany.

Average density about 380 kg/m^3.

Ramin (Gonystylus spp.)

Other names — melawis.

Distribution — Malaya, Sarawak.

Characteristics — a pale yellow, straight grained timber with no obvious figure. The wood is moderately hard and tough, works easily with sharp hand tools, but has a tendency to split when nailed, especially in smaller sections. Ramin takes glue well, and can be stained and polished satisfactorily if the large pores are first filled. Resistance to insect attack and decay are poor.

Uses — mainly in small sections for mouldings, picture framing, architraves, door lippings, etc. Suitable for good class joinery (interior).

Average density about 660 kg/m^3.

Sapele (Entandrophragma cylindricum)

Other names — sapele mahogany, aboudikro.

Distribution — West Africa.

Characteristics — a medium to dark, reddish brown timber with a very pronounced stripe when quarter sawn. Logs occasionally yield a highly decorative blister or fiddle-back grain. A lightly scented timber, sapele is of medium density for a hardwood but interlocked grain often makes working very difficult. When carefully cleaned up, sapele can be polished to a superb and attractive finish.

Uses — furniture, veneers, high class joinery and shopfitting.

Average density about 630 kg/m^3.

Sycamore (Acer pseudoplatanus)

Other names – sycamore plane, great maple, plane (Scotland).

Distribution – mainly Europe, some from West Asia.

Characteristics – a creamy white to yellowish-white timber with a very even texture and natural lustre. The grain or figure is not distinct unless the wood is polished, when it stands out as fine pinkish-brown "contours" against the pale background. Occasionally an attractive fiddle-back figure may be met with. The timber works fairly easily with sharp tools, glues well and can be brought to a fine, smooth finish, when it takes an excellent polish. In common with most other pale, whitish timbers, sycamore has little resistance to fungal attack.

Uses – veneers, cabinet making, shopfitting, decorative work (inlaying), food containers and kitchen utensils.

Average density about 610 kg/m^3.

Teak (Tectona grandis)

Distribution – India, Java, Indo-China and Thailand.

It has been grown with some success in East and West Africa.

Characteristics – a golden brown timber, often with a distinctive and attractive dark figure. The timber darkens on exposure, generally being greenish-yellow when freshly opened. Teak is in many ways an outstanding timber with excellent natural durability and resistance to insects, acids and fire. The timber has a greasy texture and is gritty, this latter feature having a pronounced blunting effect on hand tools and machine cutters. When worked the timber has a pungent, leathery smell which makes identification easy, and the fine dust produced is known to be an irritant which may adversely affect some joiners. The timber is tough and strong with a fairly even texture but apart from its blunting aspect is quite mild working. Teak takes stain and polish satisfactorily and glues well with polyvinyl acetate and urea resin adhesives.

Uses – high class joinery, furniture, veneers, shopfitting, boat building, garden furniture, laboratory fittings or any situation requiring a strong, stable, durable and attractive timber.

Average density about 640 kg/m^3.

Utile (Entandrophragma utile)

Other names – utile mahogany, sipo.

Distribution – West Africa.

Characteristics – a rich, reddish-brown timber very closely resembling sapele (of which it is a near relative) but lacking the spicy scent of the latter and less frequently exhibiting a stripe when quarter sawn. The wood is generally more even in texture and easier to work than sapele but in most other respects the two timbers are very similar.

Uses – as for sapele. (The name "utile" literally means "useful".)

Average density about 550-750 kg/m^3.

Walnut, African (Lovoa trichilioides)

Other names – Nigerian walnut, Benin walnut, sida.

Distribution – West Africa.

Characteristics – not a true walnut, an attractive golden brown to bronze timber with a distinct stripe when quarter sawn and often with a dark figure which gives it a superficial resemblance to certain types of true walnut when polished, hence its common name. Lovoa is an even, mild textured timber which works easily with sharp tools, takes nails, glue and screws well, and can be brought to an excellent finish for polishing. Moderately durable.

Uses – high class joinery, shopfitting, furniture and veneers.

Average density about 550 kg/m^3.

Walnut, European (Juglans regia)

Other names – English walnut, Italian walnut, etc.

Distribution – Great Britain, France, Italy, Turkey and Yugoslavia.

Characteristics – a purplish-brown timber with a very attractive dark figure, sometimes with a curl or burr. The working properties of walnut vary considerably from easy to very difficult depending on the straightness or otherwise of the grain. The wood can be glued, nailed and screwed fairly easily and takes an excellent polish, the grain normally being first filled. Walnut is a valuable timber, much sought after by cabinet makers and veneer manufacturers, and has quite good natural resistance to fungal decay although it is rarely used outdoors.

Uses — furniture, veneers, very high class joinery, decorative work and small turnery items.

Average density about 640 kg/m^3.

CONVERSION OF TIMBER

The breaking down or sawing of a log into boards or planks for use by the carpenter and joiner and other woodworking trades is known as conversion, and whilst this is a specialist job and not part of the craftsman's work (years ago it used to be), the way in which this conversion is carried out can have a direct effect upon the subsequent behaviour and usefulness of the timber. It is therefore important for the craftsman to have a basic understanding of the methods employed in the conversion process in order to get the best possible use from the timber.

Sawmill

The location of the sawmill is dependent to some extent upon whether the timber is a hardwood or a softwood. Softwoods are almost invariably sawn into marketable sizes in their country of origin, with the exception of certain heavier size timbers which are resawn after importation.

Hardwoods on the other hand are often imported in log or billet form and converted by the timber merchant, sometimes to the customers' requirements.

Sawing machines

Various types of sawing machines are used to cut the logs into the required sections, but in the main two types are currently in use illustrated in diagrammatic form in Fig. 2.8.

Fig. 2.8. *Cutting action of saws.*

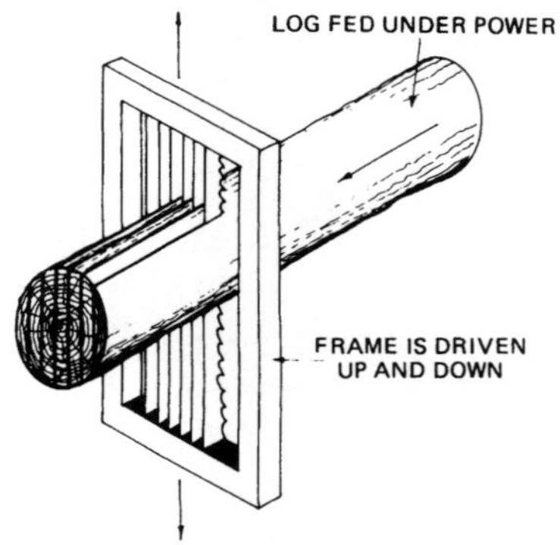

(a) Frame saw.

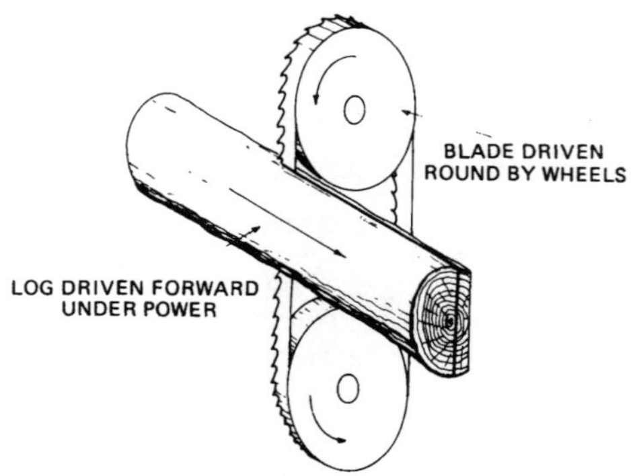

(b) Band mill saw.

Frame saw

This is a rather old fashioned and outdated type of machine but it is still met with on occasions and performs satisfactorily. The machine comprises of a number of saw blades held in tension in a very heavy vertically reciprocating frame. As the log moves forward under power, it passes through the saw blades which cut it into boards. The distance between the saw blades is adjustable to enable varying thicknesses of board to be cut. Figure 2.8*(a)* shows in diagrammatic form a log being cut into boards on the frame saw.

Band mill saw

This machine has largely superseded the frame saw and embodies a saw blade in the form of a wide endless band running on two large wheels which hold the blade in tension and drive it round in a continuous cutting action (*see* Fig. 2.8*(b)*). The log is fixed to a carriage which moves forward on rails and is power fed past the saw blade to cut a board. The carriage then moves back to its starting point when it is again run forward for a second cut and so on. The sawyer can stop the machine at any time and also vary the thickness of the board at the commencement of each cut if he so desires. It is thus possible to cut any log to the best advantage.

The band mill saw is a remarkably efficient machine with a very fast cutting action, and has thus gained in popularity over the years until, in one or other of its many forms, it has become the main instrument of log conversion. In recent years, advances in electronic technology have brought about the introduction of band sawing machines which enable the sawyer to control the cutting by the touch of a button. Some of the latest machines can be programmed to perform automatically the cutting process to the best advantage irrespective of the size or shape of the log.

Methods of conversion

Through and through
Figure 2.9*(a)* shows a log which has been sawn *through and through*. This is the most economical way in which to convert timber and gives the least possible waste. However, as will be seen later, this method of sawing does not always give the best results as regards the usefulness of the boards. Through and through sawing is known alternatively as "flat", "slab", or "slash" sawing.

Fig. 2.9. *Methods of conversion.*

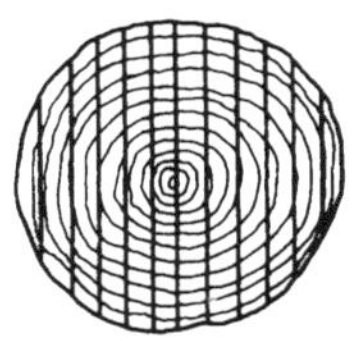

(a) Through and through sawing.

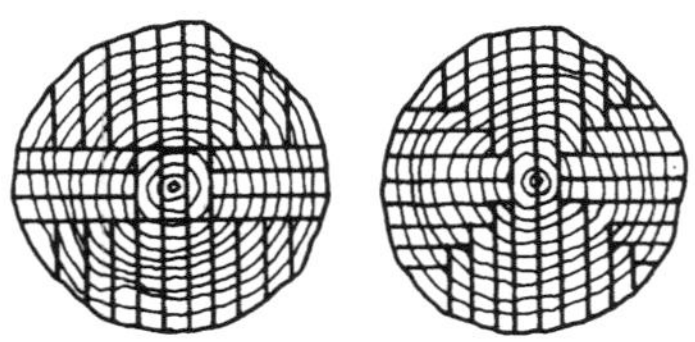

(b) Quarter sawing (2 methods).

Quarter sawing
Figure 2.9*(b)* shows logs which have been converted by quarter sawing. This method of cutting is less economical than the through and through method, both in material wastage and in time involved. It does however, produce the greatest quantity of "rift" or "radial sawn" boards, which are generally superior for joinery.

SEASONING OF TIMBER

The timber from newly felled trees is of no immediate use to the carpenter and joiner, or indeed to any woodworker requiring a workable, stable material.

At this stage, the wood is "green" or wet, since it contains a high proportion of water in the form of sap. Most of this water has to be removed from the wood by some form of drying process before it becomes useful to the craftsman, and this drying of the timber is known as "seasoning".

Various ways of seasoning or drying timber are in use today but basically they can be broken down into two main methods, namely "air drying", sometimes referred to as natural seasoning, and "kiln drying", in which mechanical means are adopted to dry the timber more rapidly than is otherwise possible.

NOTE: The term "natural seasoning" is something of a misnomer since the timber does not season naturally – rather it would decay.

Air drying
For air drying the timber is stacked in a pile in such a way as to allow a circulation of air all around each piece, as shown in Fig. 2.10. The stack should be made on level bearers the tops of which are at least 300 mm above a clear base free from plant growth. The boards are stacked lengthwise on the bearers with lateral spaces between each board, and each layer of boards is separated from that above and below by means of strips of wood or "stickers" about 25 mm thick. A point of importance here is to ensure that the stickers are placed vertically above each other in line with the bearers in order to avoid bending stress on the stacked timber. Obviously, the stickers in each layer must be the same thickness or bending is bound to occur. It is possible to some extent to vary the speed at which the timber dries by means of the thickness

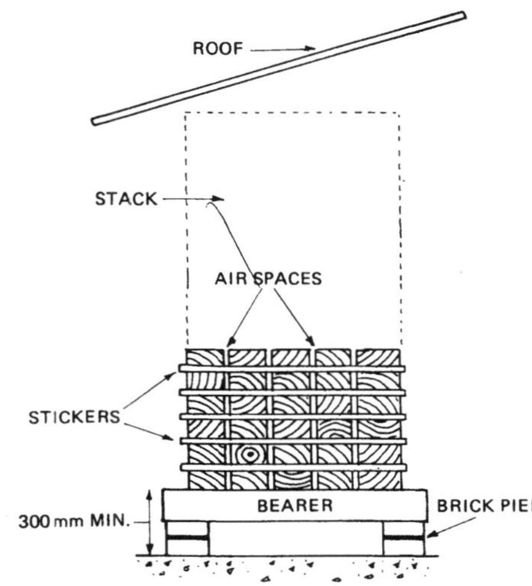

Fig. 2.10. *Air drying.*

of the stickers used, thick ones, of say, 38 mm giving greater contact with the air and thus quicker drying than thinner ones. Further to the use of stickers, it should be mentioned that these are best made of softwood, such as European redwood, since hardwood used for this purpose may stain the surface of the boards where it is in contact.

Stacking the timber

In building a stack of timber for air drying common sense dictates that the longest boards be laid first so as to avoid overhanging ends which would inevitably bend and take on a cast during drying. It is also sensible to keep one end of the stack level, so that the length of the boards can be easily determined by inspection.

Height of the stack

Generally speaking, the height of a stack of seasoning timber is governed simply by the ease with which it can be built, up to two metres being common. Above this, the difficulties of stacking the timber by hand increase proportionately.

Width of stack

No hard and fast rules apply here, much depending upon the ground area available, but the stack should always be sufficiently wide to remain stable in high winds, and not to imperil workmen in the vicinity.

Protection from the elements

It cannot be too strongly emphasised that where air drying is concerned it is the *air* circulating around the boards which causes them to dry, and it is in no way beneficial to allow strong sun, driving rain or snow to interfere with this process. For this reason, some form of roof is required in order to protect the timber. Whilst the roof may be very basic from a constructional point of view, it should be substantial and effective enough to do its job and remain stable during the rather extended drying period required.

Drying time

It is almost impossible to state precisely how long it will take to dry timber by air seasoning as there are too many variable and uncontrollable factors involved.

In the first place, it must be realised that air dried timber can never be any drier than the air which dries it. This varies considerably from season to season and even from day to day, but on the whole, air dried timber is considered "dry" when its moisture content has been reduced to about 20 per cent (*see* Table I, p. 37). Secondly, the type of wood in stack has a profound effect on the drying out period, hard dense timbers requiring up to twice the time necessary for soft porous ones. Finally there is the thickness of the planks themselves to take into account. Obviously, it will take longer for a plank 75 mm thick to lose its moisture than it will for one of 19 mm thickness.

To sum up, it can be generally taken as a rule of thumb that one year is necessary for each 25 mm of board thickness for drying to about 20 per cent moisture content. Thus, although air drying has much to recommend it in terms of simplicity and quality of the wood texture, it does mean that a tremendous amount of capital must necessarily be tied up for a year or more in the timber yard before the asset can be realised, i.e. before the timber can be used. Furthermore, as will be seen later, a moisture content of 20 per cent is still far too high for many items of woodwork.

Protecting the ends of the boards

During drying, especially air drying, where there is no control over temperature and humidity, the ends of boards in stack tend to dry more quickly than do the middle portions. This can cause end splits or checks to occur, due to the differential shrinkage. To avoid this, the board ends are often protected from too rapid drying by means of hoop iron strips, timber strips (as shown in Fig. 2.11) or simply by painting.

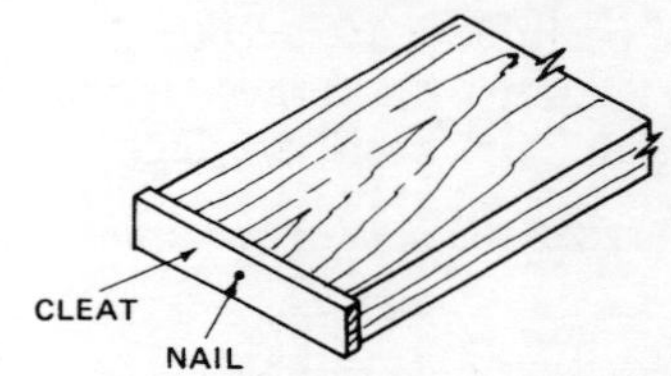

Fig. 2.11. *Protection of end of board.*

Kiln seasoning

Kiln drying of timber is a mechanical means of reducing the moisture content of wood to a level at which it will be suitable for use.

Basically, a kiln for drying timber is a large container into which the timber is placed or into which it is run on rails. Facilities exist inside the kiln to enable a forced draught to pass through the stack to carry away excess moisture, the temperature and humidity of the air being variable by the operator in order to dry the timber quickly, efficiently and safely.

Two types of kiln are in common use – the compartment kiln and the progressive kiln. In either case, the drying of the timber is similar in principle but differs somewhat in the mechanical methods employed.

The compartment kiln

With this type of kiln (*see* Fig. 2.12), the timber is stacked in much the same way as for air drying, but on a trolley mounted on rails so that the stack of timber can be pushed into the kiln which is then closed.

Inside the kiln, warm air is circulated by means of fans whilst the humidity is controlled by steam or water vapour jets, in order to prevent the timber drying too rapidly. The water removed from the timber is condensed outside the kiln so that the air inside does not become saturated to the point at which drying would be halted or slowed down to any great extent.

By this means timber can be dried out to any required moisture content, tests being made periodically to test the progress of the drying process, and this in a fraction of the time which air drying would take.

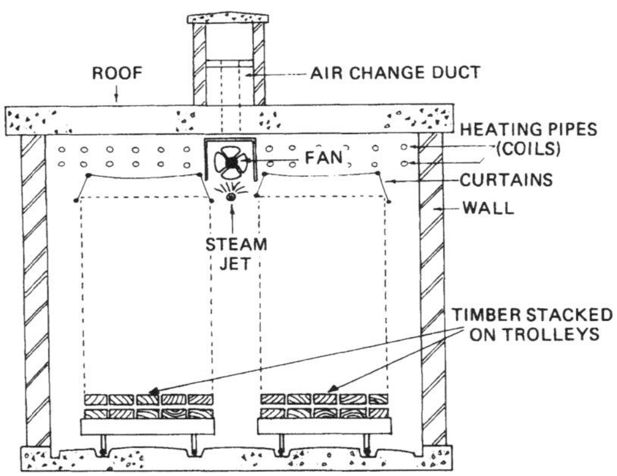

Fig. 2.12. *Compartment kiln.*

It must not be assumed, however, that kiln drying of timber is quite as simple a technique as the foregoing may lead one to believe. In fact, rapid as the process is, it must be most carefully controlled in order that the quality of the timber is not impaired. This can easily happen if the timber is dried too rapidly causing differential rates of moisture loss, i.e. if the outside were to dry leaving moisture in the centre (*see* defects due to seasoning on p. 39), or if the moisture is removed from the cells more rapidly than air can replace it.

The various species of timber previously listed, all have their own individual characteristics as regards behaviour during drying and this must be taken into account during seasoning.

Generally, kiln seasoning commences with the air being warm and humid, and finishes with the air hot and relatively dry.

Progressive kilns

In the progressive type of kiln the timber, stacked as before on railed trolleys, passes slowly and gradually through a long chamber, the change from warm and humid to hot and dry occurring in gentle stages as the timber moves from one end of the kiln to the other. It will be appreciated that whilst the timber is dried by a process basically similar to that of the compartment kiln, the more or less continuous seasoning cycle of the progressive kiln makes for greater economy, as regards the quantity of timber handled in a given time.

Advantages of kiln seasoning

The main advantages derived from kiln as opposed to air drying are threefold:

(*a*) the speed at which the seasoning is carried out;

(*b*) the facility to dry timber to any desired moisture content;

(*c*) the sterilising effect of the heated air upon fungi and insects in the timber, particularly where high proportions of sapwood are involved.

Moisture content of wood

The sap or water which is present in newly felled timber may be regarded as taking two distinctly separate forms. Firstly there is the "free water" held within the cell cavities, much like the water held in a fully saturated sponge.

Secondly there is the water which is held within the cell walls. This will remain for some time after the free water has gone, in the same way that the sponge will remain damp after it has been wrung out. Figure 2.13 shows this loss of water in diagrammatic form. During seasoning, the free water in the cells dries out relatively quickly. At this

Fig. 2.13. *Loss of water from wood cells.*

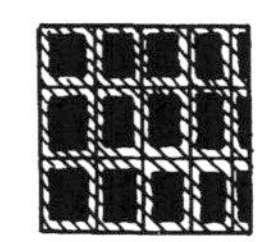

(a) Wood saturated.

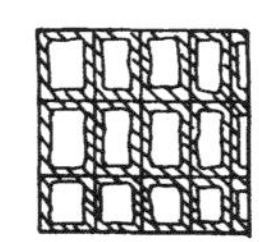

(b) Free water gone. Cell walls still saturated.

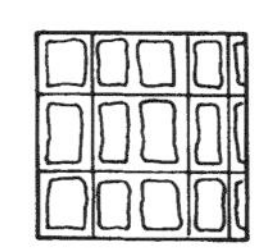

(c) Cell walls dry.

stage the timber has undergone a loss in weight, due to the removal of this water, but basically no other change has occurred.

After the free water has gone, the cell walls themselves begin to dry out — this part of the drying process taking much longer — and as this occurs, the timber not only continues to lose weight, but also commences to shrink as the mass of material forming the wood substance starts to harden and contract.

Distortion due to moisture content

When a piece of timber shrinks, due to loss of moisture, it may be thought that it will simply become smaller more or less evenly in all directions. Such, however, is not the case, and herein lies one of the most important aspects of timber behaviour with which the craftsman must contend. In stressing the importance of this area of timber technology, it would be true to say that without a sound basic knowledge of the effects of shrinkage (or swelling) on various sections of timber, it is unlikely that a really satisfactory piece of joinery will ever be produced, other than by pure chance.

Figure 2.14 shows, in diagrammatic form, a log of timber in which the natural arrangement of the cells is marked by arrows *(a)*, *(b)* and *(c)*. The arrows indicate the principal directions in which the cells are joined together to form the wood, i.e. longitudinally, tangentially and radially, respectively.

Shrinkage which takes place as timber dries varies considerably from one species to another, but on the average it is found that longitudinally it amounts to about 0.1 per cent, i.e. about 1/1000 of its length (in direction *(a)*). To all intents and purposes, therefore, longitudinal shrinkage may be ignored as it is so small.

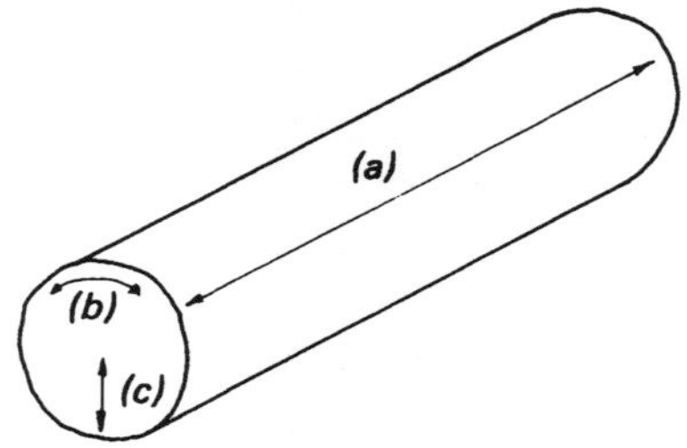

Fig. 2.14. *Disposition of cells in a log.*

Radially (direction *(c)*), timber shrinks to a considerably greater extent, up to about 5 per cent, whilst tangentially (direction *(b)*), shrinkage is greater still, up to as much as 15 per cent in some instances. Thus it will be appreciated that since this movement due to loss of moisture is by no means even, pieces of timber will distort through shrinkage in different ways, depending upon the principal direction in which the cells lie in any particular plank or section.

Figure 2.15*(a)* shows a section of a log which has been cut into planks "through and through" and also the same boards after drying (and shrinkage) has taken place. It should be noted that the least distortion has occurred in the planks nearest the centre of the log.

Figure 2.15*(b)* shows a radially or "rift" sawn board together with the same board after drying. Note that the distortion is minimal being mainly a slight loss of thickness.

Figure 2.15*(c)* shows a flat or slash sawn board before and after drying. Note that in this instance the distortion is quite marked, being apparent in both width and shape. The board has warped badly as well as having lost width. This would have serious consequences if the board had been used in a piece of joinery.

Fig. 2.15. *Distortion due to loss of moisture.*

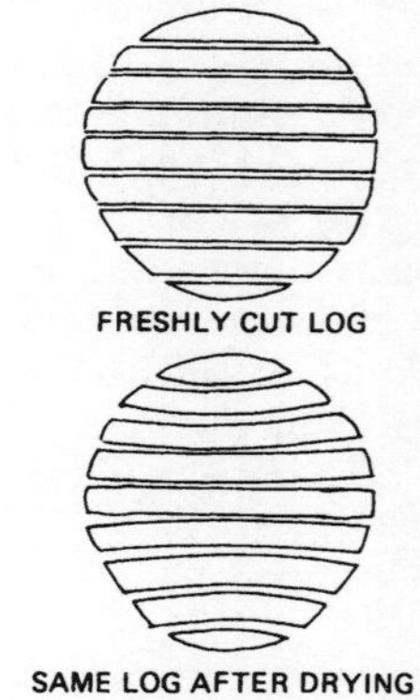

(a) Log sawn through and through.

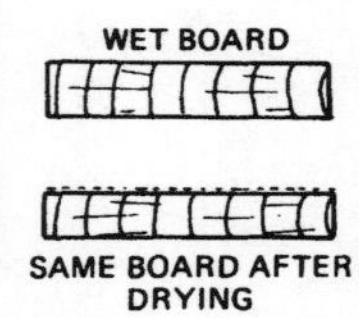

(b) Rift sawn board.

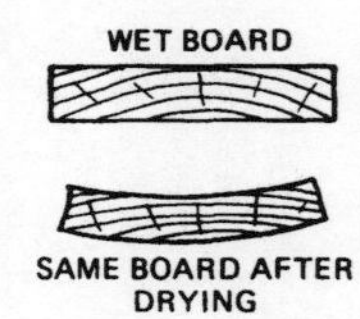

(c) Flat sawn board.

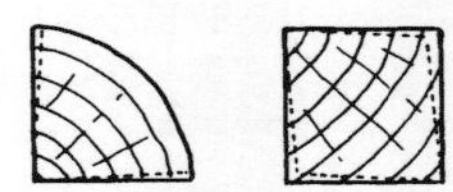

(d) Further distortion due to shrinkage.

Figure 2.15*(d)* shows two further instances of distortion due to loss of moisture.

It should be clearly understood that in most instances, rift sawn boards are the more stable and least likely to give trouble in service. It is generally accepted that a board may be termed "rift" sawn when the annual rings form an angle of 45° or more to the face, as shown in Fig. 2.16.

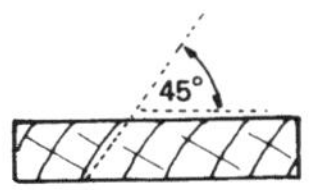

Fig. 2.16. *Rift sawn board.*

Furthermore, it must be accepted that some distortion is bound to occur when wet timber dries out; it is the inevitable result of shrinkage in the sawn board. Moreover, the woodworker should realise that a dry timber which absorbs moisture will swell and in so doing it will almost certainly distort, this time in the direction opposite to that due to shrinkage.

Moisture content

It should, by this time, be obvious that the basic raw material of the carpenter and joiner – namely timber – is by no means the stable material which may have been imagined. Wood is in fact a *hygroscopic* material, which is to say that it will lose or absorb moisture (and thus shrink or swell) if it be exposed to air that is relatively dry or wet.

Fortunately there is a good deal that can be done to prevent, or at least to minimise, the effects of moisture movement in timber.

Equilibrium moisture content

Since, as we have seen, a timber with a relatively high moisture content placed in a very dry position will shrink, and conversely a dry timber in a relatively damp situation will swell, it would seem logical to ensure that the moisture content of any article of woodwork be compatible with the relative humidity of the air in the situation in which it is to be in service. This balance between the moisture content of the timber and the dryness or otherwise of its location in a building is a most important factor in preventing moisture movement in the wood. It is known as "equilibrium moisture content" – the point in fact at which further moisture movement is least likely to occur.

Table I below gives suitable moisture contents for timbers in various situations.

TABLE I. MOISTURE CONTENT

Specific use of timber	*Suggested moisture content (%*)*
Carcassing timbers – rafters, joists, etc.	20-22
External joinery	16-18
Interior joinery – intermittently heated	13-15
Interior joinery – continuously heated; Panelling and furniture	10-12
Joinery near to heat sources	8-9

**See below.*

Calculation of moisture content

The moisture content of a piece of timber is expressed as a percentage. This figure represents the weight of water present in the timber as a percentage of the completely dry weight of the wood. The following formula will be found satisfactory in calculating the percentage moisture content of a sample of timber:

$$\frac{\text{Wet weight} - \text{dry weight}}{\text{Dry weight}} \times 100$$

where the wet weight is the initial weight of the sample of timber being tested and the dry weight is the weight of the same sample after drying out completely, or at least to a point at which no further loss in weight occurs.

The formula can thus also be expressed as:

$$\text{Percentage moisture content} = \frac{\text{Loss in weight}}{\text{Dry weight}} \times 100$$

EXAMPLE

Weight of timber sample = 45 gm
Weight after completely drying = 36 gm
∴ Loss in weight = 45 – 36 = 9 gm
(i.e. the sample contained 9 gm of water).

Using the formula:

$$\frac{\text{Loss in weight}}{\text{Dry weight}} \times 100$$

we get $\frac{9}{36} \times 100 = \frac{900}{36} = 25\%$

∴ The moisture content of the sample is 25 per cent.

Kiln drying timber to a specific moisture content

Let us assume that a quantity of green timber is to be kiln dried to a moisture content of 12 per cent. A sample of the timber would first be weighed and tested for moisture content in order to determine roughly how much water has to be removed. From records of previous seasoning the operator will now be able to estimate fairly accurately the drying schedule required for the timber in question. Having loaded the timber into the kiln, the seasoning cycle can commence, as described previously, the operator keeping a careful check on the movement, temperature and humidity of the air inside the kiln.

During the drying process samples are taken from time to time to establish the rate of drying and behaviour of the timber until the required moisture content is obtained.

Samples for testing moisture content

The samples which are taken to determine the moisture content of the timber in the kiln should be taken from a board near the centre of the stack. Figure 2.17*(a)* shows how the stickers can be cut to enable the board to be withdrawn whilst Fig. 2.17*(b)* shows how the sample is taken from the board. Obviously, since the ends or outside of the boards tend to dry in advance of the centre, this avoids a false reading being obtained.

Fig. 2.17. *Taking a sample to test for moisture content.*

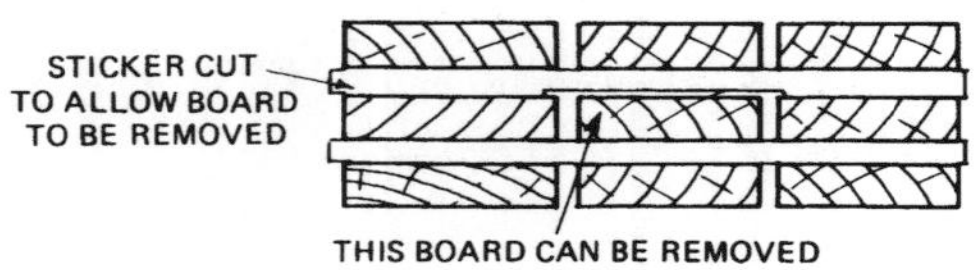

(a) Provision for board removal.

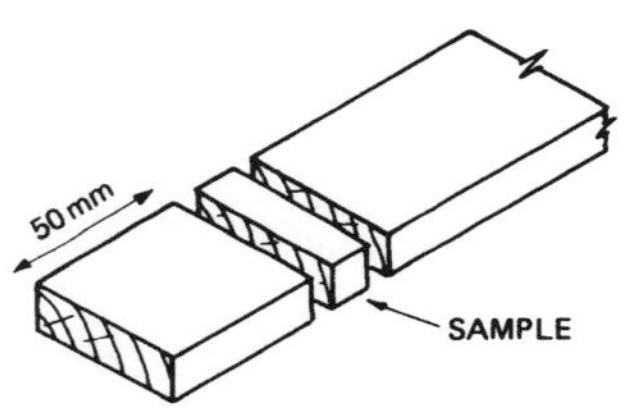

(b) Taking the sample.

Drying time

As with air drying the time required to kiln dry the timber depends largely upon the following factors:

(a) species of timber;
(b) thickness of timber;
(c) desired moisture content.

Since these are all variable factors, no hard and fast drying period can apply, but average figures for drying to about 14 to 16 per cent moisture content would be in the region of six days per 25 mm thickness of softwood (up to 75 mm thick), and about twice this for hardwoods.

Moisture meter

Figure 2.18 shows a moisture meter, an instrument which is used to determine on the spot the approximate moisture content of a piece of timber. The meter works by measuring the flow of electric current between two pointed contact pins which are pressed into the wood. The moisture content is registered on a calibrated scale by means of an arrow or by a series of light emitting diodes.

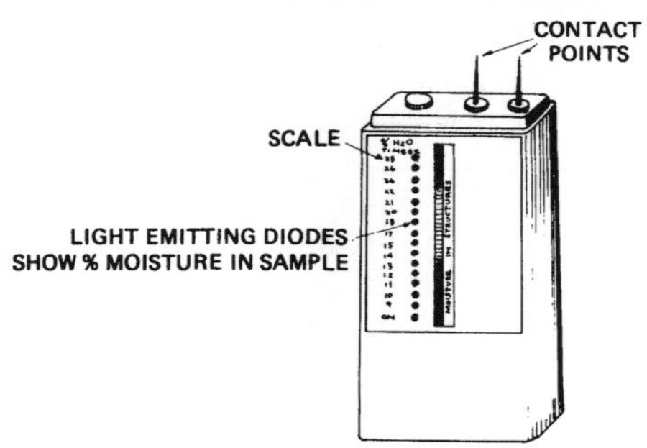

Fig. 2.18. *Moisture meter.*

TIMBER DEFECTS

Defects are faults which are found in timber, and are due to natural or other causes. They may originate in the growing tree, in the felled log, during conversion or during seasoning. Some defects present a serious structural weakness in the timber; others do little more than spoil its appearance or cause a certain amount of inconvenience.

Since most of these defects occur before the woodworker comes into possession of the timber, there is little he can do to prevent them, but he can at least recognise them for what they are, decide on their seriousness or otherwise as regards the job in hand, and possibly even reject the timber as unsuitable for its purpose because of the defect.

Star shakes

As shown in Fig. 2.19*(a)* these are radial cracks which occur around the periphery of a log. Star shakes follow the line of the medullary rays and are widest at the outside, tapering towards the centre. They are caused by shrinkage at the outside of the log whilst the middle remains more or less stable – thus the circumference decreases whilst the diameter does not. This type of shake is usually the result of leaving the log for too long before conversion.

Heart shakes
These are generally the result of disease or over-maturity of the growing tree. As shown in Fig. 2.19*(b)*, heart shakes radiate from the centre of the log following the line of the medullary rays and are caused by internal shrinkage of the log.

nutriment or to excessive bending or twisting of the tree in high winds. Bad cup shakes make economic conversion of the log very difficult.

End splits
End splits or end shakes are shown in Fig. 2.19*(d)* and are caused by too rapid drying of the board.

concave side, these rings being longer than on the convex side.

Where timbers of 50 mm or more in thickness are concerned, the reluctance of the section to move evenly often causes a crack or shake to develop on the convex side, as shown in Fig. 2.19*(f)*. Warping may sometimes occur in reverse as the result of a relatively dry board absorbing moisture as shown in Fig. 2.19*(g)*.

Fig. 2.19. *Timber defects.*

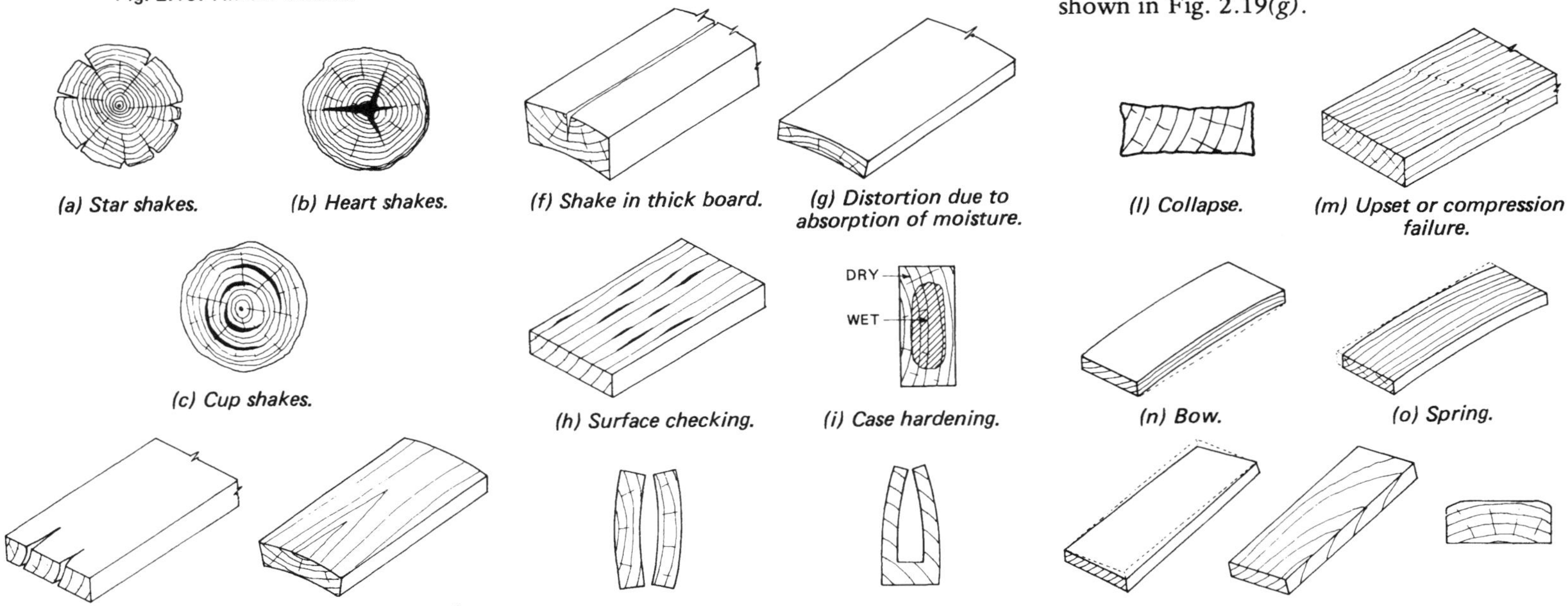

(a) Star shakes. *(b) Heart shakes.* *(c) Cup shakes.* *(d) End splits.* *(e) Warping or cupping.* *(f) Shake in thick board.* *(g) Distortion due to absorption of moisture.* *(h) Surface checking.* *(i) Case hardening.* *(j) Re-sawn boards.* *(k) Test for case hardening.* *(l) Collapse.* *(m) Upset or compression failure.* *(n) Bow.* *(o) Spring.* *(p) Twist.* *(q) Short grain.* *(r) Waney edge.*

Cup shakes
Cup shakes or ring shakes as they are sometimes known are shown in Fig. 2.19*(c)*. They take the form of separations between the annual rings and extend longitudinally down the grain of the log. They originate in the growing tree and are generally due to lack of

Warping or cupping
This defect, shown in Fig. 2.19*(e)*, is very common in flat or slash sawn boards. It occurs through shrinkage of the timber when drying, as the contraction of the cells along the line of the annual rings is greater on the

Surface checking
These are small longitudinal shakes which occur on the faces and ends of boards. They are the result of severe frost or sun on air dried timbers and, occasionally, of too rapid kiln seasoning. Surface checks, shown in

Fig. 2.19*(h)*, are not serious defects structurally, but they spoil the appearance of the wood. Some timbers are more prone to this trouble than others.

Case hardening

This defect is mainly confined to kiln dried timber of relatively heavy section, i.e. timbers 50 mm or more in thickness. It is caused by too rapid drying in the early stages of seasoning, the result being a board in which the outside cells have dried and hardened, sealing off the moisture in the central, more remote portion (*see* Fig. 2.19*(i)*). Thus there is a tendency towards differential movement setting up stresses within the board. When such a board is re-sawn as shown in Fig. 2.19*(j)* the stresses are released, causing the two newly exposed surfaces to become concave. In certain cases, this movement takes place whilst re-sawing is in progress. In others it will occur more slowly as the wetter sides become exposed to the air and dry out.

Case hardening is often present to a greater or lesser degree in most kiln dried timbers in excess of 50 mm thickness. A test for case hardening is shown in Fig. 2.19*(k)* where a short length of timber is sawn as shown and the inward movement noted after a few hours (sometimes a few minutes).

Collapse

This defect, fortunately a rare one, is shown in Fig. 2.19*(l)*. Once again it is a defect confined to kiln dried timber, being due to excessively high temperature and too rapid drying in the early stages. This results in moisture being withdrawn from the centre of the board at a rate quicker than that at which air can replace it, thus dehydrated cells collapse. Cell collapse can sometimes be rectified by re-seasoning in the kiln.

Upsets

This defect, shown in Fig. 2.19*(m)*, is sometimes referred to as "thunder shake", although there is little or no evidence to connect the fault with electric storms. The defect is almost always caused by bad felling (across a fallen log) or through the use of explosives to clear a jam when floating logs down river. It occurs mostly in tropical hardwoods, especially mahogany, and can be difficult to detect before planing the timber.

The defect, which is extremely serious, is in the form of a transverse hair line fracture and wood affected can be snapped off with little effort.

Bowing

This defect, shown in Fig. 2.19*(n)*, is quite serious as it may cause an otherwise good board to be suitable only for use in short lengths. The defect may be due to one of several reasons such as poor stacking of boards, the axis of the board not being parallel to the grain, or to internal stresses set up in the growing tree being relieved after conversion.

Springing

This is an edgeways curvature of a board as shown in Fig. 2.19*(o)*. It is nearly always due to the release of internal stresses during seasoning. Small amounts of springing can often be used to advantage in construction work to counteract a natural tendency to sag.

Twist

This defect, shown in Fig. 2.19*(p)*, can be very serious as it may restrict the use of the timber to short lengths, and even then it may be necessary to substantially reduce its thickness to obtain a flat surface. It is caused by poor seasoning, poor stacking or by the release of internal stresses during conversion and seasoning.

Short grain

This term applies to timbers in which the general run of the grain is not parallel to the longitudinal axis of the wood (*see* Fig. 2.19 *(q)*). It may be the result of poor conversion, but more often it is due to "wandering heart" in the growing tree. Obviously unless the trunk of the tree is exactly straight, some degree of short grain is inevitable. As the sketch shows, short grained wood has a tendency to break easily, so it is a defect which should be avoided in any load bearing member.

Knots

These are natural defects which occur mainly in softwoods such as European redwood and whitewood. A knot marks the origin of a branch in the growing tree, thus the axis of a knot is always radial. The various types of knot shown in Fig. 2.20*(a)* are due to the disposition of the plank in the original tree section.

Knots may be termed "dead" or "live" the distinction being due to the condition of the branch which caused it.

Small, sound, live knots are no real detriment to the timber, unless it be highly stressed. Dead knots, which are often loose, and large knots, whether live or dead, present a real structural weakness.

Wherever possible, the craftsman should

arrange his timbers in such a way that knots in structural members are in compression (*see* Fig. 2.20(*b*)). As shown, some forethought as to the position of a knot in the member can greatly reduce its weakening effect. Often a dead knot in a piece of joinery may have to be cut out and a "little joiner" inserted (*see* Fig. 2.20(*c*)).

Fig. 2.20. *Knots.*

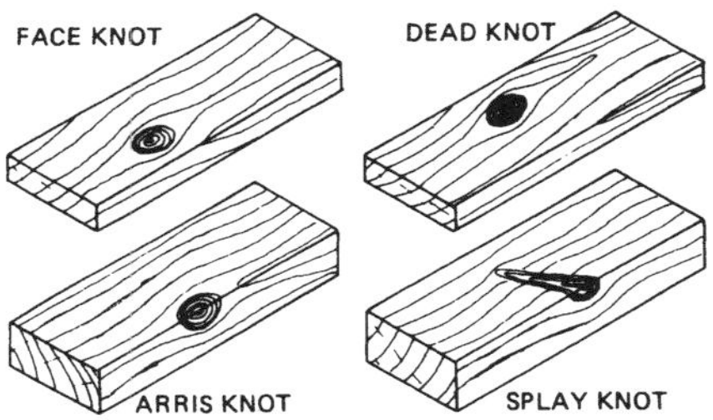

(a) Types of knot.

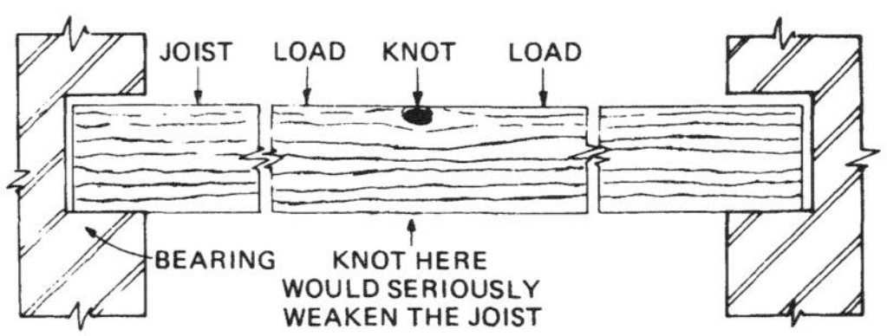

WHEN THE KNOT IS AT THE TOP OF THE JOIST AS SHOWN, THE WEAKENING EFFECT IS QUITE SMALL, AS THE KNOT IS IN COMPRESSION. HOWEVER IF THE JOIST WERE TO BE PLACED IN POSITION WITH THE KNOT AT THE LOWER SIDE IT WOULD HAVE A CONSIDERABLE WEAKENING EFFECT ON THE JOIST. THE WEAKENING EFFECT OF A KNOT IS *LESS* WHEN IT IS NEAR THE *END* OF THE JOIST.

(b) Sketch showing effect of a knot in compression.

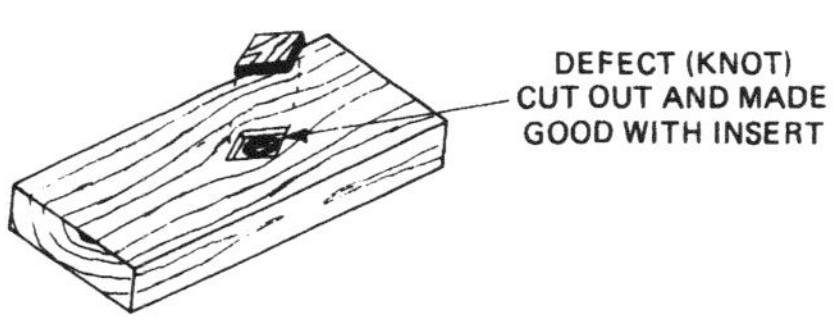

(c) Little joiner.

Waney edge

This defect (*see* Fig. 2.19(*r*)) is the result of too economical conversion of the log. Waney edges present a weakness in structural timber on account of part of the wood being absent. Moreover, the timber will contain a high proportion of sapwood, adjacent to the wane, which in itself is undesirable.

Blue stain

This is a blue-grey discolouration of the timber due to the presence of certain fungi. The fungus attacks many of the lighter coloured timbers, especially the softwoods such as European redwood and European whitewood, and lives on the food stored within the living cells. The defect is therefore largely confined to the sapwood, the heartwood only being affected marginally.

Since the fungus lives on the contents and not the cell walls there is no structural weakening of the timber so the defect is not in itself a serious one. However, blue stain, or sap stain as it is sometimes called, does indicate a high proportion of sapwood in the timber, which is always undesirable.

TIMBER DECAY

Timber, like all other living matter, is an organic material and is therefore prone to be attacked and possibly destroyed by other living organisms which in nature break down and clear the dead vegetation – including trees which would otherwise lie for ever in the forests where they fall.

These organisms, therefore, perform an important task in the natural ecology of trees, but obviously they cannot differentiate between the dead tree in the forest and the timber which we use for constructional purposes. The woodworker, therefore, must make every effort to ensure that his work is not spoilt or destroyed in this way.

In the main, constructional timbers which have crumbled or decayed in service have been attacked by either insects or fungi – or both.

FUNGAL ATTACK

There are various forms of fungi which, given suitable conditions, will attack timber until it is destroyed. These can broadly be classified into two groups:

(*a*) the wet rots;
(*b*) the dry rots.

Both types may present a serious hazard to constructional timbers, especially if the timber has a high moisture content.

Dry rot (Serpula (Merulius) lacrymans)

This is a fungus which is prone to attack damp, unventilated timber which has moisture content in excess of about 20 per cent. The fungus does most damage to softwoods but will also attack suitably damp hardwoods, especially if they are in close proximity to a softwood which has already been infected.

Dry rot is a very serious timber disease since, if undetected or untreated, it may well lead to the ultimate destruction of much of the timber in a building. Once dry rot is established, it can carry its own water supply through thick strands of fine threads known as *hyphae* in order to dampen dry timber upon which it will then feed. It can also reverse this process to dry out wet timber partially to a moisture content which is suitable for its purpose. The strands of hyphae are known to pass through plaster, brickwork and even concrete to reach the timber upon which they feed. Under good growth conditions dry rot can produce a fruiting body (sporophore) within about twelve months, and this then releases millions of tiny orange-red spores, any one of which settling in a suitable environment within the building may start up a fresh growth.

Symptoms and characteristics of dry rot

The following are the most obvious signs of an outbreak of dry rot within a building:

(a) an unpleasant, musty smell;

(b) distortion of affected wood – bulging, buckling of skirtings or floorboards, etc;

(c) softening and crumbling of the wood – test by probing with a screwdriver or penknife blade;

(d) presence of whitish grey strands of hyphae – this may sometimes be very prominent taking the form of a cotton wool-like mass (Mycelium) on the surface of the timber. On occasions, this sheet-like growth may be tinged with yellow or lilac;

(e) presence of fine, orange-red spore dust on the floorboards and other parts of the structure within the building;

(f) presence of a sporophore within the building – this is a sign of advanced decay and takes the form of a large fleshy growth 300 mm or more in diameter which is usually attached to the woodwork or plasterwork adjacent to the worst of the decay. The sporophores (often more than one) are the same colour as the mycelium sheets but with a reddish centre where spores have been released;

(g) cuboidal cracking of the wood – timber badly decayed by dry rot has deep cracks, both along and across the grain, giving it the appearance of badly charred wood. Quite frequently, the most significant symptoms of the disease (stranding, crumbling and cracking, etc.) are to be found on the underside of floorboards, the back of skirtings and architraves, etc., whilst the visible faces of these may show little sign of decay. Any investigation must always be very thorough, requiring removal and examination of suspect pieces of woodwork.

Remedial measures

Dry rot is a very difficult problem to eradicate since it may be widespread within a building. To be successful, therefore, treatment must be carried out both thoroughly and methodically; failure to do so leads inevitably to further outbreaks, damage and expense. A good system to work to would be as follows.

(a) Make a careful and probing survey of the building or suspect area to determine:

(i) the initial cause of the attack (blocked ventilators, defective dpc, leaking pipes, etc.);

(ii) the extent of the damage.

(b) Cut away all infected timber. Affected timbers should be cut back at least 600 mm beyond the last signs of decay. Timber removed should be destroyed – preferably by burning on site, or as close as possible.

(c) Clear away all rubble, debris and decayed matter in the vicinity, cut back plasterwork adjacent to signs of decay, remove all obvious stranding from brickwork or masonry and clean down with a wire brush.

(d) Rectify if at all possible, the defective structure responsible for the onset of the trouble, e.g. faulty dpc, leaking pipes, etc.

(e) Treat all adjacent brickwork, masonry, concrete, etc., with an effective fungicide and/or flame gun. Fungicides may be applied by brush or spraying but is best done by irrigating under pressure. (The aim is to surround the site of the outbreak with a fungicidal barrier.)

(f) Treat any remaining timbers with two good flowing coats of fungicide.

(g) Replace rotted timber with new material which *must* have been treated against decay, preferably under pressure.

(h) Keep a watch on the completed job for several weeks to make sure the work has been successful.

NOTE: Dry rot is so called because of the dry, crumbly appearance of timber that has been badly affected. It should, however, be understood that excess moisture in the wood is the main reason for the start of the decay, and that where wood is kept dry and well ventilated (below 20 per cent moisture content) there is little likelihood of any such trouble occurring.

Wet rot

This is a general name given to several types of wood destroying fungi, the two main forms of which are cellar fungus (Coniophora cerebella) and pore fungus (Poria vaporaria). Both of these require more moisture than dry rot in order to survive (hence the name "wet rot") and are capable of withstanding greater extremes of temperature. Wet rots, although capable of destroying wood completely, do not present such a serious problem as dry rot, since their progress is much slower and, more important, they cannot survive in dry conditions. Thus if the source of wetness is removed the disease is halted. Also, it is very unusual to find decay of this sort which has travelled far from the original source.

Hazard areas

In a building, the most likely places to find wet rot are:

(a) external doors and windows where rain has penetrated ill fitting, old or badly maintained joints (the joint between mullion and cill on bay and bow windows is particularly susceptible), and failing putties;

(b) fascia boards, behind cracked gutters;

(c) roof members where water has penetrated through faulty or cracked slates, tiles or flashings. Flat roofs and lead lined timber gutters are danger spots;

(d) fences and gate posts, especially at ground level;

(e) any external woodwork which has been insufficiently weathered allowing water to stand or collect;

(f) the ends of ground floor joists where a dpc is non-existent or not effective, allowing the wood in contact with the wall to become saturated.

Identification of wet rot

The following description is typical of damage caused by wet rot:

(a) musty smell;

(b) distortion of the face of the wood;

(c) darkening, softening and crumbling of affected timber;

(d) cracking of the wood surface – cracks are mainly along the length of the grain but a limited amount of cuboidal fracturing may be evident, though this is always less pronounced than is the case with dry rot;

(e) localised pockets of decay close to obvious wetness;

(f) presence of hypahe – very fine dark brown threadlike strands in the case of Coniophora; white, delicate fan shaped mycelium in the case of Poria.

NOTE: Wet rots do not produce heavy stranding, thick mycelium sheets or colourful fruiting bodies as does dry rot.

Treatment of wet rot

As the disease is not nearly so virulent as dry rot less extreme measures are called for. It is generally sufficient to remove the rotted timber, treat remaining timber with a fungicide and ensure that replacement timber is similarly treated. (Impregnation under pressure is alway advisable where new wood is to replace decayed.) It is also a sensible precaution to burn all the infected timber removed, on site if possible. Finally, rectify the cause of the wetness where this is feasible.

Simple as these remedial measures are there is, however, one very important measure which must never be overlooked; always positively ensure that the decay in question is *wet rot*, and that no dry rot is present. Where there is any doubt on this score, the trouble must be investigated very thoroughly, even if this means calling on the services of a company which specialises in timber decay.

Avoidance of fungal decay in timber

The following are what may be termed "good building practices" and are essential if decay is to be prevented.

(a) Timber should be of good quality, with as little sapwood present as possible.

(b) Timber structures should be well ventilated to avoid the build-up of damp or humid air pockets.

(c) Damp proof courses and membranes must be fully effective.

(d) No timber must be used below the dpc unless it has been treated (under pressure) with a preservative.

(e) Exterior softwood joinery (doors, windows, etc.) should be well constructed with close fitting joints, and preferably treated with a suitable preservative before assembly.

(f) Joinery which is eventually to be painted should be primed before it leaves the workshop and delivered to the site at the correct time – not too soon.

(g) All exterior woodwork should be well designed with good weatherings, drips, throatings and anti-capillary grooves.

(h) Maintenance of exterior painted surfaces must be carried out at reasonable intervals.

(i) Special attention should be given to glazed frames where timber movement, rainwater and condensation might lead to early putty failure.

(j) The timber selected for a particular job must be sufficiently durable for its purpose.

(k) The ingress of water through any part of the fabric of the building must be prevented.

(l) As far as is practically possible timber should be kept dry.

INSECT ATTACK

There are several species of insect, principally beetles, which lay their eggs in timber. In due course, the eggs turn into grubs or larvae which are responsible for the destruction of the timber. They include: *(a)* the common furniture beetle; *(b)* the powder post beetle; *(c)* the death watch beetle; and *(d)* the house longhorn beetle.

The common furniture beetle

This is a small insect up to about 5 mm in length the female generally being the larger of the two sexes. The eggs are laid in cracks, crevices and on rough surfaces of either hardwoods or softwoods and within a few weeks they hatch into tiny grubs properly known as larvae.

The larvae live within the wood, feeding upon it until the interior is a mass of minute tunnels. The larval stage of the insect lasts from one to three years (depending upon the conditions prevailing), after which the insect forms a small chamber just below the surface of the wood where it changes, first into a pupa, and then a few weeks later into an adult beetle. The adult then bites its way to the surface, leaving the charactistic flight hole about 1.5 mm in diameter, and proceeds to look for a mate. The cycle then starts all over again. Emergence and mating take place during the summer, usually June or July.

The powder post beetle

This is about the same size as the common furniture beetle (up to 5 mm in length) and has a life cycle – egg, larva, pupa, adult – of up to two years, although this may be as short as ten months in a well heated building.

The insect derives its name from the very fine, talcum powder-like bore dust which tends to accumulate on the floor around an infected hardwood post or beam.

The female lays her eggs within the pores or vessels of a hardwood; timbers with large pores being more vulnerable than those with small. Softwoods, being non-porous, are more or less immune.

Since it is the starchy cell contents which provide the larvae with food, it is the sapwood which is the more vulnerable to attack by the powder post beetle, although emergence may be through the heartwood leaving a flight hole about 1.5 mm in diameter. Although any starchy hardwood is liable to be attacked, probably the greatest amount of damage caused by this insect is suffered by timbers such as oak, ash, elm and walnut during the period when they are stacked for air seasoning.

Death watch beetle

This is related to the common furniture beetle, but is much larger, approximately twice the size, and therefore the damage it causes can be more severe.

The life cycle of this insect is similar to that of those previously described, but may last up to ten or twelve years in sound dry wood (less than half this in partly decayed, damp wood).

Generally speaking, the death watch beetle confines its activities to hardwoods such as oak, although softwood in the vicinity of an already infested timber is likely to be attacked. The insect has a distinct preference for old and partly decayed wood and is responsible for much of the damage to the woodwork in many fine old buildings, notably Westminster Hall in London where the great hammer beam roof has required extensive repairs.

Death watch beetles leave a flight hole about 3 mm in diameter, nearly always in some dark or inconspicuous position, and tend to leave visible faces of the wood unmarked although this may be little more than a thin veneer over an otherwise eaten-away timber.

House longhorn beetle

This insect attacks seasoned softwoods and is capable of doing enormous damage to structural timber on account of its size – a full grown beetle may be up to 25 mm long – its long life cycle which may span as much as seven years, and the ability of the larva to eat its own length through the wood every 24 hours. The beetle takes its name from the long, horn-like antennae which protrude and curve backwards from its head.

Longhorn beetles bore a tunnel about 6 mm in diameter and characteristically leave an oval flight hole about 10 mm in length. Like the death watch beetle, the insect often leaves a veneer of semi-sound wood on the visible surfaces of an infested timber. Fortunately in the United Kingdom the insect is found only in the southern part of England, mainly in Surrey and parts of Hampshire. Inside these areas, it is necessary to treat all structural softwoods with an insecticide as a measure of defence against its depredations.

PRESERVATIVE TREATMENTS

The organisms which prey upon timber, e.g. fungi and insects, require much the same essentials for life as does any other form of life (including ourselves), namely air, water, food and a suitable temperature. If the fungi or insects can be deprived of any one of these four main criteria, then our timber would become safe from attack.

At the present moment, the most effective measures that can be taken to combat the aforementioned organisms depend upon the following:

(a) deprivation of water – keeping the timber dry. This is obviously the first line of defence against fungi, but is less effective against insects.

(b) deprivation of food – poisoning or otherwise treating the timber so that it is no longer a source of food. This is effective against both insects and fungi and is perhaps the only sure and certain way to prevent attack from either source. Thus the chemical preservative becomes the most important weapon that can be used to combat decay in timber.

Types of preservative

Timber preservatives may be divided into three main groups:

(a) tar oils;
(b) water-borne preservatives;
(c) organic solvent preservatives.

Each type has its own peculiar advantages and disadvantages over the others.

Tar oils (creosotes)
These are very effective all round preservatives and are relatively cheap but they may be difficult to paint over, have a tendency to creep into adjoining brick or plasterwork and give off a strong odour which becomes ultimately unpleasant and may contaminate other materials (including foodstuffs).

Water-borne preservatives
These consist mainly of salts of copper, zinc, mercury or chromium, dissolved in water.

They are highly toxic to fungi and insects, have good powers of penetration into timber, have no adverse effect on the subsequent application of paint or glues, and are relatively inexpensive. They do, however, cause a certain amount of distortion to the wood due to the re-introduction of water and are therefore more suited to use with structural carcassing timbers than joinery.

Organic solvent preservatives
These are probably the most effective preservatives, but also the most costly. They are composed of various chemicals, dissolved in a spirit base (often volatile), have excellent penetrating qualities, little or no adverse effect on the application of paint or glues, dry out rapidly and having a non-water base do not cause distortion when applied to finished articles of joinery.

Methods of application

Preservatives can be applied or introduced to timber in any of the following ways.

(a) Non-pressure methods:
(i) brushing – applied liberally;
(ii) spraying – using a coarse spray;
(iii) dipping – immersion in an open tank.

(b) Pressure methods:
(i) empty cell method – preservative is forced into the timber under pressure. This compresses the air within the wood cells; when the pressure is released the air within the cells expands and blows out much of the preservative for re-use, having effectively rendered the timber toxic to some depth. This method is particularly suitable for application of water-borne and organic solvent preservatives;
(ii) full cell method – similar to *(i)* but a vacuum is applied to the timber *before* the introduction of the preservative; the wood cells thus retain much of the preservative. Useful with tar oils and water-borne preservatives;
(iii) double vacuum method – a vacuum is first applied to the timber to extract air from the cells, then the preservative is introduced, the vacuum released, and pressure applied. Finally, the pressure is released and a second vacuum applied to recover as much preservative as possible. This method is mainly used in conjunction with an organic solvent type preservative.

General information

In connection with timber preservation, the following points are of importance.

(a) The liquids used in water-borne and organic solvent preservatives are merely the "vehicles" – the means of carrying the chemicals into the timber. Subsequently they evaporate leaving the toxic chemicals *in situ*.

(b) Some timbers are more easily impregnated with preservative than others, and even the most easy timbers in this respect have a limit to the depth to which penetration is possible. It should be noted, therefore,

that where impregnated timbers are subsequently cut, the cut ends, thus exposed, require treatment by brush or spray to ensure 100 per cent effectiveness.

(c) The effectiveness of modern preservatives depends more upon their depth of penetration than the type of chemical used.

(d) It is possible to incorporate a fire retardant into the preservative. By doing so the timber not only becomes protected against insects and fungi, but also has a greatly increased resistance to fire.

VENEERS AND MAN-MADE BOARDS

In its natural form timber is one of the most useful and versatile of the materials used by man. It has, however, certain inherent shortcomings which tend to limit its usefulness in its solid form, i.e. as planks, boards, etc.

These shortcomings include the following.

(a) Width of boards – in its natural form a board is limited by the diameter of the log from which it is cut. This can vary from 100 mm to a metre or more depending upon the species of timber, or possibly, where very large trees are concerned, the maximum size the saw can accommodate.

(b) Stability – timber, especially in large sizes, is likely to warp, shrink or swell as described on pp. 39 and 40.

(c) Strength – although timber is remarkably strong longitudinally, i.e. with the grain, its strength across the grain is relatively poor. Moreover, even in the direction of its maximum strength, a defect such as a large knot can seriously weaken it.

Man-made boards of various types have been in use for many years. These boards are an attempt to overcome the shortcomings of solid timber and, in most cases, have done so satisfactorily and economically.

Plywood (BS 565:1963 and BS 3493:1962)

Plywood is available in large, standard size sheets in a range of thicknesses from 1 mm upwards. It comprises of a series of veneers, cemented together so that the grain of each veneer is at right angles to its neighbour. The number of veneers varies with the thickness of the board, but is normally an odd number, 3, 5, 7, etc., so that the sheet is balanced, i.e. the two outside veneers have the grain running in the same direction.

Where more than three veneers are employed in the construction, the material is known generally as "multi-ply". Figure 2.21 shows sections of various types of plywood.

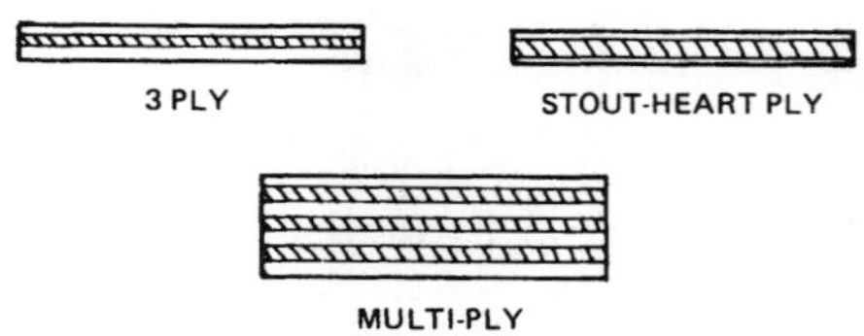

Fig. 2.21. *Types of plywood.*

Timbers used in plywood

Many timbers are used for the manufacture of plywood, the number of these having increased greatly in the past few years as some of the lesser known species have proved successful for this purpose. Some of the better known timbers include: redwood, parana pine, Douglas fir, alder, gaboon, birch, afara, beech, African mahogany, mengkulang and obeche.

Apart from the aforementioned species, many of the most beautiful and attractive timbers are used as decorative face veneers on less exotic base timbers.

NOTE: Where a single decorative face veneer is applied to a standard plywood sheet, the construction becomes less stable due to the fact that it is no longer a balanced material, i.e. the veneers total up to an even number. The better qualities of plywood so treated have a balancing veneer applied to the rear face.

Adhesives

The "plies" or veneers are bonded together with one or other of the adhesives described in Chapter 3, some of which are waterproof and others not. Thus when purchasing plywood care should be taken to ensure that it is suitably bonded for the purpose required. Plywoods for exterior and marine use are generally stamped 'WBP" (*see* Classification of adhesives, Chapter 3).

Peeled and sliced veneers

The veneers for plywood manufacture are produced by one of the following methods which are illustrated in Fig. 2.22.

Peeling

The great majority of plywood veneer is produced by this method, it being the most economical. For peeling, the log is first softened by boiling or steaming and is then mounted on a large lathe or peeling machine as shown in Fig. 2.22*(a)*, and a long knife peels off a continuous sheet of veneer.

Whilst this method of cutting the veneer is both fast and economical, the figure obtained in this way is not particularly attractive, so peeled veneers are mainly used for the production of non-decorative plywood.

Fig. 2.22. *Production of veneers.*

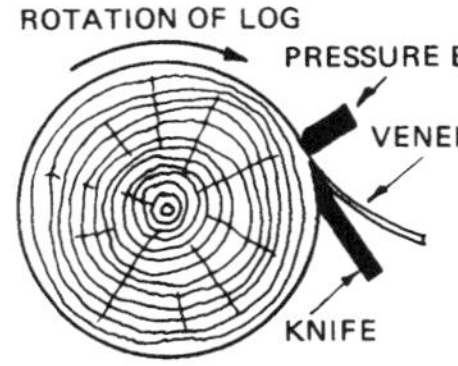

(a) Peeling veneer.

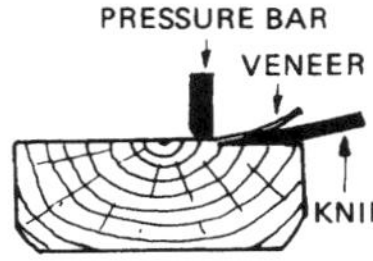

(b) Slicing veneer.

Slicing

This process, shown in Fig. 2.22*(b)* gives a true wood grain on the veneer. The slab from which the veneer is cut is carefully selected and sawn to give the best possible figure. As with peeling, the timber first requires softening.

Veneers cut by slicing are used for decorative faces on peeled plywood, and for application either by hand or machine to other types of board in the manufacture of high class joinery and furniture.

Blockboard (BS 3444:1961)

This is a very useful sheet material made by sandwiching a core of narrow strips of wood between two face veneers as shown in Fig. 2.23*(a)*. The best grades of blockboard have double face veneers so that the core strips and the grain of the face veneers run in the same direction.

Blockboard is available in the usual standard sized sheets in thicknesses from 12 to 48 mm, and has core strips up to 25 mm wide.

The inherent stability of blockboard makes it ideally suited for table tops, working tops, counters, doors, large items of specially made joinery and furniture, and as a base for veneering and the application of laminated plastic sheet.

Fig. 2.23. *Other types of man-made board.*

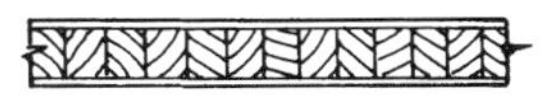

(a) Blockboard.

(b) Laminboard.

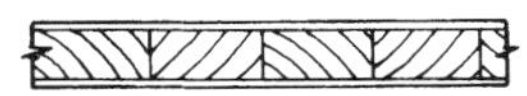

(c) Battenboard.

(d) Chipboard (pressed).

Laminboard (BS 3444:1961)

Laminboard is essentially similar in construction to blockboard as shown in Fig. 2.23*(b)* but is a generally superior and more stable material owing to the narrowness of the core strips which have a maximum width of 7 mm. Also, the core strips in laminboard are always glued edge to edge, unlike some of the cheaper grades of blockboard.

Used for similar purposes to blockboard and available in boards from 12 to 25 mm in thickness, it is probably the best possible base for the application of veneer.

Battenboard

This material, shown in Fig. 2.23*(c)* is a coarser version of blockboard and has core strips up to 75 mm in width. It is obviously a less stable material than blockboard and it is best suited for jobs where overall strength and tendency to "ripple" are less important.

Particle boards (BS1811:1961 and BS2604:1963)

Generally referred to as chipboards, these are made of wood chips and flax chives which are bonded together with a synthetic resin adhesive. The boards are available in lengths of up to 5.300 m, widths up to 1.700 m and thicknesses from 8 to 40 mm.

Chipboards are made either by pressing between platens or by extrusion, the pressed boards generally being stronger and having a finer surface. Platen pressed chip boards are available in various densities and grades, the denser boards having a finer, smoother face and being suitable for flooring panels. Lower grade, less dense boards are used for roof decking, ceiling and wall panels and general joinery, whilst extruded boards are used mainly as a base for veneers and laminated plastic sheets (the strength of these being improved by the applied outer skin).

Flooring grade chipboard

This is generally 19 mm thick and has a relatively fine smooth face due to the smaller, finer particles on the outside and the tendency of the chips to lay more or less parallel to the surface as shown in Fig. 2.23 *(d)*.

Suitably sealed and supported, chipboard can be used for concrete formwork.

Good quality chipboard can be worked satisfactorily by hand and machine, but tends to dull cutters rapidly due to the abrasive action of the synthetic resin binder. Fixings should not be too near the edge of the sheet, and screws tend to split the board when driven into "end grain".

Chipboards generally have less resistance to bending stresses than solid timber or plywood.

Fibre boards (BS1142:1961)

Hardboard

This is available in standard sized sheets from 3 to 6 mm in thickness and is used for the same purpose as plywood, being somewhat cheaper, heavier and inferior.

It is made from timber or sugar cane pulp with a certain amount of adhesive "binder", the pulp or "wet lap" being pressed to the required thickness.

Oil tempered hardboard is available and has a good degree of weather resistance through being impregnated with certain oils and resins.

Other variations include perforated hardboard (pegboard), enamelled hardboard, plastic faced hardboard, moulded hardboard (reeded, fluted, etc.) and double faced hardboard.

Insulation boards

These are low density boards made from the same material as hardboards, but are not pressed as are the latter.

Insulation board or "softboard" is available in various textures and finishes to suit the wide variety of work for which it can be used.

Basically a very efficient insulating material, it is used for wall and ceiling coverings in either sheet form or tile form in thicknesses from 6 to 20 mm.

Medium hardboard

This is a medium density fibre board which is basically a compromise between hardboard and softboard. It has better heat insulation qualities than hardboard whilst it retains greater mechanical strength than softboard.

Available in standard sized sheets, it is normally 9.5 mm in thickness. Its main use is for wall and ceiling panels though it is also an excellent material for pin boards.

Conditioning of fibre boards

Fibre boards being composed of timber pulp tend to swell and contract with variations in moisture content and it is good practice to "condition" them before use in order to minimise this fluctuation.

Hardboard sheets which are to be rigidly fixed should first be laid face down and brushed or sprayed with water on the back surface. This will ensure that when the sheet is fixed – fifteen minutes or so later – it will be in an expanded state and will thus pull tightly on drying, reducing the risk of buckling at a later date due to swelling.

Medium hardboards and softboards, on the other hand, being relatively soft, should be unpacked and stored on edge so as to permit air to circulate round them for forty-eight hours or so in the place where they are to be used, to allow them to adjust to the conditions prevailing in normal service. Oil tempered boards do not require conditioning.

FURTHER READING

British Standards and Codes of Practice

BS 565:1972	Glossary of terms relating to timber and woodwork
BS1142	Specification for fibre building boards
	Part 2:1971 Medium board and hardboard
	Part 3:1972 Insulating board (softboard)
BS1186	Quality of timber and workmanship in joinery
	Part 1:1971 Quality of timber
	Part 2:1971 Quality of workmanship
BS3444:1972	Specification for blockboard and laminboard
BS4471	Specification for dimensions for softwood
	Part 1:1978 Sizes of sawn and planed timber
BS5450:1977	Specification for sizes of hardwoods and methods of measurement
CP5589:1978	Preservation of timber

Building Research Establishment Digests

No. 72	Prevention of decay in external joinery
No.156	Specifying timber

Trade literature

Wood Preservation: The Facts. Protim Ltd.

SELF-TESTING QUESTIONS

All the information required to answer the following questions is contained within this chapter. Attempt each section *as fully or as briefly* as the question demands, and then check your answers against the information given in the chapter.

1. Explain briefly, the functions of the following parts of a tree: *(a)* roots; *(b)* trunk; *(c)* crown; *(d)* heartwood; *(e)* sapwood; *(f)* cambium.

2. *(a)* Sketch a cross section through a piece of wood to show: *(i)* spring wood; *(ii)* summer wood; *(iii)* rays.

(b) Explain briefly why the annual rings are more clearly defined in some timbers than in others.

(c) State the basic differences between a hardwood and a softwood.

(d) Name the types of cell found in a softwood and state their particular function in the living tree.

3. State the country of origin, approximate density, classification (hardwood or softwood) and specific uses of each of the following timbers: *(a)* Douglas fir; *(b)* parana pine; *(c)* redwood; *(d)* Western red cedar; *(e)* ash; *(f)* iroko; *(g)* mahogany (any species); *(h)* oak (any species); *(i)* teak; *(j)* utile.

4. *(a)* State what is meant by the "conversion" of timber.

(b) Sketch sections to show what is meant by: *(i)* through and through sawing, and *(ii)* quarter sawing.

(c) State which of the two methods of sawing in *(b)* is: *(i)* the most economical; *(ii)* likely to produce the most stable boards.

(d) Give reasons for your answer to *(c)*.

5. *(a)* Briefly outline the process of seasoning timber by: *(i)* air drying; *(ii)* kiln drying.

(b) State the advantages the latter has over the former.

6. *(a)* Sketch sections to show what is meant by:

(i) a "rift" or radially sawn board;

(ii) a "slash" or tangentially sawn board.

(b) Indicate on your sketches of *(i)* and *(ii)* the probable shape the boards would take up during drying.

7. *(a)* Calculate the percentage moisture content of a stack of partially seasoned timber, a sample of which has a wet weight of 45 g and a dry weight of 36 g.

(b) State what is meant by "equilibrium moisture content".

(c) Specify suitable moisture contents for timber to be used for the following purposes:

(i) roof members;

(ii) external door or window frames;

(iii) floorboards in a centrally heated building.

8. Explain with sketches, the following timber defects: *(a)* star shake; *(b)* heart shake; *(c)* warping or cupping; *(d)* waney edge; *(e)* short grain; *(f)* splay knot.

9. Describe as briefly as possible the symptoms of timber decay caused by: *(a)* dry rot fungus; *(b)* wet rot fungus.

10. Briefly outline the measures to be taken when dealing with decay caused by: *(a)* dry rot fungus; *(b)* wet rot fungus.

11. *(a)* Name four types of insect which are known to attack seasoned timbers, stating their preference (if any) for a particular type of timber.

(b) Outline the procedures adopted when preservative is applied to timber by: *(i)* a non-pressure method (give three examples); *(ii)* the "double vacuum" method.

(c) Name the three basic groups of wood preservative.

12. *(a)* Sketch diagrams to illustrate the composition of the following "man-made" boards: *(i)* "stout heart" plywood; *(ii)* blockboard; *(iii)* laminboard; *(iv)* "multi-ply".

(b) State the advantages that "man-made" boards have over solid timber.

(c) Describe two methods of cutting veneers for use in plywood manufacture.

3. Basic Joints and Adhesives

After completing this chapter the student should be able to:

1. State the functions of a woodworking joint.
2. Describe the requirements of a woodworking joint.
3. Sketch the basic woodworking joints and indicate their proper proportions.
4. Differentiate between scribing and mitring a moulding and state the factors which determine the method to be used.
5. Select the best joint for a particular situation.
6. Name the commonly used woodworking adhesives and describe their properties.
7. Select a suitable adhesive for a particular situation.

JOINING WOOD

When we consider the nature of timber as discussed in Chapter 2, and also give some thought to the sizes in which these timbers are available, it becomes obvious that much of the craftsman's skill and expertise is involved directly with the joining together of pieces of timber to form a complete unit. The unit may well be a small one such as a nail box, or it may be a large structure such as a roof or a floor, but whatever the size of unit involved, it will almost invariably be made up of several pieces of timber joined together.

There are, of course, sound reasons why this is so. Even if it were possible to obtain a solid slab of timber large enough to make, say, the entrance door to a house, it would not be practical to do so because:

(a) the cost would be prohibitive;

(b) the weight would be excessive;

(c) it would be weak because of the considerable amount of short grain;

(d) it would be unstable – that is to say, it would be likely to shrink, swell, warp, twist and crack.

Thus we have to consider joining together smaller sections of timber in ways which rule out, or at least minimise, these inherent shortcomings in the material itself.

Functions of a woodworking joint

Depending upon the nature of the work under construction the joint may be required to fulfil one or more of the following functions:

(a) to increase the size of the material, i.e. to make a wide board from two or more narrower ones or to extend the length of a member such as a ridgeboard or wall plate;

(b) to form the angle between two members as in the case of the head and jambs of door and window frames, or between the rails and stiles of a door or sash;

(c) to form vertical or horizontal corners on three dimensional work such as kitchen units, wardrobes, deep chests or boxes, and internal and external angles in wall panelling;

(d) to allow a certain amount of movement to occur between close fitting parts without cracking or distortion occurring as in the case of door panels and the fixing down of solid timber work tops;

(e) to provide a neat finish to intersections in mouldings and to the exposed edges of plywood, blockboard and the various other man-made boards now available.

Requirements of a joint

Strength

The joint must be so designed as to give adequate strength in service.

Simplicity
The joint should be as simple as possible for the circumstances prevailing. There is no virtue whatsoever in complicating a joint without good reason.

Appearance
This requirement is met with more often in joinery than in carpentry where much of the work is ultimately out of sight. Generally, in joinery work, the joints should have clean lines and be close fitting. They should not detract from the appearance of the job.

Weather resistance
This is most important when considering items of joinery exposed to the elements, e.g. doors and sashes. Every effort should be made to prevent the penetration of water into the interior of the joint. In particular, exposed end grain should be kept to an absolute minimum.

Securing a joint
It should be noted that in most instances good design of a woodworking joint is simply that which provides the most satisfactory way of *fitting* the two or more members together, and as often as not it will be found necessary to provide some means of *holding* the joint together in service. This can be done by means of adhesives, wedges, dowels, bolts, nails, or screws. In some cases, paint may be used in place of an adhesive.

EDGE JOINTS

These are used for joining boards together edge to edge to give increased width. They may be secured with adhesive, or left dry thus allowing the individual pieces to move independently of each other due to absorption or loss of moisture. A little consideration as to the function of the joint will denote whether or not an adhesive is required.

Butt joint
This is the simplest of the edge joints (*see* Fig. 3.1) and is used for jointing solid timber for door panels, working tops, etc. The boards are held in the vice, singly or in pairs, and the edges "shot" with a trying plane. The "shooting" must be done carefully to produce an accurate joint with no gaps. When placed together, no light should be visible through the joint, and the faces of the boards should be flat and free from twist. When long boards are to be joined, most joiners would plane the edges very slightly concave so that the ends meet tightly when the middle is cramped, but this is not necessary for boards less than 1.500 m long.

After the edges have been shot, the boards should be marked on the face as shown in Fig. 3.1*(a)* so as to make assembly straightforward, and then with one piece held in the vice, the adhesive should be applied simultaneously to both edges as shown in Fig. 3.1*(b)*. The glued edges are then rubbed together as shown in Fig. 3.1*(c)*, the ends being brought carefully into alignment during the final movement and care being taken to keep the faces of the two boards flush. Cramps are not normally required unless the boards are fairly long and have been planed concave. It is normal practice simply to remove the jointed boards carefully from the vice and to stand them against a wall to dry.

Fig. 3.1. *Butt joint.*

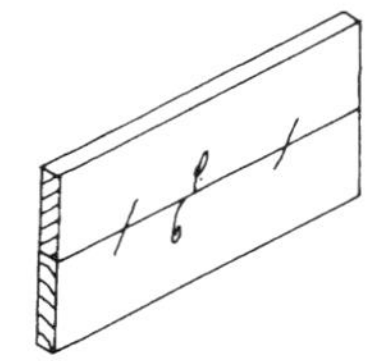

(a) Showing marking.

(b) Gluing the edges.

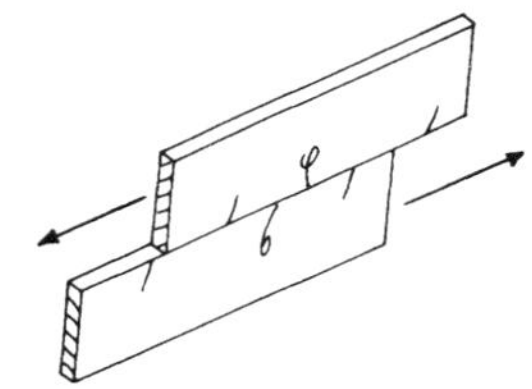

(c) Rubbing the joint.

Where more than two boards are necessary to obtain the required width however, it is easier to glue and assemble them on bearers placed across the bench using cramps. The rubbing tends to squeeze the glue into the pores of the timber, thus increasing the strength of the joint and giving rise to the name "rubbed joint" by which it is often known.

Cross tongued joint
This joint is basically similar to the butt joint, but has increased strength by virtue

of the cross tongue ploughed into the shot edges of the joint (*see* Fig. 3.2(*a*)). The cross tongue is normally a strip of plywood of suitable dimensions (*see* Fig. 3.2(*b*)) but can also be made from the same timber as the boards themselves. In this case, the cross tongue is made up in short sections cut with the grain at about 45° (*see* Fig. 3.2(*c*)). The increase in strength is derived from the stiffness of the cross grained or plywood tongue

Fig. 3.2. *Cross tongued joint.*

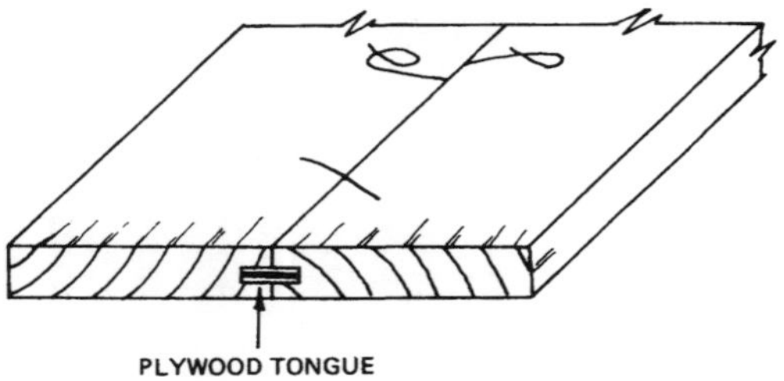

(a) Showing marking and tongue.

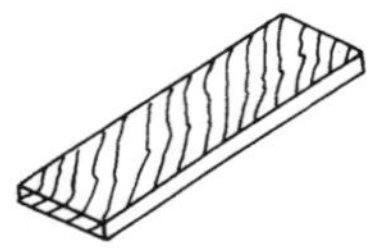

(b) Plywood tongue. *(c) Cross grained tongue.*

which reinforces the wood adjacent to the glue line, and also from the increase in gluing surface — in the region of 100 per cent in boards 22 mm thick. The tongue also provides an accurate registration for the face of the boards during assembly which is normally carried out on bearers laid across the bench. Rubbing is not a practical proposition with this joint, and the boards are generally held together with sash cramps and dogs until the glue has set. When making this joint, which is used mainly on exterior work, it is essential to plough the grooves accurately from the face side of the boards.

Slot screwed joint

This is really a variation of the rubbed or butt joint utilising concealed screws within the joint itself to give both a cramping action and a degree of alignment during assembly (*see* Fig. 3.3(*a*)). The countersunk headed screws also give a certain amount of added strength to the joint making it possible to work on the jointed boards within a short time of gluing up. It also does away with the need for a great many sash cramps — useful when a number of joints are to be made in a relatively short space of time.

When making the joint, the boards to be joined are shot in the same way as for a rubbed joint, the position of the screws and slots then being marked carefully with a marking gauge. (Note that the ends of the boards should be offset when marking so that they come into alignment when the protruding screw heads are forced along the slot — *see* Fig. 3.3(*b*).) The screws are then inserted in one edge, and the slots formed on the other.

When the two edges are brought together the screw head enters its respective hole and the ends of the boards can then be tapped or forced into alignment causing the shank of the screw to slide along its slot, the screw head cutting a recess in the sides. Before the

Fig. 3.3. *Slot screwed edge joint.*

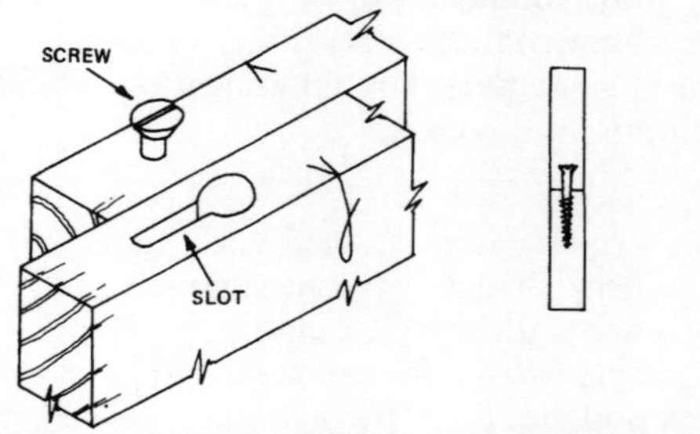

(a) Showing screw and slot. *(b) Joint complete.*

joint is glued the screws are given a further half turn to ensure a tight joint assembly.

This joint is not confined to the edge jointing of boards, but has many applications in joinery and fixings generally.

Tongued and grooved joint

This is an extremely useful and versatile joint with wide applications throughout the woodworking industry. When used as a glued edge joint, as shown in Fig. 3.4(*a*) its function is mainly to provide an accurate location of one piece of timber into another, there being only a minimal gain in strength over a butt joint.

Tongues and grooves also serve very effectively as a means of preventing unsightly or unwanted gaps occurring due to shrinkage between parallel members such as floor or match boards. Figure 3.4(*b*) shows a tongue and groove joint applied to floor boarding. Its purpose here is to hold the faces of the boards flush and in alignment with each other. Also, of course, when the floor boards shrink, as they almost invariably do, the tongue prevents a gap occurring right through the joint. Figure 3.4*(c)* shows a similar tongue and groove joint applied to match boarding, its purpose being the same as that for floor boards. The small chamfer forming the vee joint between the boards is simply to emphasise the joint which is almost certain to open to a greater or lesser degree, thus making the joint obvious and thereby acceptable.

Fig. 3.4. *Tongued and grooved joint.*

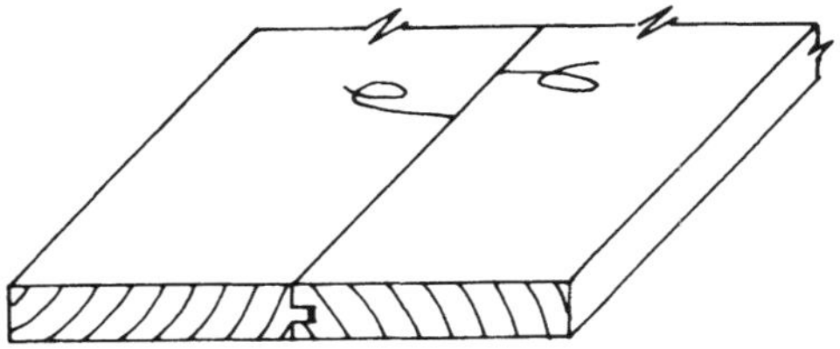

(a) Completed joint.

(b) Floor board. *(c) Match board.*

Neither of the latter two tongued and grooved joints would be glued; indeed to do so would defeat the whole purpose of the joint since the wall or floor surface so treated would thereby become a single wide board with all the associated problems of shrinkage, cracking and distortion. Figure 3.5*(a)* shows a tongued and grooved joint used to form a right angle as might be required between the side and front frames of a cupboard or wardrobe. Its function in this case is to provide an accurate location between front and side to make exact assembly easy, either in the work-shop or on site. The joint would almost

Fig. 3.5. *Tongued and grooved corner joints.*

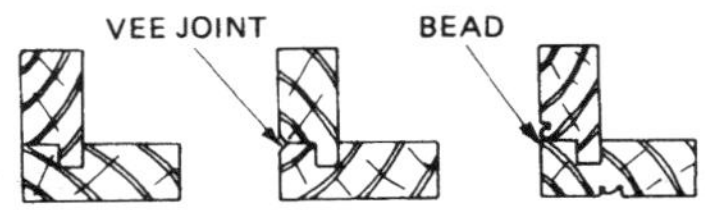

(a) *(b)* *(c)*

certainly be glued or nailed – or both. Since the members are relatively narrow, shrinkage does not present any real problem but, in any case, the principle of accentuation can be applied if necessary as shown in Fig. 3.5 *(b)* and *(c)*.

LENGTHENING JOINTS

Joints of this type are used to extend the length of a piece of timber as may be necessary from time to time when working on jobs requiring extra long members, e.g. skirting boards, ridge boards, purlins, etc. The main considerations to be given to the design of these joints are:

(a) whether or not the joint is under stress;

(b) the type and direction of any applied stress;

(c) whether or not the joint will be reinforced by other members within the structure.

Scarf joints

The scarf joints described below are illustrated in Fig. 3.6.

Plain scarfed joint

The simple scarfed joint shown in Fig. 3.6*(a)* is often used for joining members where there is little or no applied stress, as in the case of a ridge board. The joint is usually spiked together through the feather edges and has no great strength, but is nevertheless an improvement on a plain butt joint.

Fig. 3.6. *Scarf joints.*

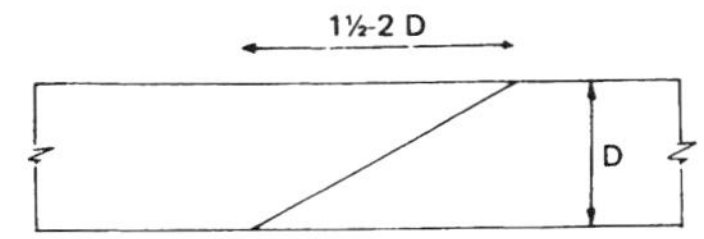

(a) Plain scarfed joint.

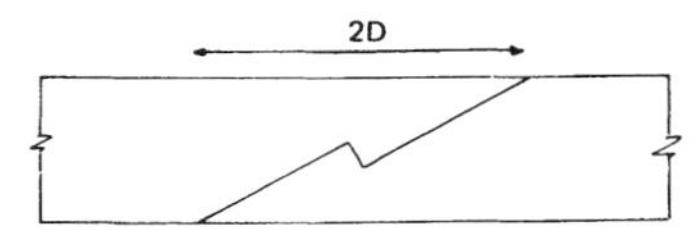

(b) Notched scarf joint.

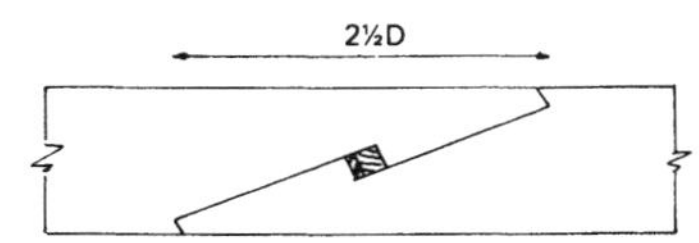

(c) Locked scarf joint.

Notched scarf joint

The notched scarf shown in Fig. 3.6*(b)* is a variation of the previous joint and is used in similar situations. The notch provides a limited degree of resistance to tensile stress (i.e. being pulled apart) and forms a positive location when spiking the joint together. This joint is commonly used for joining purlins, for which purpose it is quite suitable, especially when there is some support by means of a prop or brick pier under the centre of the joint.

Locked scarf joint

This joint (*see* Fig. 3.6*(c)*) is a further development of the plain scarfed joint and gives a reasonable resistance to tensile and compressive stresses. The joint is tightened by driving in a pair of folding wedges, which force the undercut ends into close contact. The construction of this joint is somewhat time-consuming, and therefore it would not be used where a simple joint would serve.

Fish plated joint

The fish plated joint shown in Fig. 3.7 is both strong and simple, and has very good resistance to compressive, tensile, and bending stresses. The members to be joined are butted together, reinforced with fish plates on either side, and then fastened by means of coach bolts, with timber connectors inserted between the members. The timber connectors bite into the wood on each side, thus increasing the area under direct stress from the bolts, and reducing the risk of failure due to shear or splitting. Toothed or split ring connectors as shown in Fig. 3.7*(b)* are suitable for this purpose. Fish plates are rather unsightly, but where strength is the prime consideration – as in structural joints – this would be of secondary importance.

Fig. 3.7. *Fish plated joint.*

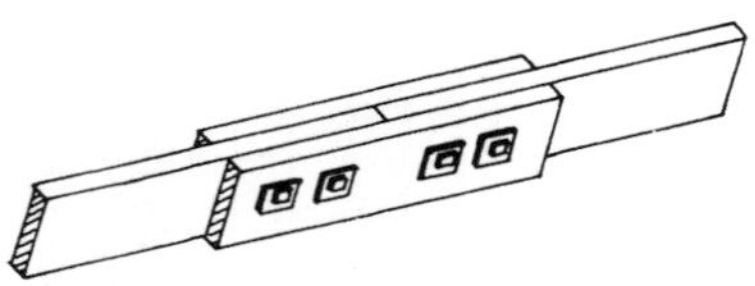

(a) Showing bolts, washers and timber connectors.

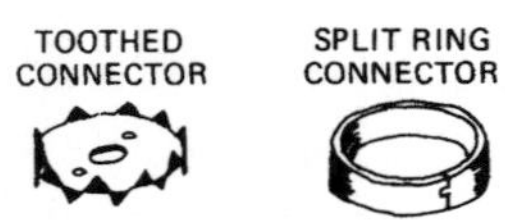

(b) Timber connectors.

Lapped joints

Various forms of lapped joint are used, both in joinery and for constructional purposes, and are illustrated in Fig. 3.8. They have the advantages of being quick and easy to make, retain the line of the member, and have a good resistance to compressive stress. Resistance to tensile and bending stresses is only fair. Figure 3.8*(a)* shows a half lapped joint as used for lengthening a wall plate.

Fig. 3.8. *Lapped joints.*

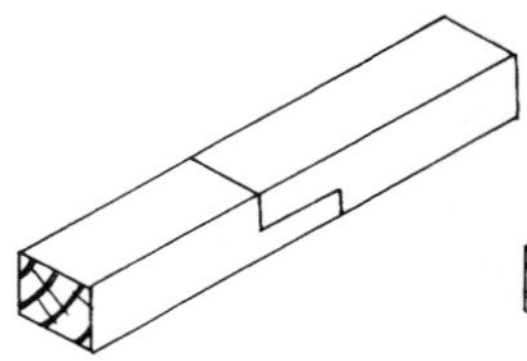

(a) Half lap joint.

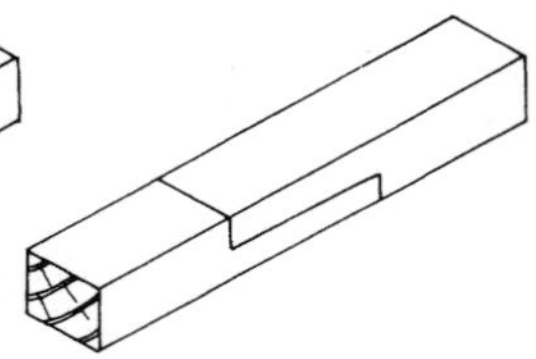

(b) Long lap joint.

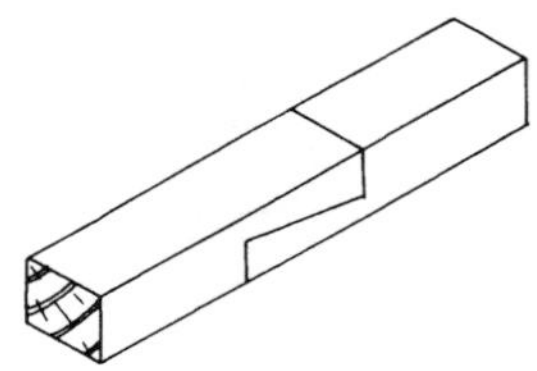

(c) Bevelled lap joint.

The joint at Fig. 3.8*(b)* is a variation of the half lap previously mentioned, the extra length of lap giving more space for securing the joint, thereby increasing its resistance to tensile stress.

The bevelled lap joint shown in Fig. 3.8*(c)* is a way of further increasing resistance to tensile stress. This joint is most effective when sandwiched between other members, or in any situation where spreading of the joint can be prevented by the cramping action of other surfaces. Lapped joints are normally secured by means of nails or screws, but where strength is important bolts may be used. In joinery the joint is usually glued and screwed.

FRAMING JOINTS

Framing joints are those generally associated with the construction of items of joinery such as doors, windows, cupboards, furniture, etc. They are not wholly confined to joiner's work, however, and occur in various forms in all types of constructional woodwork.

In fairness to the young craftsman, it should be understood that in today's industry, the great majority of the joints described in this section would be produced by machines, often on a mass production basis in large joinery works and factories. There remains, however, a significant proportion of work which for one reason or another is still done by hand. Where specialised and one-off jobs are concerned it may still be economical to produce the work on a hand production basis. Even so, it is almost certain that as much work as possible will be done by machine, either by a wood machinist or by the joiner himself. Many joiners are highly skilled in the use of the basic woodworking machines, and in the smaller workshops might well be expected to use them as a matter of course.

Whatever the circumstances prevailing in individual situations, the fact remains that if good quality work is to be produced economically a sound understanding of the design and construction of woodworking joints is highly desirable, whether the operative be apprentice, foreman, carpenter, joiner, shopfitter or wood machinist. It should be realised that the purpose and function of a joint and the need for its reliability in service are prime considerations by whatever means it is produced.

Halving joints

These are very common joints which take several forms, each suited to a specific purpose, and are illustrated in Fig. 3.9.

Fig. 3.9. *Halving joints.*

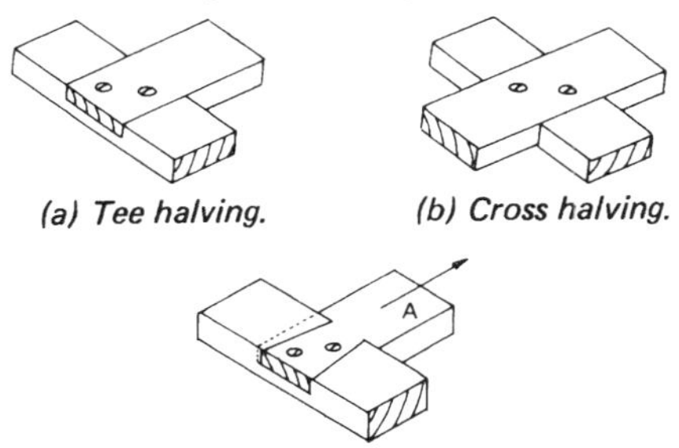

(a) Tee halving. *(b) Cross halving.*

(c) Dovetailed halving.

Tee halving

Figure 3.9*(a)* shows a tee halving, a joint commonly used in light joinery framework. It is normally secured with an adhesive or screws, or both. Generally, the joint has equal proportions, half the thickness of the material being cut away from each part, although this is not invariably the case and does not matter unduly provided that the setting out has been done correctly and both parts gauged from the face side. The joint has a fair resistance to twisting and swivelling, and is ideal for a framework which is to be covered with a plywood or hardboard skin.

Cross halving

Figure 3.9*(b)* shows a cross halving, this being the simplest way of obtaining a flush finish to two members which cross each other. It is generally fastened with glue and screws, and is used mainly in the construction of light frameworks and carcasses.

Dovetailed halving

Figure 3.9*(c)* shows a dovetailed halving. As can be seen its purpose and form are similar to those of a tee halving but the dovetailed portion prevents withdrawal (in the direction of the arrow). This is a useful joint where part A is in tension. The dotted line shows a simplified version of the joint where only one edge of the lap is dovetailed, thus saving some work at the expense of a little strength. This form of the joint is used mainly where the members joined are not at right angles to each other.

Mortice and tenon joints

This joint in one or other of its many forms illustrated in Figs. 3.10 to 3.21 is undoubtedly one of the most useful and effective ways of forming a framework. It provides for a more accurate location of members than the halving joint and has much greater resistance to twisting. Also, since the tenon is enclosed on all four faces, there is good resistance to withdrawal, and when used in conjunction with wedges and glue or is draw-bored, the joint is very strong indeed.

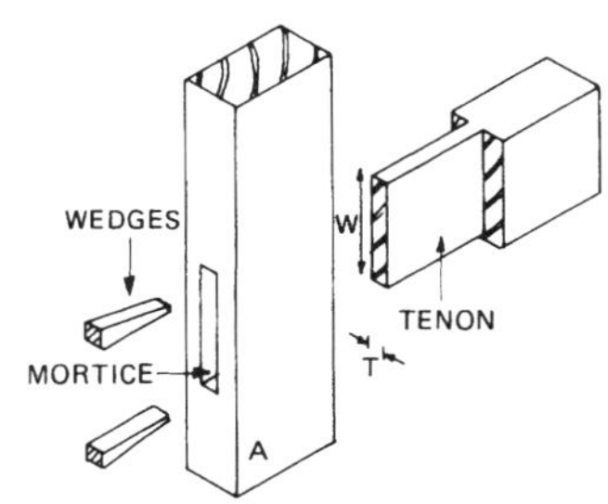

Fig. 3.10. *Through mortice and tenon.*

Through mortice and tenon

Figure 3.10 shows a "through" mortice and tenon which is designed to be secured with glue and wedges, thus making the tenon into a dovetail and preventing withdrawal. Ideally, the thickness of the tenon (T) should be one-third the thickness of the material. This proportion gives the best strength ratio to both parts of the joint, any variation from this tending to strengthen one part at the expense of the other. It is generally acknowledged that the width of the tenon (W) should be limited to five times its thickness, thus keeping to a minimum any tendency for the tenon to shrink and become loose. Also, of course, an over long mortice would tend to weaken the stile (A). It should be understood that these proportions are not inviolate – they are used for guidance rather than as invariable rules.

Haunched mortice and tenon

Figure. 3.11 shows a haunched mortice and tenon which is used when the horn (A) is eventually to be cut off. The haunch keeps the tenon shoulders in alignment with the stile, reducing any tendency to twist, whilst the cut away section of the tenon, shown by dotted lines, prevents the joint becoming open ended when the horn is removed. It will be appreciated that a haunch can be used as a means of reducing the width of the tenon to limit shrinkage as mentioned earlier.

Mortice and tenon joints as illustrated in Fig. 3.10 and 3.11 are used in the construction of frames, the stiles and rails of which have planed square edges (PSE), i.e. frames which have no rebates, grooves or mouldings on the inside edges.

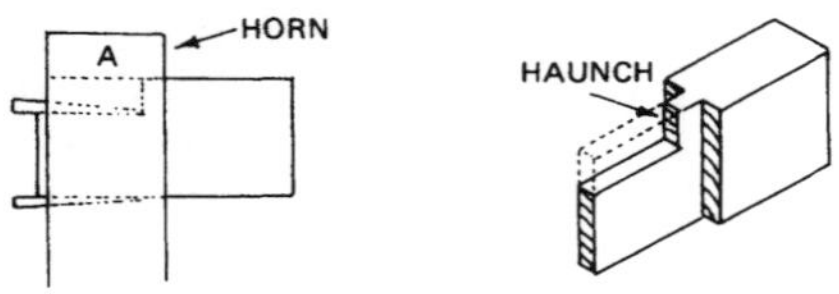

Fig. 3.11. *Haunched mortice and tenon.*

Grooved framing

Figure 3.12*(a)* and *(b)* illustrates tenons on rails which are grooved to receive panels on one or both edges respectively. It will be readily seen from Fig. 3.12*(c)* that where such a groove occurs, the width of the tenon is reduced by the depth of the groove. This part of the tenon will in fact be unavoidably planed away when the groove is ploughed, and there must therefore be a corresponding allowance made when setting out and chopping the mortice or a loose ill-fitting joint will result.

Where grooved framing is to be morticed and tenoned together, the haunch serves a secondary purpose of filling up the end of the groove which would otherwise show on the end of the stile when the horn is removed. These joints are typical of those used for making panelled doors.

Fig. 3.12. *Tenons for grooved framing.*

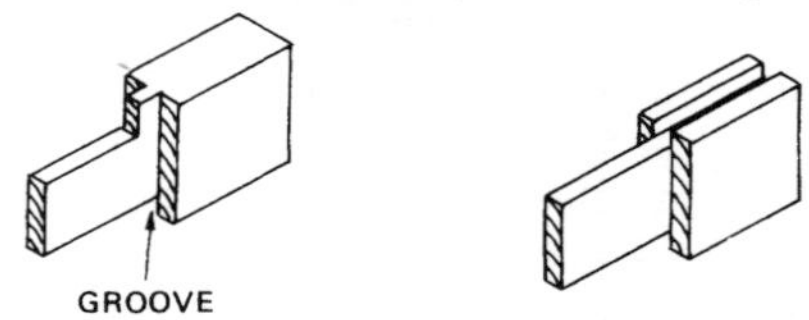

(a) Haunched tenon with groove. *(b) Tenon with groove.*

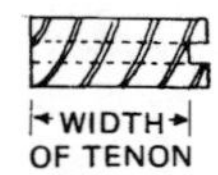

(c) Effect of a groove on a tenon.

Tenons for wide rails

Figure 3.13*(a)* and *(b)* shows tenons designed for use on wide material such as the middle and bottom rails of a door. Note that Fig. 3.13*(a)* shows a pair of tenons separated by a single haunch and is a suitable joint for any wide intermediate rail. The grooves shown on the top and bottom edges are only required when a panel is to be fitted. Note also that Fig. 3.13*(b)* shows a pair of tenons and *two* haunches, the lower one of which serves to contain the tenon within the stile thus preventing the joint becoming "open".

Fig. 3.13. *Tenons for wide rails.*

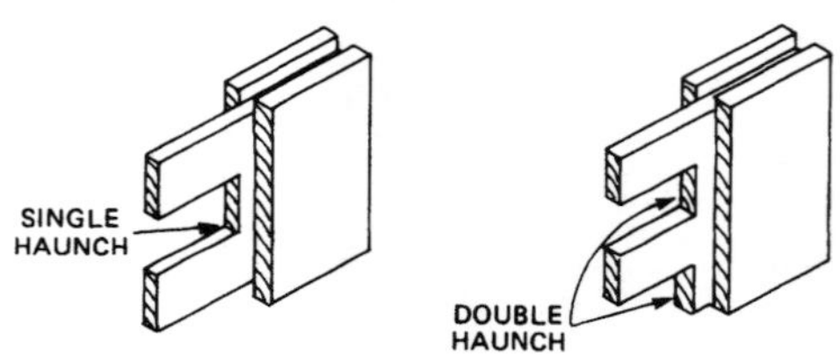

(a) For wide middle rail. *(b) For wide bottom rail.*

Blind tenon

Figure 3.14*(a)* shows a "blind" mortice and tenon for use where it is unnecessary or undesirable for the tenon to pass right through the stile. The joint is weaker than the more normal through mortice and tenon and would be used where strength is not the prime criterion, or where the visible end grain of the tenon would be objectionable. It is possible to give the joint greater resistance to withdrawal by the use of fox-wedges as shown in Fig. 3.14*(b)*).

Fig. 3.14. *Blind mortice and tenon joint.*

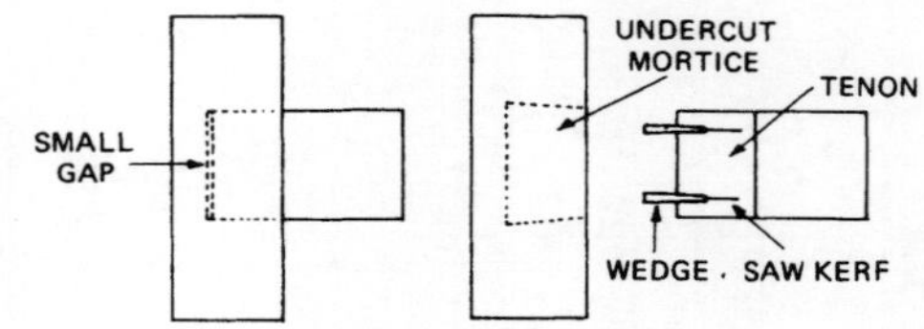

(a) Blind mortice. *(b) Fox wedged tenon.*

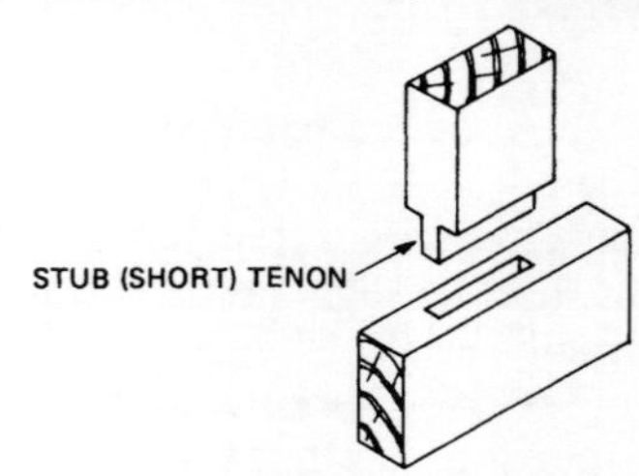

(c) Stub tenon.

In fox-wedging, the mortice is undercut at the ends in order to accommodate the spreading of the tenon when the joint is cramped up. Great care must be taken when fox-wedging a blind mortice and tenon. All parts must be measured and cut with extreme accuracy, since if the completed joint is either too loose or too tight a fit it is difficult to take the joint apart for re-fitting.

Stub tenon

This joint has a short tenon, usually referred to as a stub tenon, and is used only where the member having the tenon or tenons (there is often one at either end) is in compression, since there is little resistance to withdrawal (*see* Fig. 3.14(*c*)). Stub tenons serve mainly as a means of positive location between members at right angles to each other and of preventing lateral (sideways) movement of the rail.

It should be noted that with any form of blind mortice there must be a slight gap between the end of the tenon and the bottom of the mortice, otherwise any shrinkage of the stile will cause the shoulders to open up.

Rebated framing

Figure 3.15 shows a mortice and tenon joint where the tenon has long and short shoulders, i.e. the shoulders on the face and back of the rail are not in line. Joints of this type become necessary where there is a rebate or a moulding on the edge of the framework. Reference to Fig. 3.15(*b*) will make it clear that the shoulder on the rebated side of the work must be extended by the depth of the rebate in order to avoid a gap. The same principle applies where there is a moulding on the inside of the framing.

Fig. 3.15. *Mortice and tenon joint for rebated framing.*

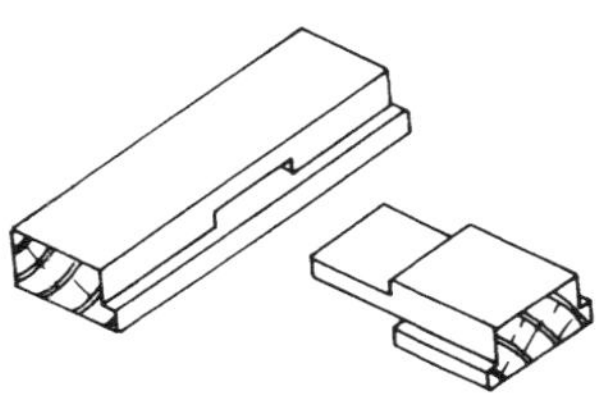

(a) Long and short shoulders on tenon due to rebate.

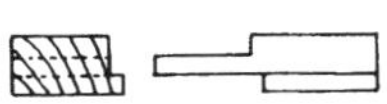

(b) Use of long and short shoulders to close rebate.

Barefaced tenon

Figure 3.16(*a*) shows a tenon which has only one shoulder. This is known as a barefaced tenon and is used where the rail is thinner than the stile, to bring the barefaced side of the rail in line with a rebate or groove as shown in Fig. 3.16(*b*). Some care is required when cramping up frames containing barefaced rails as the stiles have a tendency to twist over towards the side of the rail without a shoulder. This can be prevented by cutting a piece of scrap timber to the shoulder width of the rail and placing it loosely on the barefaced side of the rail to form a temporary shoulder. After cramping and wedging it is removed.

Barefaced rails are used in the construction of roof lights, framed, ledged and braced doors, tables, and various other items of joinery.

Fig. 3.16. *Barefaced tenon.*

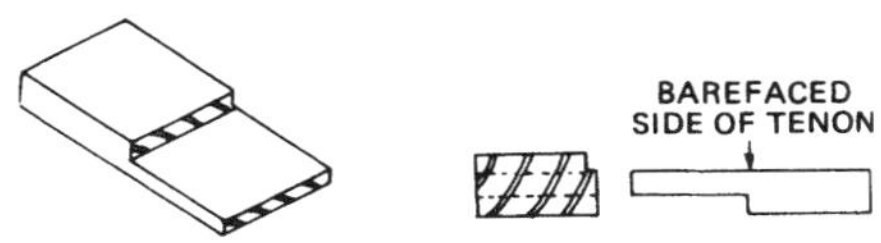

(a) *(b) Use of barefaced tenon.*

Mortice and tenon joints with stuck mouldings

Figure 3.17(*a*) to (*d*) shows sections of framing which are commonly joined with mortices and tenons. Such sections occur in doors, door frames, sashes, window frames, etc. These illustrations represent sections taken through the actual joints in the framework, the position of the mortices being indicated by dotted lines.

Fig. 3.17. *Mortice and tenon joints with stuck mouldings. (Sections (c) and (d) show mouldings which should be mitred.)*

(a) Bevel (scribe).

(b) Ovolo (scribe).

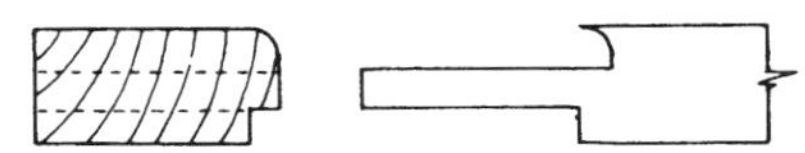

(c) Pencil round (scribe difficult).

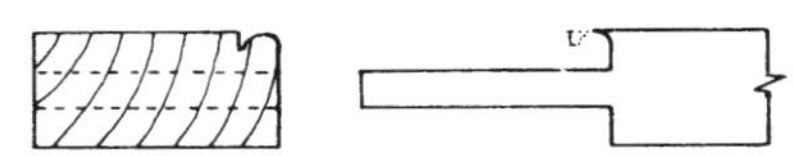

(d) Corner bead (scribe impossible).

Intersection of mouldings

The problem facing the craftsman in these instances is how to form a good neat intersection of the mouldings. The first point to note is that the moulding should be on the face/face edge corner of the timber, so that any slight variation in the thickness of the material will show up on the back of the frame and will not affect the intersection of the mouldings. Secondly, the shoulder on the moulded side of the rail must be extended by the depth of the moulding so as to avoid unsightly gaps. Where there is a rebate on the edge opposite to the moulding, as shown in Fig. 3.17*(a)*, *(b)* and *(c)* both shoulders must be extended. It is good practice, wherever possible, to make the depth of the rebate and moulding equal so as to bring the extended shoulders back into line. This simplifies the marking out.

Mitring and scribing

The intersection of mouldings is accomplished generally in one of two ways: *(a)* by mitring; *(b)* by scribing.

A mitre involves cutting the moulding with a sharp chisel to form a bisection (half the angle formed by the joint), usually – but not necessarily – at 45° as shown in Fig. 3.18*(a)*.

A scribe involves cutting the reverse shape of the moulding across – or partly across – the tenon shoulder, thus allowing it to fit over the moulding as shown in Fig. 3.18*(c)*.

An important point for consideration is whether a moulding should be mitred or scribed. Fortunately, the decision is an easy one to make. The scribed intersection is less prone to open through shrinkage, is more resistant to water penetration, and lends itself more favourably to making a close fit when the job is cramped up. The mitre is less effective in these instances, and the answer, therefore, is "scribe whenever possible". A brief study of the sections in Fig. 3.17*(a)* and *(b)* reveals that the scribings required to

Fig. 3.18. *Mitring and scribing.*

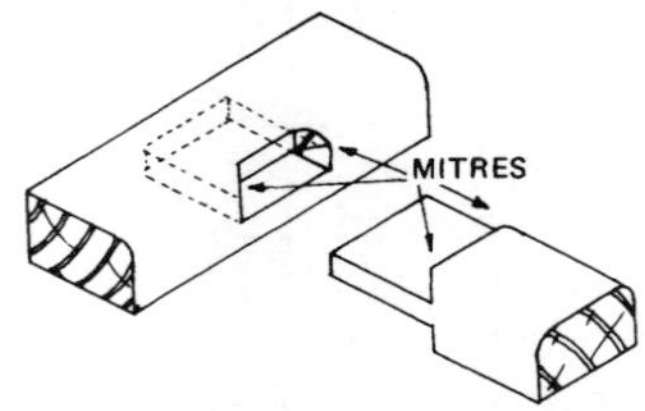

(a) Mitring a pencil round moulding.

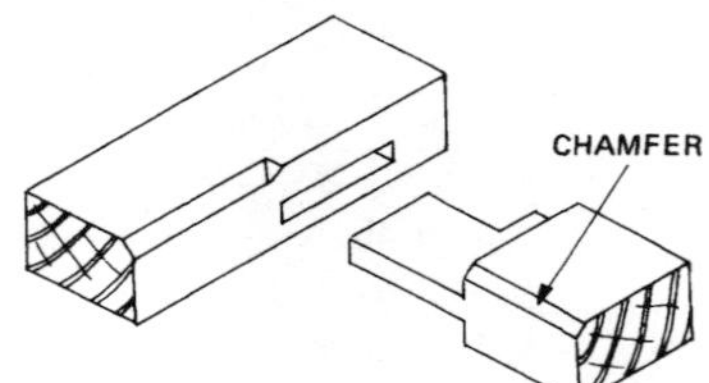

THIS JOINT IS ALSO IDEAL FOR A PENCIL ROUND MOULDING

(b) Mason's mitre.

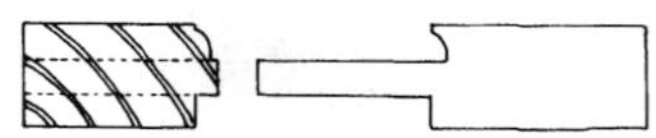

(c) Scribed ovolo mould.

fit the bevel and ovolo mould respectively are simple and straightforward, whereas the scribing required to fit over the section in Fig. 3.17*(c)* is difficult and results in a fine feather edge which tends to pick up when the job is cleaned off. The corner bead shown in Fig. 3.17*(d)* turns back on itself making scribing impossible. The latter two mouldings, therefore, would be better mitred. The woodworker would do well to realise (with respect to the foregoing examples) that mouldings which require to be mitred are best avoided unless there is a good reason for using them. (Mouldings on external angles are invariably mitred.)

Hand and machine scribing

When stuck mouldings are scribed by hand it is common practice to fit the shoulder over only part of the moulding, leaving the rest of the shoulder intact, and cutting away the moulding on the stile to accommodate it as shown in Fig. 3.19*(a)*. This reduces the

Fig. 3.19. *Hand and machine scribing.*

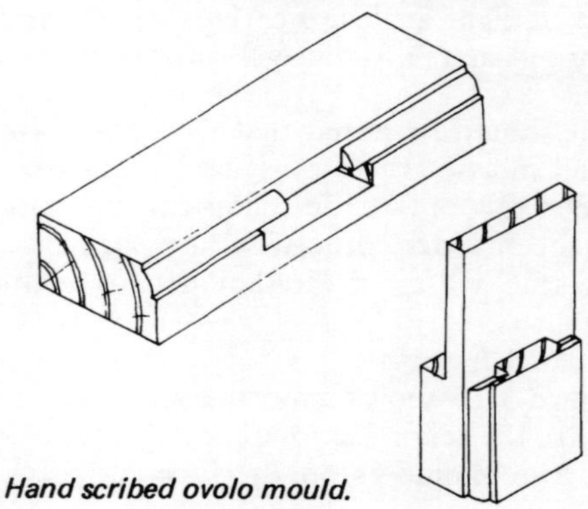

(a) Hand scribed ovolo mould.

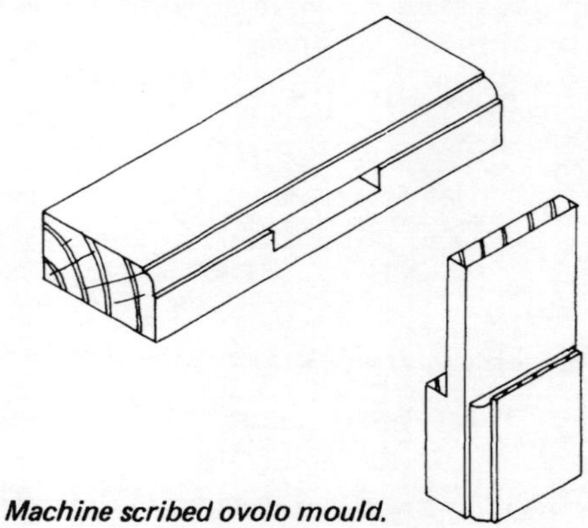

(b) Machine scribed ovolo mould.

amount of work involved and gives a better abutment when cramping up. Machine scribes, on the other hand, are carried right across the full width of the shoulder as shown in Fig. 3.19*(b)*. This is done simply because it is quicker and easier to scribe in this manner when the work is carried out by machine.

Dowelling and draw boring of tenons

Tenons are occasionally locked into their mortices by means of dowels or draw boring.

Dowels

When dowelling a tenon, the frame is cramped up tightly and a suitably sized hole is bored straight through the joint which is then locked by driving a dowel through it as shown in Fig. 3.20*(a)*.

Draw boring

Draw boring is a more effective method of securing the tenon since it pulls the shoulders up tightly and holds them in position, without the use of cramps.

In draw boring the hole is first bored through the mortice, the tenon is inserted, and the position of the hole marked on it with the point of the twist bit. The joint is then taken apart and the hole bored through the tenon, offsetting the centre of the hole slightly towards the shoulder. The hole through the joint is, therefore, slightly out of alignment as shown in Fig. 3.20*(b)*. Obviously, when the pin is driven through, the holes will be forced into line thus pulling the joint together tightly. The tip of the dowel should be sharpened as shown in Fig. 3.20*(c)* so as to avoid spelching out on the underside of the joint when it is driven through.

Fig. 3.20. *Locking a mortice and tenon joint.*

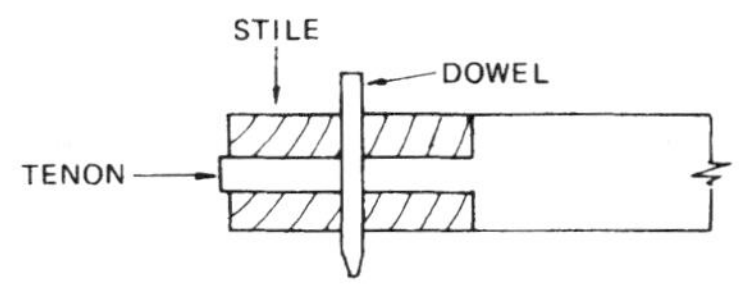

(a) Joint locked with dowel.

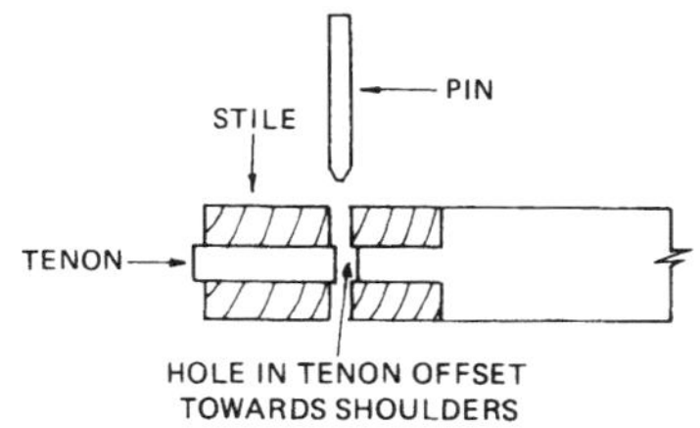

(b) Joint draw-bored.

(c) Hardwood pin for draw-boring.

Bridle and combed joints

The bridle joint shown in Fig. 3.21*(a)* is basically an open mortice and tenon. It is used for work requiring accurate alignment of the members but no great strength. It is generally secured with glue in conjunction with nails or screws.

The combed joint shown in Fig. 3.21*(b)* is a variation of the bridle joint deriving its strength from the increase in gluing area due to the double tenons. For maximum effect, the tenons and slots should be equal in thickness. This joint is used mainly in the machine production of sashes for storm-proof windows where it is secured with an adhesive in conjunction with a star-dowel.

Fig. 3.21. *Bridle and combed joints.*

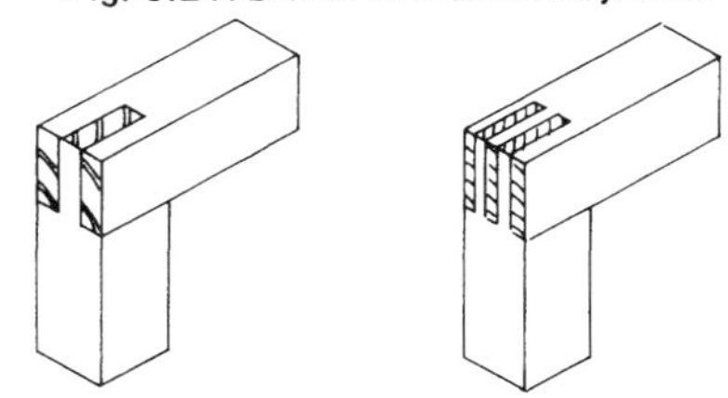

(a) Bridle joint. *(b) Combed joint.*

DOWELLED JOINTS

Dowelled joints are used mainly as alternatives to the mortice and tenon. Holes are bored into the stile and into the end of the rail to receive the dowels. These are glued into their holes during the assembly of the job. To be effective the dowels must be a tight fit in the holes, and it is good practice to cut a fine groove along the length of the dowel to allow air and surplus glue to escape. A typical dowelled joint is shown in Fig. 3.22, and is commonly employed in the mass production of panelled doors. A dowelled joint is not as effective as a mortice and tenon, either in resisting withdrawal or in keeping the shoulders in alignment with the stile. Although the joint looks simple, considerable care is required when setting out and boring the holes if the joint is to be really effective.

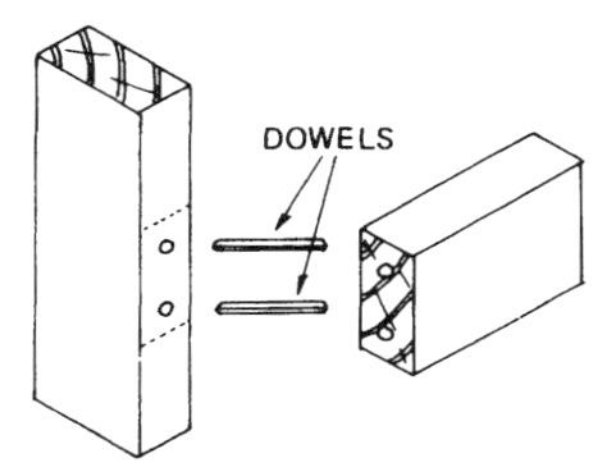

Fig. 3.22. *Dowelled joint.*

DOVETAILED JOINTS

These joints illustrated in Fig. 3.23 can be produced by hand or machine, and are generally employed for joining the corners of drawers, boxes and similar articles which have deep sides. When properly made, the joint is a strong one, almost invariably being secured with an adhesive. A little consideration will make it clear that the joint is more resistant to being pulled apart in one direction than in the other. This is a point to consider when deciding which part is to have the "pin" and which the "tail".

Fig. 3.23. *Dovetailed joints.*

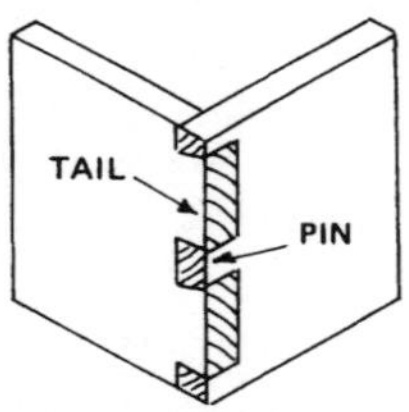

(a) Common or through dovetail.

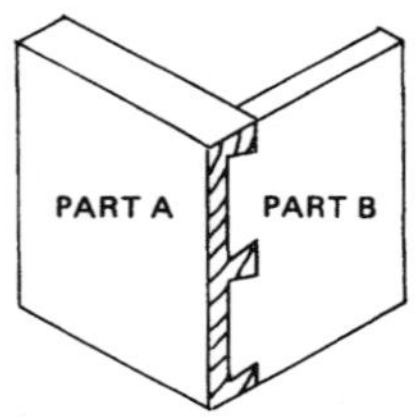

(b) Lapped dovetail.

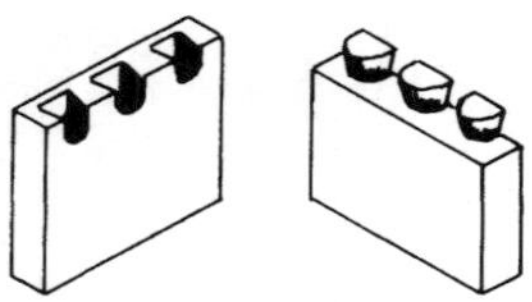

(c) Machine made dovetails.

Through dovetail
Figure 3.23*(a)* shows a common or through dovetail such as might be used when making a box. The angle of the dovetail is generally taken to be 1 in 6 or 1 in 7 (about 80°), and the proportions of pin to tail may be varied to suit the job in hand. Generally, equally proportioned parts give the strongest joint, whilst a fine pin with a wider tail gives a more pleasing appearance. When making the joint by hand, it is normal to cut the "tails" first, using them to mark the "pins" or "sockets". When made by machine, the two parts are cut simultaneously.

Lapped dovetail
Figure 3.23*(b)* shows a lapped dovetail as used for joining the sides and front of a drawer. In most instances the side (part B) would be thinner than the front (part A), thus maintaining maximum strength without unduly reducing the interior dimensions of the unit. The lapped dovetail is a very useful joint in situations where it is undesirable to show end grain on the face or front of the work.

Machined dovetails
Figure 3.23*(c)* shows the shape of machine-made dovetails.

HOUSED JOINTS

These simple but effective joints illustrated in Fig. 3.24 are used for joining wide boards at right angles as, for example, when making book cases and units containing shelves.

Plain housing
This joint is made by cutting a trench across the width of one piece of timber to receive the end of another as shown in Fig. 3.24*(a)*. The joint is secured by means of glue and nails.

Stopped housing
The stopped housing shown in Fig. 3.24*(b)* serves the same purposes as the through housing, but gives a neater appearance to the front edge of the work, the shoulder on part A concealing the joint and giving a clean uninterrupted line to the edge of part B.

Fig. 3.24. *Housed joints.*

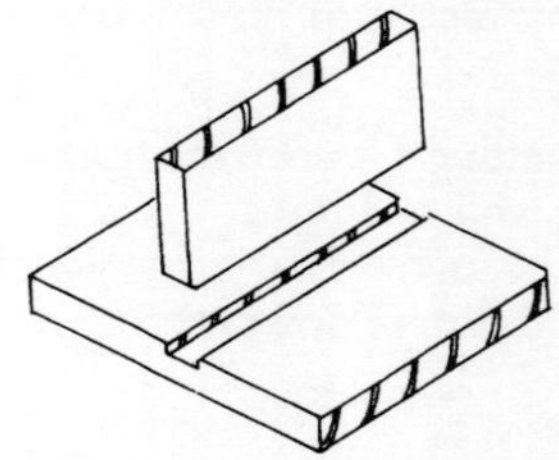

(a) Plain housing.

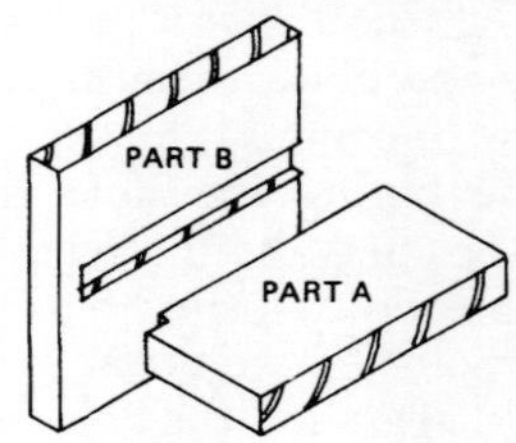

(b) Stopped housing.

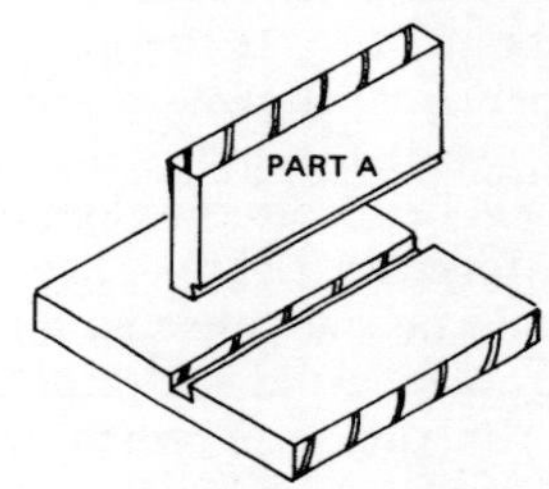

(c) Dovetailed housing.

Dovetailed housing

A further variation of the joint is the dovetailed housing shown in Fig. 3.24*(c)*. This joint would be used when the shelf or division (part A) is in tension, e.g. acting as a tie across the sides of the carcase. Only one side of the housed portion is normally dovetailed, this being more than sufficient to hold the parts in close contact. During assembly, glue is applied to the housing and the shelf or division is slid into position. The dovetailing renders nails or screws unnecessary – a useful feature of the joint when the item under construction is to be polished. Dovetailed housings can also be used in the "stopped" form shown in Fig. 3.24*(b)* if the joint is to be concealed on the front edge.

SPECIALISED JOINTS

Apart from the joints that have been considered in this chapter so far, there are many others which the craftsman is likely to utilise from time to time. Often these will be special joints, developed to suit a specific situation or requirement, and could therefore hardly be called basic. These will be dealt with in later chapters as the areas of work involving them arise.

ADHESIVES FOR TIMBER

Adhesives (or glues) are so widely used and commonplace throughout the construction industry and everyday life that they are often taken for granted, without much thought as to how they work or why they very occasionally fail. We tend to say simply that they stick things together. This complacency is due largely to the fact that modern adhesives are remarkably efficient, and are available in so many varieties, each to suit a particular purpose. In other words, the thinking has been done for us. However, to get the best use from an adhesive, and in order to avoid the problems associated with misuse, the craftsman should have a basic knowledge of what it does and how it does it.

Physical principles

When we consider the principles on which a glue works, we are concerned immediately with two natural phenomena; *(a)* cohesion, and *(b)* adhesion.

Cohesion

This is the force of attraction which exists between similar molecules in a substance. This cohesive force may be relatively weak or very strong. It is cohesion which holds together the molecules in a piece of steel to give it a particular shape. In this case the cohesive force is extremely powerful.

Similarly, it is cohesion which causes a water droplet to form on a dripping tap. The force here is weak – sufficient only to hold the liquid together in a very limited mass. (When water freezes cohesion obviously increases.)

Adhesion

On the other hand, adhesion is the force of attraction which exists between unlike molecules. It is the force of adhesion which causes the print to "adhere" to this page; it is the force of cohesion which holds the paper together, retaining its size and shape. Thus, here, both of these forces are at work together.

Properties of an adhesive

Where glues are concerned, adhesion itself takes two forms:

(a) specific adhesion – the molecular attraction we have just discussed, and

(b) mechanical adhesion – which for many years was considered to be the only way in which glue could form a strong bond.

When we make a glued joint in timber – such as a rubbed joint, for example – the glue penetrates the pores and cells of the wood to act like millions of microscopic dowels – glue dowels – which, having set, hold the joint together. Thus such a joint is really a mechanical joint in the same sense as that made with a screw or a nut and bolt. In the majority of cases, specific adhesion will also occur to help give a strong bond.

The properties required in a wood glue, therefore, are:

(a) low initial cohesion making it easy to spread and able to penetrate the surfaces;

(b) high ultimate cohesive strength so that when the glue has set, it is at least as strong as the material to be joined;

(c) good adhesion, both mechanical and specific, so that it will not easily release its hold on the wood.

Other properties, less fundamental but still important, are:

(a) water resistance;

(b) heat resistance;

(c) resistance to oils and spirits;

(d) resistance to bacteria and micro-organisms;

(e) suitable speed of set – preferably controllable;

(f) economy in use (cost);

(g) reasonable pot and shelf life (the keep-

ing properties of the glue when prepared and when stored respectively);

(*h*) simplicity in use;

(*i*) safety in use.

The foregoing list represents a formidable set of requirements which can rarely, if ever, be met in a single adhesive.

Selection

In view of all the possible requirements in selecting an adhesive for a particular job, it becomes necessary to define the main criteria for the joint and make the choice on that basis, achieving the best compromise possible with the rest.

Durability classification

Under the relevant British Standards Specifications, the durability of adhesives used in the construction industry is classified by their resistance to water as shown below:

INT — Not waterproof — suitable for interior use only.
MR — Moderately weather resistant; moisture resistant.
BR — Boil resistant.
WBP — Weather and boil proof.

In considering these classifications, it should not be automatically assumed that one adhesive is inferior to another by reason of its BS classification, but rather that it is more suited to a different type of work.

TYPES OF ADHESIVE IN COMMON USE

Animal glue

(BS classification: INT)

Best known as "scotch glue", this is made from the bones and hides of animals, and is available in jelly, cake and pearl forms. The latter two varieties require soaking in water and heating in a special glue pot before they can be used; the former generally merely needs warming.

Animal glues develop high strength, are relatively harmless and non-toxic, but have little resistance to moisture. They should therefore be used only in situations where they will be kept dry and free from damp. They are non-staining, and surplus glue is easily removed from a surface by swabbing with hot water — qualities which make them very suitable for furniture construction and for hand veneering. Frequent heating for re-use should be avoided as this leads to deterioration in strength.

Shelf life is practically unlimited provided the glue is stored under good conditions.

In use animal glues chill very rapidly, especially in cold weather, and therefore assembly of the components from the application of the glue to cramping up should be done as speedily as possible.

Casein glue

(BS classification: INT)

Casein glues are normally supplied as powders which require mixing with water for use. To obtain the best and most consistent results, the mixing should be done accurately, preferably by weight rather than by volume.

Casein is derived from milk and is made soluble by the addition of an alkali. Thus the glue is alkaline by nature and tends to stain timbers to which it is applied — often so badly as to preclude its use. Less alkaline types are available, but non-staining properties can only be achieved at the cost of some loss in strength.

Casein glues are used cold, have quite good gap-filling properties, and have better resistance to moisture than animal glues. However, they are not waterproof adhesives, and are liable to attack by bacteria and micro-organisms.

Properly stored, the shelf life is good and the pot life at normal room temperatures is in the region of 6 to 8 hours. When mixing, the use of non-ferrous metal containers should be avoided — glass, earthenware, plastic and iron all being satisfactory.

Whilst casein glue gives a very strong bond, its properties generally restrict its use to interior softwood joinery and similar work which is to be painted. It is used commonly in the construction of flush doors.

Polyvinyl acetate

(BS classification: INT)

This adhesive has found increasing use in the woodworking trades over the past few years, mainly due to the fact that it is so quick and clean in use. The adhesive is a creamy white liquid with a syrupy consistency and is supplied ready for use generally in a plastic container. Woodworking adhesives of this type are water based emulsions and, although relatively costly, their simplicity in use and wide range of application make them both reliable and economical. Most types may be diluted with water to give increased coverage on large areas, making them ideal bonding agents on surfaces where adhesion might otherwise be a problem (plastering, rendering, etc.).

Polyvinyl acetates contain no harmful ingredients and are thus quite safe in general use, surplus or spilt glue being easily removed from the work or from the hands with water. Generally, these glues retain a fair amount of flexibility and some have good gap-filling qualities, but since they are "thermoplastic" resins, they have a tendency to creep under prolonged load, and should not be considered as constructional adhesives for use in stressed components. Resistance to heat and water is poor.

One of the most versatile and widely used adhesives, its main use is undoubtedly in the assembly of furniture and interior joinery.

NOTE: A good many modern adhesives, including polyvinyl acetates, urea resins and resorcinol resins, are what are known as "plastics". They are in fact similar in behaviour to the more commonly known plastic materials such as nylon, polythene, bakelite, etc.

Plastics fall naturally into two main groups:

(a) thermoplastic materials – materials which soften with the application of heat, e.g. nylon, polythene, etc., and

(b) thermosetting materials – materials which harden with the application of heat, e.g. bakelite, melamine, etc.

Since a certain amount of heat is developed when the molecules of a substance are severely stressed – either by squeezing or stretching – *thermoplastic* adhesives tend to allow a stressed joint to creep and are therefore unsuitable for highly stressed situations. *Thermosetting* resins, on the other hand, are ideal for this.

Urea resins
(BS classification: MR)

These thermosetting synthetic resins are produced by reacting urea resin with formaldehyde, and are available either as syrups or as powders, generally requiring the addition of a hardener – usually an acid – to cause the adhesive to set. Powdered resins have a longer shelf life than syrups but have to be mixed with water to form a syrup before they can be used.

Hardeners are usually supplied with the resin but in a separate container, and the manufacturer's instructions for mixing should be carefully followed if good, consistent results are to be obtained. Certain urea resins, supplied in powder form, have the hardener already added and require only the addition of water – in the correct proportion – to make them ready for use.

Method of application

The hardeners used with urea resins are themselves available in both powder and liquid form thus giving the user a choice as to the method of application.

(a) Mixed application. In this method, the hardener is added to the liquid resin allowing the resultant glue to be applied to the joints with a brush, stick or glue spreader in a single operation. It should be realised that once the hardener is added to the liquid resin the glue has a strictly limited pot life which will vary in accordance with room temperature – the higher the temperature, the shorter the time available.

(b) Separate application. This is an extremely useful method of application, especially where large components are concerned and assembly time is prolonged. The resin and hardener are kept separate – the resin being applied to one part and the hardener to the other – the two parts being brought together when all is ready. Thus the set does not commence before the components are assembled, and the pot life of both resin and hardener is greatly extended with consequent reduction in wastage.

Uses

Urea resins are non-staining, have a high resistance to water (cold), and are thus very suitable for use in all types of interior or exterior joinery and cabinet making, their setting time often being greatly reduced by raising the temperature of the glue line by means of high and low frequency radio waves or low voltage strip heating.

Handling

Urea resin adhesives are skin irritants and should be used and handled with care. Splashes on the skin should be removed with soap and water as soon as possible, and, where contact with glue is prolonged, the use of rubber gloves and/or barrier creams is advisable.

Resorcinol resins
(BS classification: WBP)

These adhesives have been developed during the last two decades and are supplied in the form of dark, reddish-brown liquids with either liquid or powder hardeners which are added to the resin to cause it to set. They are thus two-part adhesives, but are normally used by the mixed application method, having a slower set than the urea resins which gives the operative more time during assembly. As with urea resins, accurate mixing of the component parts is essential.

Uses

Undoubtedly, the most useful feature of the resorcinol adhesives is their outstanding strength and durability, properties which make them the obvious choice for any timber structures likely to be subjected to high stresses and/or severe exposure to the weather, such as portal frames, box beams, laminated trusses and boat building. They also have good resistance to oils, spirits, heat, fungi, bacteria and micro-organisms. In considering the possible uses of resorcinol resin, it should be borne in mind that the material is very expensive and, therefore, it would not be economical to use it where a urea resin could prove satisfactory. Because of the cost of the basic raw material, resorcinol resins often contain some proportion of less expensive resin such as phenol. They are sometimes also added, during manufacture, to other less powerful adhesives, to serve as a reinforcement.

Handling

As is the case with most types of thermosetting synthetic resin, resorcinol is a skin irritant and requires the same care and hygiene in use as urea. It is, however, far more readily removed from the skin with soap and water than are urea resins, and the risk of damage to the skin is therefore reduced.

Rubber based adhesives

The only adhesives of this type with which the carpenter and joiner is likely to be involved are the "impact" or "contact" adhesives used for the bonding of laminated plastic sheets. They consist of mixtures of natural and synthetic rubber and resin in a highly volatile solvent, such as naphtha or acetone. They have no great structural strength and little resistance to heat or water, being formulated strictly for the purpose stated.

Application

Depending upon the actual brand used, the adhesive is applied to both surfaces by means of a spray gun, brush or comb, and then left for the solvent to evaporate. The two surfaces are then brought together. Since the bond is instant on contact, it is essential that care is taken to ensure correct alignment when the surfaces are brought together, there being little or no "slide". Solvents are available for removing unwanted adhesive.

NOTE: Rubber based impact adhesives give off a heavy inflammable vapour which may be harmful if inhaled. The user should refrain from smoking, make sure there are no naked lights in the vicinity, and ensure that the work area is well ventilated.

FURTHER READING

British Standards and Codes of Practice

CP 112	Parts 1 and 2:1973 The structural use of timber.
BS 565:1972	Glossary of terms relating to timber and woodwork
BS 745:1969	Animal glues for wood
BS 1186	Quality of timber and workmanship in joinery
	Part 1:1971 Quality of timber
	Part 2:1971 Quality of workmanship
BS 1204	Synthetic resin adhesives (phenolic and aminoplastic) for wood
	Part 1:1979 Specification for gap filling adhesives
	Part 2:1979 Specification for close contact adhesives
BS 4071:1976	Specification for polyvinyl acetate (PVA) emulsion adhesives for wood
BS 5442	Part 3:1979 Adhesives for use with wood

Building Research Establishment Digests

No. 175	Site use of adhesives: Part 1
No. 211	Site use of adhesives: Part 2
No. 212	Choice of glues for wood

SELF-TESTING QUESTIONS

All the information required to answer the following questions is contained within this chapter. Attempt each section *as fully or as briefly* as the question demands, and then check your answers against the information given in the chapter.

1. *(a)* State four functions of a woodworking joint.
(b) State three requirements of a woodworking joint.

2. *(a)* Sketch sections to illustrate the following edge joints: *(i)* rubbed butt joint; *(ii)* cross tongued joint; *(iii)* slot screwed joint.
(b) State the main advantages in using: *(i)* cross tongued joints, and *(ii)* slot screwed joints, in preference to rubbed butt joints.

3. *(a)* Name and sketch three types of

joint used to lengthen a timber member, giving a situation where each might be used.

(b) Sketch two types of timber connector used where timbers are bolted together and state how they give added strength to the joint.

4. *(a)* Sketch the following types of joint and give an instance where each might be used: *(i)* tee halving joint; and *(ii)* dovetailed halving joint.

(b) Sketch a typical "through" mortice and tenon joint and explain briefly:

(i) how the joint would normally be secured;

(ii) the proportions to be borne in mind when setting out the joint.

5. *(a)* State three functions of the "haunch" in a haunched mortice and tenon joint.

(b) Explain, with a sketch, how a groove on the inside edges of a framework affects the mortice and tenon joints.

(c) Explain, with a sketch, how a rebate on the inside edges of a framework affects the tenons of the mortice and tenon joints.

6. *(a)* Explain, with simple sketches, the main factor which determines whether a "stuck" moulding on the inside edges of a mortice and tenon jointed framework should be mitred or scribed.

(b) Sketch details to show the essential difference between: *(i)* hand scribing, and *(ii)* machine scribing, of an ovolo mould on the inside edges of a mortice and tenon jointed frame.

(c) Explain with sketches, what is meant by: *(i)* dowelling, and *(ii)* draw-boring, a mortice and tenon joint.

7. *(a)* Sketch and name joints suitable for:

(i) joining the corners of a tool box;

(ii) joining the front and side of a drawer.

(b) Sketch two types of "housed" joint, giving instances of where each might be used.

8. *(a)* State the meaning of the following abbreviations in respect to adhesives: *(i)* INT, *(ii)* MR, *(iii)* BR, *(iv)* WBP.

(b) *(i)* Name two types of adhesive which are generally classified INT.

(ii) Name a type of adhesive which is generally classified WBP.

(c) Name adhesives suitable for the following purposes:

(i) assembly of furniture and interior joinery;

(ii) exterior softwood joinery;

(iii) laminated portal frames.

9. *(a)* Explain briefly what is meant by: *(i)* mixed application, and *(ii)* separate application, as applied to synthetic resin adhesives.

(b) State the precautions which must be observed when using: *(i)* synthetic resin adhesives, and *(ii)* rubber based contact adhesives.

4. Workshop Drawings, Calculations and Geometry

After completing this chapter the student should be able to:

1. State the function of a setting out rod.
2. Prepare a setting out rod for a four-light casement or a four-panel door.
3. Prepare cutting lists for casement windows and panelled doors.
4. Calculate the cost of the materials for simple items of carpentry and joinery.
5. Prepare simple working drawings in conformity with BS1192.
6. Draw the plan and elevation of a simple solid object.
7. Provide graphic solutions to problems involving basic plane geometry.

The craftsman in wood, whether he be a carpenter and joiner, cabinet maker or wood machinist requires far more ability than the skilful manipulation of the tools of his trade. Indeed it would be true to say that the ability to cut accurately to a line using a saw, plane, chisel or machine is a skill which is quite readily acquired by anyone. The real craftsman, however, is the person who can not only cut and finish wood expertly and accurately but can also see the finished job in his "mind's eye" before he starts work upon it. In other words the good craftsman must be able to measure his timber and so place his marks upon it that upon completion the job is both properly constructed and of the correct size – in conformity with the specification.

SETTING OUT

Whilst many of the jobs undertaken by the craftsman may be sufficiently simple as to enable him to merely take up his tools and materials and commence work, safe in the knowledge that he knows exactly what to do and how to do it, other items of work may be of such complexity that to set about the work in this way would almost certainly be courting disaster. Indeed, it is sometimes difficult or even well nigh impossible to assess accurately the quantities and exact dimensions of the timbers required for a particular job, let alone to expect another craftsman or several other craftsmen to construct the item or items involved without some definite form of reference which can be read and understood by all.

Setting out rod

This standard reference containing all the details required by all persons working on the job is really a form of working drawing,

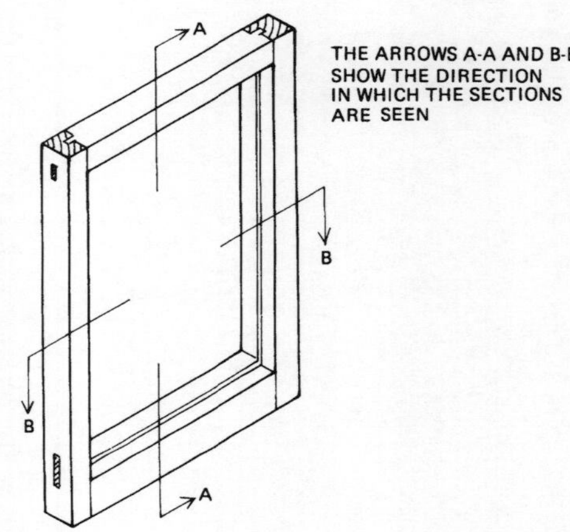

Fig. 4.1. *Sketch of a sash for a casement window.*

drawn full size, to show the shapes, sections, sizes and other essential details. It is known as a "setting out" or "setting out rod" since it is often drawn on a thin board.

A setting out rod most often shows full size vertical and horizontal sections through the job in question, but may also, especially where shaped work is concerned, include plans and/or elevations.

Figure 4.1 shows a pictorial sketch of a simple sash for a casement window, vertical and horizontal sections being identified by A-A and B-B.

Figure 4.2 shows a setting out rod for this sash, sections A-A and B-B being referred to as height and width rods respectively. Reference to Fig. 4.2 should make it plain that all the essential information necessary for the construction of the sash can be obtained from the setting out rod, and that any number of joiners making the sash *from this rod* would produce basically similar articles.

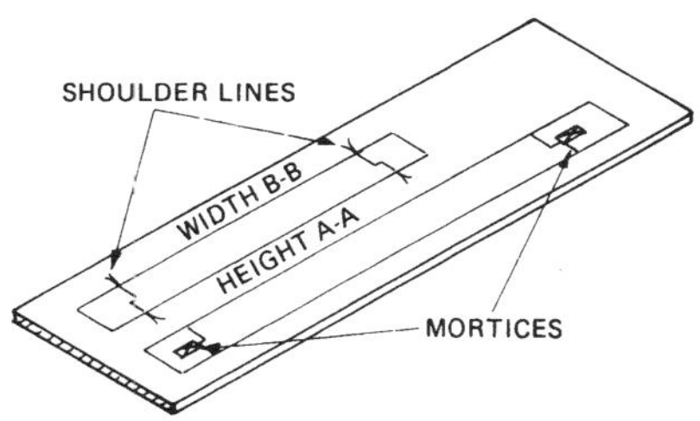

Fig. 4.2. *Setting out rod for sash in Fig. 4.1.*

The functions of a setting out rod

These can therefore be listed as follows:

(*a*) to provide a standard reference;

(*b*) to indicate overall dimensions;

(*c*) to indicate sectional shapes and dimensions;

(*d*) to enable a cutting list to be quickly and accurately prepared;

(*e*) to show the position and proportions of joints;

(*f*) to simplify the "marking out";

(*g*) to "iron out" the problems associated with the intersection of shaped members;

(*h*) to enable sections of a large piece of work to be undertaken separately.

In addition to the functions listed, the setting out rod is a valuable means of assessing the general proportions of a complicated piece of work (height of rails, width of panels, etc.), and may even show, by the use of coloured pencils, whether the timber is a hardwood or a softwood. Fixing details and allowances for joints between separate components may also be shown.

NOTE: To be useful, a setting out rod must be clearly and accurately drawn, and should be carefully checked over before use as any error will inevitably be reflected in the finished work.

CUTTING LISTS

A cutting list is an accurately prepared itemised list of all the timber required for a particular job. Its purpose is simple and straightforward – to enable the woodworker to prepare all the various pieces of timber required for the construction of the job with an absolute minimum of waste. It should, therefore, give not only the lengths, widths and thicknesses of the material, but also the names of the members and the number of pieces required. Since defects in one form or another are common in most timbers (*see* Chapter 2) it is always useful to know when selecting the material exactly which component one is dealing with, as a defect which might be a serious drawback in one situation may be minimised or completely eliminated in another.

A further point of importance is the inclusion in the list of both nominal (sawn) and finished sizes. This not only eliminates any possible misunderstandings in respect of the actual finished sizes required but also simplifies the sawing and costing of the job. It is worthwhile to indicate in the cutting list the type of material to be used for each component, i.e. hardwood or softwood, and any special instructions or treatment required, i.e. "machined to pattern" or "to cut four", etc.

Figure 4.3 shows a pictorial sketch of a

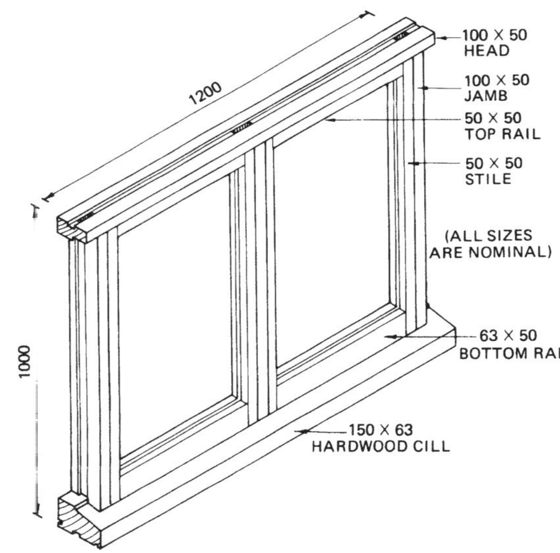

Fig. 4.3. *Sketch of a two light casement frame and sashes.*

two-light traditional casement frame fitted with opening sashes with the overall dimensions and nominal sizes of the members indicated. Figure 4.4 shows a setting out rod for the window and gives the sectional shapes of all the members and positions of all the mortice and tenon joints.

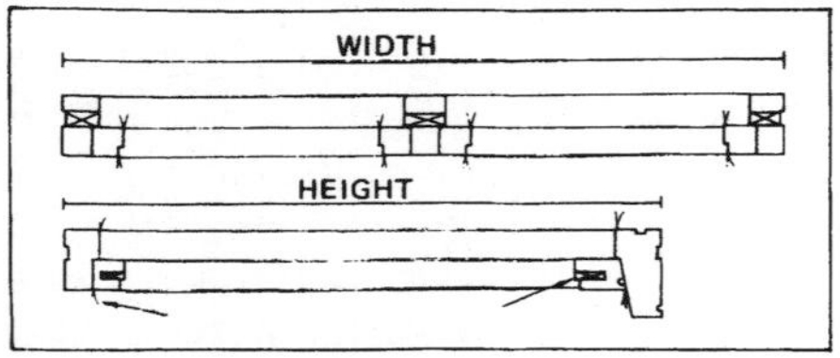

Fig. 4.4. *Setting out rod for casement window in Fig. 4.3.*

A cutting list for four of these casement windows is shown in Table II, prepared directly from the setting out rod as previously stated. The list should be studied carefully in relation to Figs. 4.3 and 4.4 until the reader is satisfied that the various components can be readily identified in each of the three illustrations.

Points to note

(a) An allowance of 60 mm has been made for each horn on the head and cill.

(b) An allowance of 20 mm has been made for each horn on the sash stiles.

(c) An allowance of 5 mm has been made on the lengths of the top and bottom rails of the sashes.

(d) No allowance has been made on the length of the two jambs and the mullion since the mortar key on the outside edges of the frame caters adequately for this.

TABLE II. CUTTING LIST FOR FOUR NO. CASEMENT WINDOWS

Member	*No.*	*Length*	*Breadth*	*Thickness*	*Finished size*		*Material*	*Remarks*
Head	4	1,320	100	50	95	45	Baltic red	
Cill	4	1,320	150	63	145	58	English oak	Machined
Jambs	8	1,000	100	50	95	45	Baltic red	
Mullions	4	1,000	100	63	95	58	Baltic red	to
Stiles	16	964	50	50	45	45	Baltic red	
Top rails	8	553	50	50	45	45	Baltic red	pattern
Bottom rails	8	553	63	50	58	45	Baltic red	

(e) The reader is urged to prepare a full size setting out rod of the casement frame in question in order to interpret better the value and function of both setting out rod and cutting list.

MARKING OUT

Marking out, sometimes loosely referred to as "setting out", is the actual marking on the material of all the lines necessary for the cutting, shaping and jointing of the various members involved in the construction of a piece of work. It is therefore a most important aspect of the craftsman's work, since any errors or inaccuracies at this stage of the job will inevitably result in similar discrepancies in the finished article.

It is difficult to over-emphasise the importance of good marking out, no matter what type of job is involved, because not only does the usefulness or otherwise of the job depend to a great extent upon its correctness, but frequently one craftsman may have to work to another's marking out, and what may be obvious to the marker out, may be quite obscure to the person cutting the joints.

It should be obvious that marking out should therefore be as standard a procedure as possible — at least within any one workshop — if errors and misunderstandings are to be avoided. Admittedly, variations in methods of marking out are bound to occur from one workshop to another, but such variations are likely to be minimal — the broad principles involved being more or less universally adopted.

Main requirements

The essential features of good marking out can be summarised in the following points which are illustrated in Fig. 4.5.

(a) Face and face edge marks as shown in Fig. 4.5*(a)* must be clearly applied to all relevant parts and used as references — datums — from which much of the marking out and working is done.

(b) Face and face edge marks should be applied only after careful inspection of the material, bearing in mind that the two faces so marked are usually the most important of the four surfaces and are normally therefore the "best" two adjacent sides of the wood. The position and severity of any defects present in the wood should also be considered

Fig. 4.5. *Marking out.*

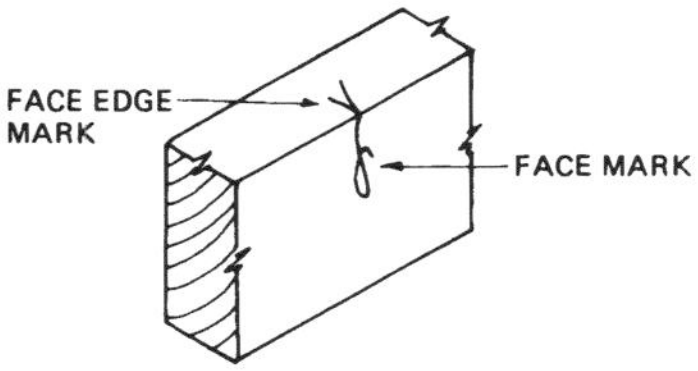

(a) Face and face edge marks.

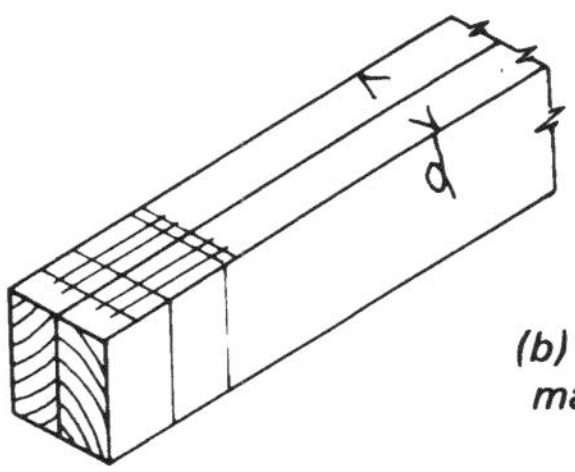

(b) Pairing members when marking out (door stiles).

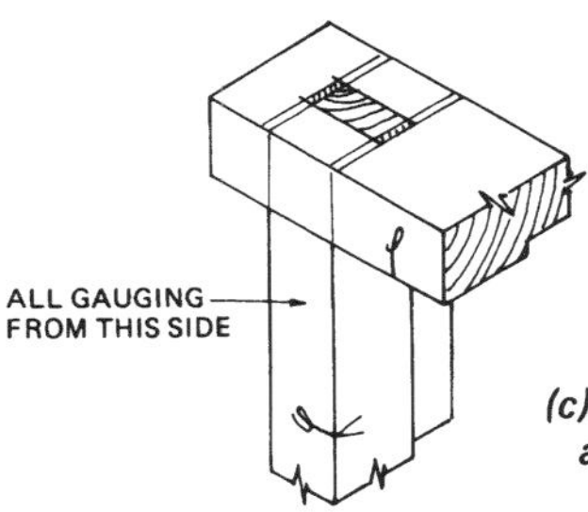

(c) Application of face and face edge marks.

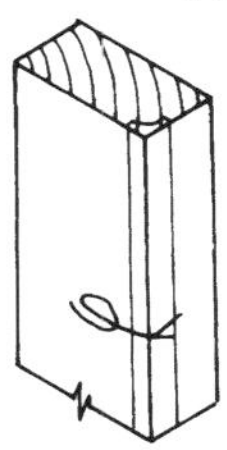

(d) Face marks on moulded frame.

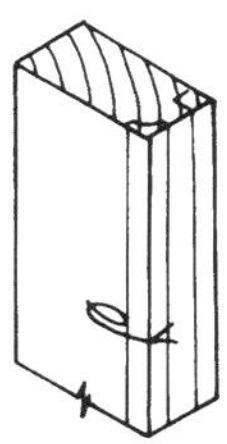

(e) Face marks on moulded and rebated frame.

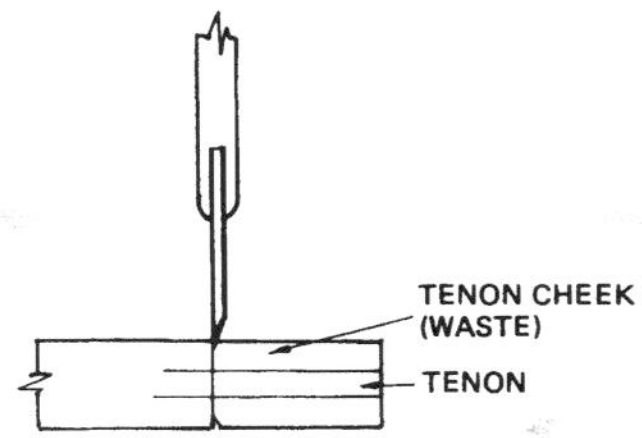

(f) Cutting shoulders with marking knife. Note bevel in waste.

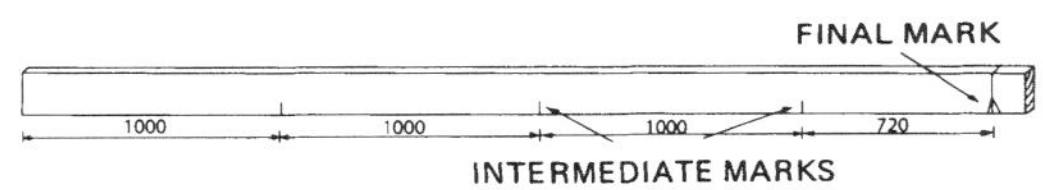

(g) Stepping off with a one metre rule. Note final mark.

at this stage. Often a damaged corner, a small knot or shake can be removed at a later stage when grooves, rebates or mouldings are cut if the face and face edge marks have been applied sensibly.

(c) "Paired members" such as door stiles and jambs which are "right and left handed" must be marked out as pairs (*see* Fig. 4.5*(b)*) since they are not reversible.

(d) The face side and face edge of the material are almost invariably the front and inside edges of a framework. Note that the face side is not necessarily the widest face of the timber but rather the face from which the gauging is done. This fact will be readily appreciated if the door frame joint in Fig. 4.5*(c)* is studied.

(e) All gauging for mortices, tenons, rebates, etc., should be done *from the face side* unless the actual gauge line is to be applied *to* the face, in which case the gauging is done from the face edge.

(f) Where a framework is to be moulded on the inside edge the face mark should be applied to the moulded side as shown in Fig.4.5*(d)*. This rule applies whether the back edge be plain or rebated as shown in Fig. 4.5*(e)*.

(g) Where one side of the framework is flush and the reverse side is not (as is the case of the casement frame shown in Fig. 4.4) the flush side should be chosen for the face side.

(h) Marking out should be clear and as simple as possible with no unnecessary lines. Pencil lines should be clear and sharp, made with the chisel edge point of a 2H pencil (*see* Fig. 4.18*(f)*), the well-known flat, soft, carpenter's pencil being of very limited use in the workshop and best avoided for anything other than carcassing work.

(i) Shoulder lines on joints, or indeed any other line which will eventually be cut with a saw or chisel and therefore requires a clean sharp edge, should be cut with a marking knife as shown in Fig. 4.5*(f)*. Obviously, the marking knife is not essential for marking out carcassing timber for roofs, floors, etc.

(j) As far as is possible, *all* the marking out for a particular job should be carried out as a single operation at the commencement of the job. This practice, which requires a certain amount of care, understanding of the job and a good deal of confidence in one's ability, is the hallmark of the competent woodworker and generally results in accurate work and saving in time.

(k) Errors in marking out, wrongly placed or double lines should be removed immediately and corrected. If left, it is almost inevitable that sooner or later, the wrong line will be cut.

Fig. 4.6. *Marking out a four-panel door.*

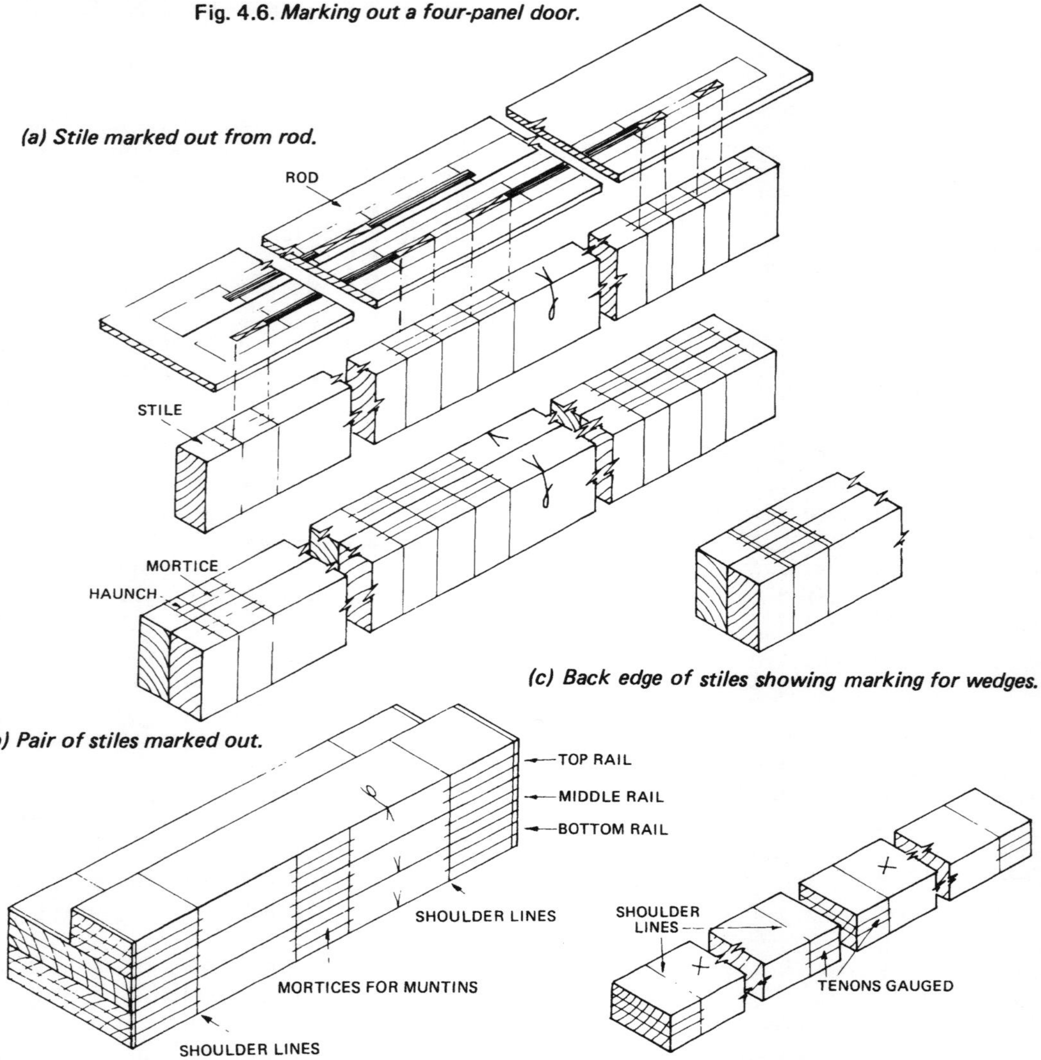

(a) Stile marked out from rod.

(b) Pair of stiles marked out.

(c) Back edge of stiles showing marking for wedges.

(d) Top, middle and bottom rails marked out.

(e) Muntins marked out (from rod).

(*l*) When using the rule to "step off" long dimensions, a different type or mark should be made for the final measurement as shown in Fig. 4.5(*g*) as it is all too easy to square off at the wrong place, especially if the last measurement is a small one.

Marking out from the rod

Figure 4.6(*a*) shows a door stile for a four-panel door marked out directly from the setting out rod. Note how the mortices have been taken from the position indicated on the rod. Figure 4.6(*b*) shows the two stiles for the same door, the second having been marked from the first. Note how the stiles have been "paired". Figure 4.6(*c*) shows the back edges of the same stiles. Note how the *wedge spaces* have been marked on the *outside* of the mortices.

Figure 4.6(*d*) shows how the rails are marked out, the positions of the shoulder lines once again being taken directly from the setting out rod, as are also the mortices for the muntins. The shoulder lines should, of course, be cut with the marking knife.

Figure 4.6(*e*) shows the marking out for the two muntins, the shoulder lines once again being taken directly from the setting out rod.

Figure 4.7 shows how door stiles (or similar components) are marked out in quantity when a number of identical items are to be made. In this case, one stile has been marked out from the rod and its "pair" marked from this as previously described. They have then been place one on either side of the remaining pairs of stiles, the whole squared up and held together with cramps, and the mortices marked across using a pencil and short straightedge.

Note how the intermediate stiles have been paired up.

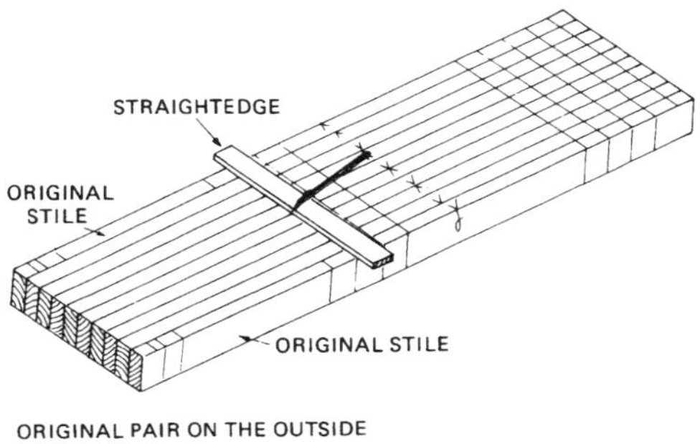

Fig. 4.7. *Marking out "batches" of stiles.*

Marking out for machine production

Where joinery is to be produced by the use of woodworking machinery, the procedure for marking out is somewhat different to that required for working by hand.

In the first instance the marking out is generally only required on one of each of the various components, since the machine, once "set up" to the pattern member, will reproduce the same cut repeatedly without further guidance. Where the chopping of mortices is concerned, much depends upon the sophistication of the machine being used – it is possible to set up a gang morticer to cut the mortices semi-automatically by the use of stops. For hand morticing with either a chain or hollow square chisel machine (*see* Chapter 12) it is usually necessary to mark out all the mortices – although gauge lines may be dispensed with where a large number of mortices are to be chopped to the same setting. It is nevertheless worthwhile to keep a check on the accuracy of mortices cut this way as worn guides on the machine may cause the chisel to wander.

Shoulder lines

Since well maintained woodworking machines cut extremely cleanly and accurately, it is not necessary to cut shoulder lines with the marking knife. In fact such lines tend to be troublesome rather than helpful. A good clean pencil line to indicate the position of the shoulders on the pattern for each batch of similar rails is all that is required, the remainder being set up to "stops".

Gauge lines

Such lines for grooves, rebates or mouldings are not required for machine woodworking although, as before, one completely marked out member is useful when setting up the machine. A gauge line can be useful from time to time where a difficult timber is being machined and the grain shows a tendency to splinter out. In such cases, a good clean line made with a cutting gauge can prove helpful.

Face and face edge marks

These are always necessary, whether the work is to be carried out by hand or machine.

Marking out on pre-machined stock

Marking out of joinery is normally carried out on timber which has been trued up and planed on all four faces (par) it being easier to handle, gauge and square over on such material. Most joiners' workshops, however, do carry quantities of standard section stock material for such items as door and window frames and sashes which are possibly required in small quantities at fairly regular intervals. The marking out for items made from stock materials is basically the same as has been previously described, for both hand and machine methods of construction, but with two subtle differences.

(*a*) The proportion of the joints, thickness and widths of tenons, etc., is predetermined by the section of the stock material.

(*b*) Squaring over on machined components such as sash glazing bars is usually done with the aid of a squaring templet as shown in Fig. 4.8. This little tool, usually made in hardwood by the joiner himself, is much easier to use for this purpose than the try square.

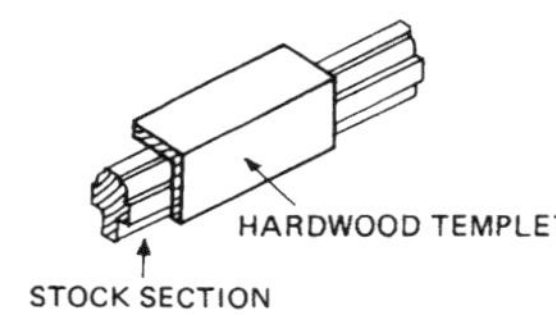

Fig. 4.8. *Squaring templet used on moulded stock.*

TAKING OFF AND COSTING OF MATERIALS

Many woodworkers, notably those self-employed or working in a smaller type firm, are expected to calculate and possibly cost the amount of timber and other materials used in a particular job. It is unlikely this responsibility will fall upon the novice or young apprentice, but almost certainly the necessity will arise sooner or later, and in any case, the usefulness of being able to do so needs no stressing.

Taking off

This is the term applied to the process of measuring or calculating from a plan, detail drawing or setting out rod the quantities of material to be used in a particular job. Obviously, since most of the components in a piece of woodwork are of different sections and sizes, separate calculations are required for each. Nominal (sawn) sizes are used in taking off.

Costing

Costing of the material involves simple arithmetic, i.e. multiplying the various amounts of the material worked out, by their *unit cost.*

Metric units (timber)

Timber is bought and sold, and therefore costed, in units which may vary in accordance with the type and quantity of the material involved.

Running metre (m)

The running or linear metre is the unit used when dealing with timber which is most generally and conveniently sold by length, i.e. mouldings, skirting boards, pre-machined stock and small quantities of standard section material.

Square metre (m^2)

The square metre is used in calculating and costing sheet materials such as plywood, chipboard, etc., and for strip materials where areas are involved such as flooring, match boarding, wood floor blocks, and cladding boards. Where very large quantities are involved, prices for such materials may be quoted in units of ten or even one hundred square metres, i.e. per 10 m^2 or per 100 m^2 respectively.

Certain hardwoods which are imported in random lengths, widths and thicknesses are often sold by the square metre, the price per m^2 being dependent on the type of timber, its quality and its thickness. Long wide boards generally cost more per square metre than shorter, narrower ones.

Cubic metre (m^3)

The cubic metre is the unit used where large quantities of timber are involved, i.e. for the bulk buying and selling of heavy section timbers such as roofing, joisting and general carcassing materials. Large quantities of joinery timber are also bought and sold by the cubic metre.

The cubic metre (m^3) is in fact merely a bulk quantity of timber (calculated) and has no direct significance as to the lengths, widths and thicknesses of the pieces of timber involved. For example, 1 cubic metre of softwood contains 100 metres of 200 × 50, 200 metres of 100 × 50 and 800 metres of 25 × 50. In the same way 300 metres of 100 × 50 = 300 × 0.1 × 0.05 m^3 = 1.5 cubic metres.

Examples of calculations

Example 1

Calculate the cost of eight shelves, each 4.000 m long by 200 mm wide by 25 mm thick, if the material is charged at £1.28 per metre run.

Total length of shelving = 8 × 4.000
= 32 m

∴ Cost @ £1.28 per metre
= 32 × £1.28
= £40.96

Note that in this instance linear quantities only are involved, the sectional size having no relevance other than the determination of the unit cost (£1.28 per metre run).

Example 2

Calculate the total cost of the following list of timbers:

	No. of pieces	*L*	*B*	*T*
1.	6	4.500 m	100	25
2.	6	1.800 m	150	38
3.	10	2.100 m	50	75
4.	4	3.600 m	225	50

given that the unit cost of each is as shown below:

1.	100 × 25	–	£0.48 per metre
2.	150 × 38	–	£0.86 per metre
3.	50 × 75	–	£0.74 per metre
4.	225 × 50	–	£2.26 per metre

Cost of Item 1
= 6 × 4.500 × £0.48 = £12.96

Cost of Item 2*
= 6 × 1.800 × £0.86 = £9.29

Cost of Item 3
= 10 × 2.100 × £0.74 = £15.54

Cost of Item 4*
= 4 × 3.600 × £2.26 = £32.54

∴ Total cost = £70.33

*Items 2 and 4 are calculated to the nearest penny.

Example 3
Calculate the number of square metres of flooring required for a room measuring 5.600 m long by 4.250 m wide, allowing 8 per cent extra for waste and cutting.

Area of floor = 5.600×4.250
= $23.800\ m^2$

∴ Area + 8 per cent

$$= \frac{23.800 \times 108}{100}$$
$$= \underline{\underline{25.704\ m^2}}$$

NOTE: Where a percentage is to be added to the calculated quantity, divide by 100 to find 1 per cent and multiply by 100 *plus* the required percentage. In this case we have 100 + 8 = 108 per cent.

Example 4
Calculate the number of square metres of matchboarding required to cover both sides of a partition 2.400 m high by 4.28 m long, allowing 5 per cent extra for waste and cutting.

Area to cover = $2.400 \times 4.280 \times 2$ (for both sides)
= $20.544\ m^2$

∴ Area plus 5 per cent

$$= \frac{20.544 \times 105}{100}$$
$$= \underline{\underline{21.57\ m^2}}$$

Example 5
How many running metres of match board are required for the partition in Example 4 if the match boards have a covering width of 92 mm? (To find this we must divide the calculated area by the width of the board.)

No. of running metres

$$= \frac{21.571}{0.092}$$
$$= 234.46\ m \text{ — say } \underline{235} \text{ metres.}$$

NOTE: The width of the board (92 mm) must be expressed as *0.092 m* since all the units of measurement in a calculation must be the same, i.e. all in mm, or all in m, the latter unit being the more convenient.

Example 6
Calculate the total cost of 34 floor joists, each 4.1 m long by 200 mm deep by 50 mm thick, given that the timber is charged at £110 per m^3.

Volume of 34 joists
= $34 \times 4.100 \times 0.2 \times 0.05\ m^3$
= $1.394\ m^3$

(Note that mm have again been expressed as m.)

∴ Cost @ £110 per m^3
= $1.394 \times £110$
= $\underline{\underline{£153.34}}$

Example 7
How many running metres of 100 X 50 mm are contained in one cubic metre? (This involves dividing *one square metre* by the sectional area of the timber involved — in this case 100 X 50 (*see* Fig. 4.9).

No. of running metres of 100 X 50 in 1 m^3

$$= \frac{1.000}{0.1 \times 0.05}$$
$$= \underline{\underline{200\ m}}$$

NOTE: Though the units used in this calculation are in m the problem could have been conveniently solved using mm, thus:

$$\frac{1{,}000{,}000}{100 \times 50} = \frac{1{,}000}{5} = \underline{\underline{200\ m}}$$

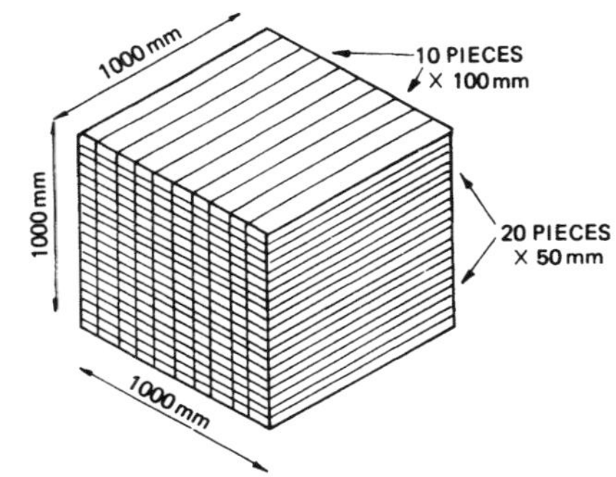

Fig. 4.9. *One cubic metre of 100 X 50 (200 pieces, 1 metre long).*

Useful formulae

Area of a triangle
$= \frac{B \times H}{2}$ (B = Base) (H = Vertical height)

Area of a trapezium
$= \frac{L + l}{2} \times H$ (L = Long side) (l = Short side)

Area of a parallelogram
$= H \times L$ (H = Perpendicular height) (L = Length)

Circumference of a circle
$= \pi \times D = 3.142 \times$ Diameter

Area of a circle
$= \pi r^2 = 3.142 \times$ Radius $\times$ Radius

Perimeter of an ellipse
$= \pi \times (R + r)$ (R = Semi-major axis) (r = Semi-minor axis)

Area of an ellipse
$= \pi \times R \times r$
Surface area of a sphere
$= 4\pi r^2$
$= 4 \times 3.142 \times \text{Radius} \times \text{Radius}$
Volume of a sphere

$= \frac{4}{3}\pi r^3$
$= \frac{4 \times 3.142 \times \text{Radius} \times \text{Radius} \times \text{Radius}}{3}$

DRAWING PRACTICE

Much of the information from which the carpenter and joiner works is given in the form of drawings, either *scale drawings* of sufficient accuracy for dimensions to be taken from them, or *sketches* – rule assisted or freehand – to explain a particular aspect of the construction. Drawings, in one form or another, are therefore of great importance to all who have to design, build or otherwise carry out construction work, and save vast amounts of verbal or written instructions.

A drawing should be completely unambiguous, giving the necessary information in a way that is clear and easy to understand by all concerned. Written instructions in the form of a *specification* may be used where necessary to clarify further the drawing or some particular aspect of it. Freehand or rule assisted sketches are one of the best ways of receiving or passing on information at craft level – from one craftsman to another – and therefore the simple art of producing clear and proportional sketches is one which should be practised by all would-be craftsmen.

British Standard 1192

British Standard 1192 (Building Drawing Practice) lays down recommendations with reference to the production of working drawings used in the construction industry in an endeavour to achieve some degree of uniformity (and therefore less chance of misunderstanding) in the methods of graphical representation of building projects. It is obviously good policy to adhere to these recommendations whenever drawings have to be produced in order to ensure that the recipient can extract from them exactly the same information as the draughtsman had in mind during their production. The main points of BS 1192 which are likely to influence the carpenter and joiner are outlined, in a simplified form, as follows.

Layout
A simple but effective layout of the drawing sheet should be adopted as a standard, a typical layout being shown in Fig. 4.10. The essential features of a good layout are:

(a) a 12 mm margin at the top, bottom and right hand side, and a 20 mm margin for filing purposes on the left;

(b) the information panel at the bottom of the sheet should contain the date, the drawing title, the scale used and the draughtsman's name. Other information such as drawing number, client's name and address, etc., can be included where necessary.

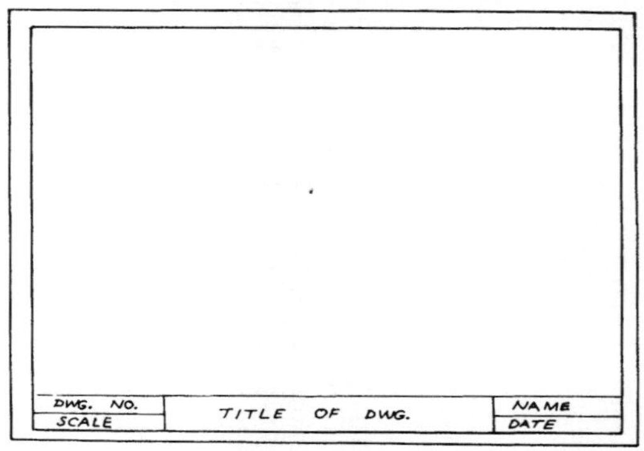

Fig. 4.10. *Simple layout of drawing sheet.*

Solid lines
These should be drawn in varying thicknesses, depending upon their specific purpose. The various types and thicknesses of lines are shown in Fig. 4.11.

(a) Thick lines are used for the site outline of new buildings on block and site plans, the section outlines of main constructional elements or any outline needing emphasis.

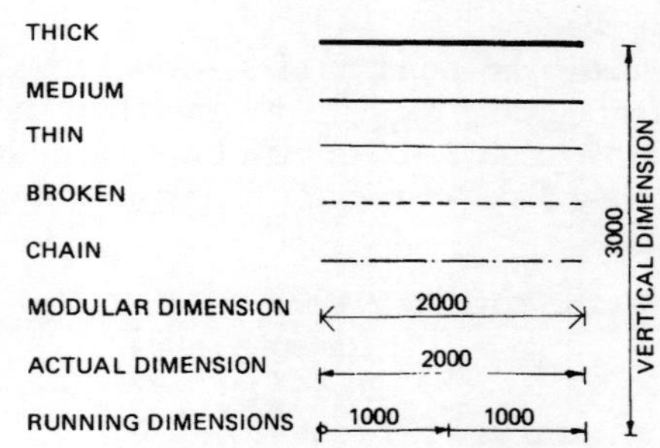

Fig. 4.11. *Types of drawn line.*

(b) Medium lines should be used for showing the position of existing buildings, outlines of secondary elements and general detailing.

(c) Thin lines are used for reference grids, dimension lines, leader lines and hatching.

(d) Broken lines of medium thickness are used to show the position of hidden details and work which is to be removed.

(*e*) *Chain lines (thin)* are used to indicate centre lines.

(*f*) *Dimension lines* are thin solid lines and should have the dimension concerned printed directly above and in the centre. Open arrowheads are used to indicate basic modular dimensions, including tolerances. Solid arrowheads are used to indicate actual dimensions. Running dimensions are indicated by arrows pointing in the same direction and are taken from the small circle which represents the terminal point. Dimension lines should always be drawn to be read either from the bottom or from the left hand side of the drawing as shown.

Dimensions

On a drawing these may be given in either metres or millimetres, although it is good policy to keep to one or the other on a single drawing as far as possible. Whole numbers on a dimension line indicate millimetres thus:

400 ⟷ = 400 mm or 1,200 ⟷ = 1,200mm

Dimensions taken to three decimal places indicate metres, thus:

1.200 ⟷ = 1.200 m

It will be seen therefore that either of the basic units (m or mm) may be used on a drawing without the need to use a symbol, since there is unlikely to be any confusion if this method of notation is used. Where dimensions of less than one metre are expressed in metres, a zero should precede the decimal point, thus:

0.400 ⟷

Where fractions of a millimetre are shown it is as well to use the symbol concerned to avoid any possible confusion, thus:

12.5 mm ⟷

Graphical symbols

These are often used on sections to indicate specific materials. Figure 4.12 shows the hatching used to represent some of the more common materials.

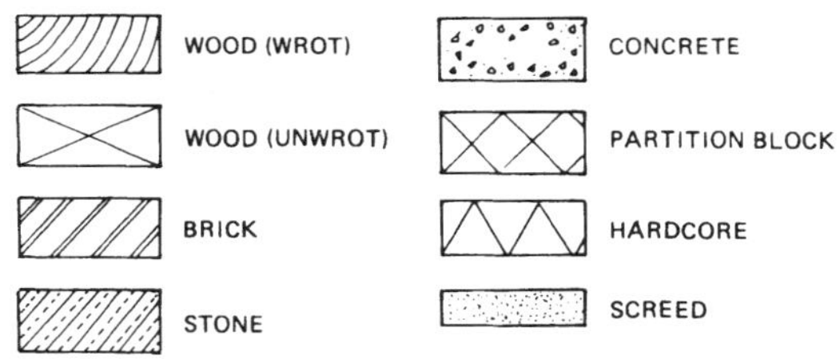

Fig. 4.12. *Graphic symbols.*

Scales

These are used in drawing to enable large objects (buildings, components, etc.) to be drawn to a convenient size while still maintaining strict and accurate proportions which can be drawn or measured as required. British Standard 1192 recommends the following scales for use as shown:

Block plans	1 : 2500	or	1 : 1250
Site plans	1 : 500	or	1 : 200
General location	1 : 200	or	1 : 100
		or	1 : 50
Detail drawings	1 : 10	or	1 : 5
	(full size)	or	1 : 1

NOTE: The adopted scale or scales should always be indicated on the drawing.

Lettering

Much use is made of lettering on a drawing in the form of headings, descriptions and notes, the main considerations of such lettering being that it should be clear, legible and neat. Since most of the lettering on drawings is done freehand, the style adopted varies from individual to individual, but students are advised to adopt and perfect a simple form of block lettering as shown in Fig. 4.13, and to print between feint parallel lines using a fairly soft pencil (HB or B).

British Standard 1192 recommends the use of 1.5 mm to 4 mm high letters for notes and descriptions and 5 mm to 8 mm high letters for headings.

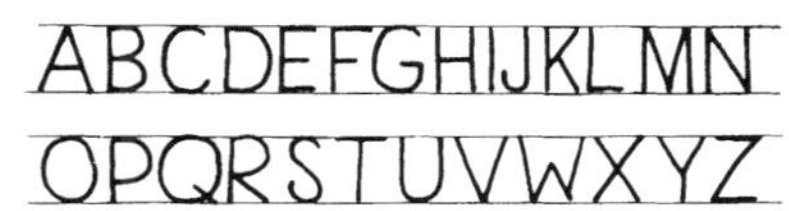

Fig. 4.13. *Lettering.*

Graphical representation of solids

When a three dimensional object such as a building or a piece of woodwork is represented by a drawing on a flat sheet of paper it becomes necessary to draw it in such a way that the third dimension – depth – can be taken into account and understood by the person reading the drawing. This can be accomplished most simply by drawing several views of the same object, a different viewpoint being taken for each one.

Orthographic (first angle) projection

This is the method most commonly used in the construction industry to represent solid objects, and involves the use of "plans" and "elevations", these being views from above and from the side respectively. Figure 4.14 shows the plan, elevation and two auxiliary elevations of a mortice and tenon joint. Note that the two auxilliary elevations A_1 and A_2 show views taken from the right and left respectively. Figure 4.15 shows a pictorial sketch of the two "coordinate" planes—the horizontal and vertical planes – onto which the views of the joint are projected, the x - y line representing the junction between the two planes.

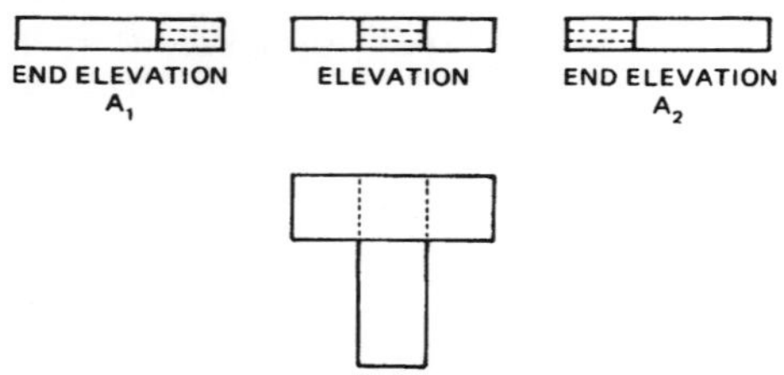

Fig. 4.14. *First angle projection.*

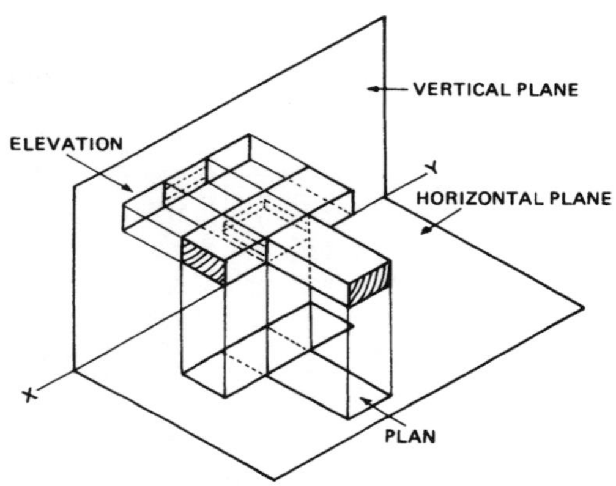

Fig. 4.15. *Sketch showing vertical and horizontal planes.*

Isometric projection

Figure 4.16 shows an isometric view of the same mortice and tenon joint. Isometric projections are useful in showing three dimensions simultaneously, thus making for a more realistic and pictorial representation of the object. As can be seen isometric projections are based on axes sloping at 30° to the horizontal plane. Measurements may be made vertically and along the two sloping axes.

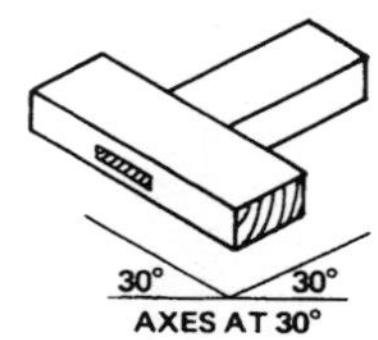

Fig. 4.16. *Isometric drawing of a joint.*

Sections

These show the exact sectional shape of an object which is cut by an imaginary plane, usually at right angles to the member or surface involved. They are therefore extremely useful, if not actually indispensable to the woodworker who invariably needs to know the sectional shapes of the pieces of timber which go together to make up a piece of joinery. Setting out rods discussed earlier in the chapter are in fact vertical and horizontal sections taken through a piece of joinery.

Figure 4.17 shows the elevation of a simple rebated frame, vertical and horizontal sections being taken through the frame at A-A and B-B respectively, the arrows on the section lines indicating the direction of the view-point. Note that on most sections some part of the object may not be cut by the section line and therefore shows up as part plan or part elevation in the background.

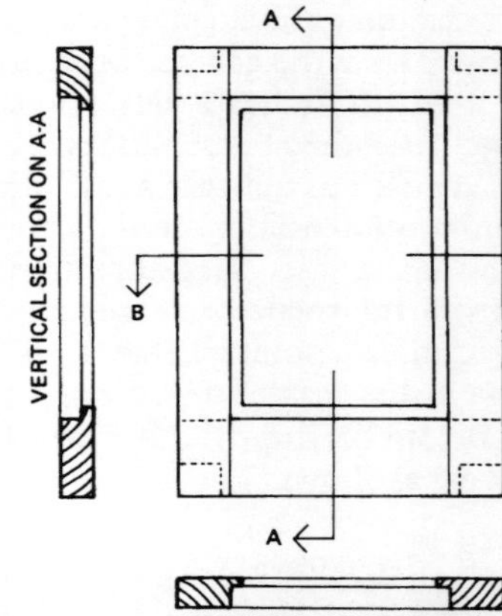

Fig. 4.17. *Elevation and sections of a rebated frame.*

Drawing equipment

Much of the woodworker's job is concerned with drawings in one form or another and the ability to produce a clear accurate and easily understood drawing is an essential part of his training. Woodworkers, carpenters and joiners in particular should realise that in order to gain an Advanced Craft Certificate of the City and Guilds of London Institute the ability to produce working drawings, details of construction and solutions to geometrical problems is more or less mandatory. The student is urged therefore to practise the art of technical drawing at every opportunity in order to acquire the necessary degree of competence. Such competence is unlikely to be developed solely through the limited experience that time allows when studying at a Technical College. Students should possess their own set of drawing equipment so that they may have the opportunity of developing drawing skills at home – they should regard drawing equipment as being of equal importance to other tools of the trade, and equally necessary to their progress.

A satisfactory basic minimum of the drawing equipment required is given below and illustrated in Fig. 4.18.

Drawing board

(*See* Fig. 4.18*(a)*.) This should be large enough to take an A2 size sheet of drawing paper, the main requirements being that it should be smooth, rigid and stable. Most woodworkers are quite capable of making their own drawing boards. A piece of soft faced 19 mm plywood or blockboard with corners lightly radiused and arrises removed is satisfactory. Good quality boards of redwood, spruce or yellow pine suitably jointed and cleated (slot screwed cleats) make an excellent drawing board.

Tee square

(*See* Fig. 4.18*(a)*.) This is another item which the woodworker is fortunate in being able to make for himself, any reasonably hard and stable hardwood being suitable. The main requirements are that it should be long enough in the blade to suit the drawing board and that the blade should be straight, smooth and *firmly fixed to the stock*. Students considering making their own tee square are advised to copy a professionally made article. Two coats of French polish or polyurethane varnish, well rubbed down, will seal it against finger prints and make it easy to wipe clean.

Tee squares are used solely for drawing *horizontal lines*, (the stock of the square being kept in contact with the left hand side of the drawing board) and as supports for set squares.

Drawing board clips

(*See* Fig. 4.18*(b)*.) These are used to hold the paper firmly in place on the drawing board and are generally more convenient than drawing pins as they do not damage the drawing board. Two clips are sufficient for most work.

Set squares

(*See* Fig. 4.18*(a)*.) These are made of plastic or celluloid and are used in conjunction with the tee square to draw vertical and angled lines. Two set squares are required, a 45° and a 60°/30° type respectively. The student is advised to purchase good quality, *large* set squares, the small ones supplied in mathematical sets being very limited in use.

Fig. 4.18. *Drawing equipment.*

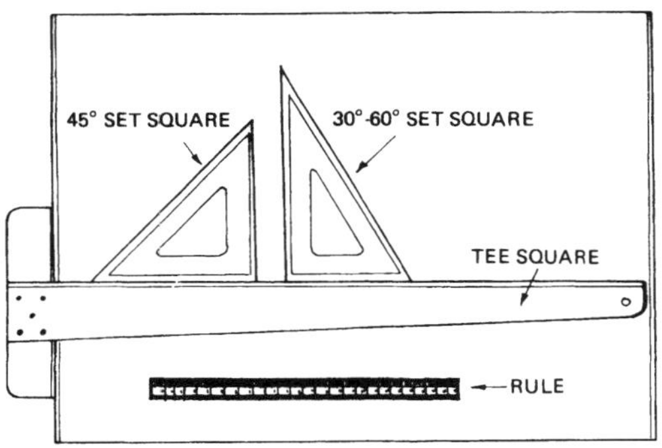

(a) Drawing board, tee square, set squares, rule.

(b) Drawing board clip.

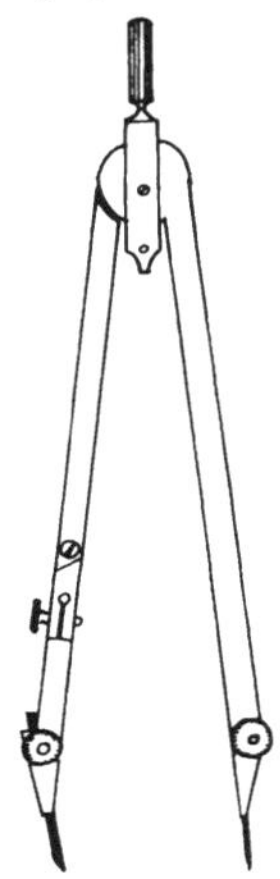

(c) Compasses.

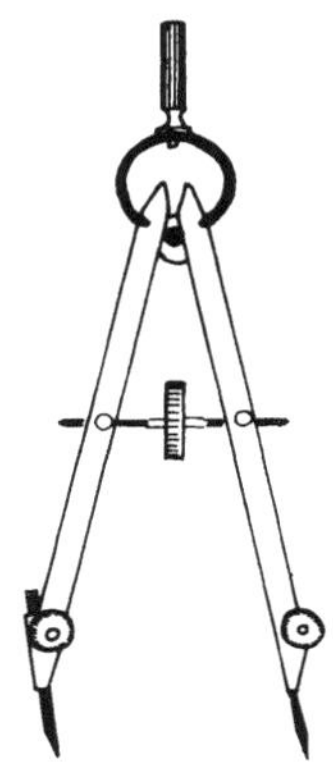

(d) Spring bow compasses.

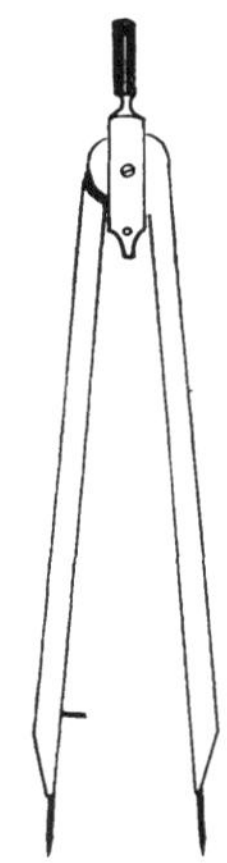

(e) Dividers.

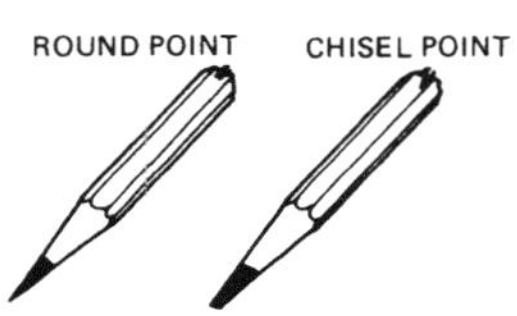

(f) Pencil points.

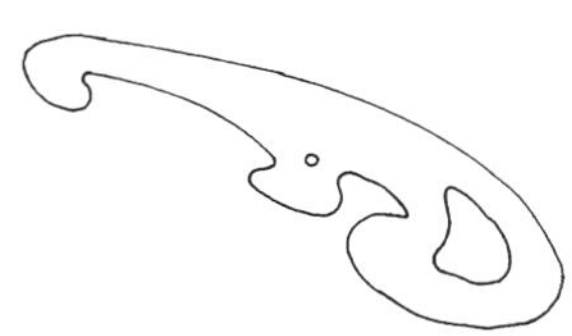

(g) French curve.

(h) Flexi-curve.

Set squares should be wiped clean occasionally to remove finger marks and pencil lead.

Rules

(*See* Fig. 4.18*(a)*.) A good quality plastic rule should be purchased for drawing and kept for that purpose. The introduction of the metric system (SI) has made it easy to use any rule graduated in mm, as scales can readily be calculated mentally. However, metric scale rules are very useful, and are both quick and convenient in use.

Compasses

(*See* Fig. 4.18*(c)*.) These are available in a wide range of qualities and sizes, some, as illustrated having various useful refinements. The student is advised to purchase the largest and best quality compasses he can afford. A pair of spring bows (*see* Fig. 4.18*(d)*) is a useful addition to the drawing equipment, not strictly essential, but convenient for drawing small circles and arcs.

Dividers

(*See* Fig. 4.18*(e)*.) These are used for "stepping off" and transferring measurements and can be regarded as essential items.

Erasers

Erasers, or "rubbers" are cheap but useful items of equipment for removing unwanted pencil lines and marks. Soft erasers are the better choice for general pencil drawings, but in any case the eraser should be used only when really necessary, as undue use – due to careless work – soon leads to rough paper thus worsening the drawing's appearance.

Pencils

Only good quality drawing pencils should be used for this purpose, sharpened to round or chisel points as shown in Fig. 4.18*(f)*. H or 2H grades will generally be found most satisfactory for drawing and geometry (lines) whilst HB or B are better for lettering. The student should possess a collection of pencils of various grades, and keep them for drawing purposes only, discarding them for other uses as they become shortened. Pencils should be handled and used with care. Absent minded tapping and accidental dropping causes the "lead" to break up inside, makes sharpening difficult, and greatly reduces their life.

French curves and flexi-curves

Illustrated in Figs. 4.18*(g)* and 4.18*(h)* respectively, these are used for drawing curves through previously plotted points. Whilst not strictly essential items, they are worthwhile acquisitions in that they save considerable time and help to produce clean accurate curved lines.

BASIC PLANE GEOMETRY

The following geometric constructions are all very simple but are, nevertheless, important fundamental disciplines which the carpenter and joiner has to put to use more and more frequently as he progresses in his chosen career. The beginner, the apprentice or novice to woodworking should not be misled into thinking that the many aspects of plane and solid geometry which he is bound to encounter during his course of study are purely academic and have no real value as a means of assisting him in practical woodworking. Such may, at first, seem to be the case, but as he comes to understand his craft in greater and greater depth, the real worth and value of having a geometrical solution at his finger tips will become apparent. To state the case more forcibly, a real lack of knowledge in basic, practical, geometrical principles may well prove a stumbling block beyond which further progress is difficult or wellnigh impossible.

To bisect a straight line

(*See* Fig. 4.19.) Let A-B be the straight line. Set the compass to rather more than half the length of A-B, and with A and B as centres strike the two arcs C-D and E-F. A line drawn through the intersections will bisect A-B at G. The bisector is perpendicular to A-B.

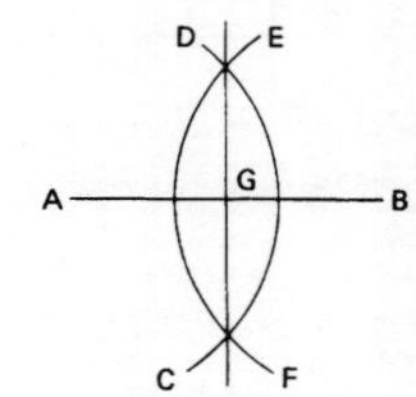

Fig. 4.19. *Bisecting a line.*

To divide a line into any number of equal parts

(*See* Fig. 4.20.) Let A-B be the line which is to be divided into a number of equal parts – say five. Draw a line from A at any convenient angle, and mark off along it five equal divisions (the exact size of the divisions is not important). Join point 5 to B. Draw through the remaining points with lines parallel to B5 to divide A-B into five equal parts.

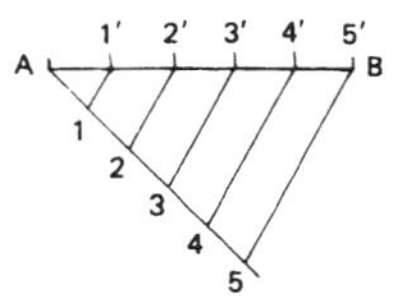

Fig. 4.20. *Dividing a line into a number of parts.*

To draw a perpendicular at a given point on a line

(*See* Fig. 4.21.) Let A-B be the line and O the point of contact of the perpendicular. With point O as centre, describe a semi-circle to give points C and D. Extend the radius of the compass and with points C and D as centres, describe the two arcs to give intersection at P. A line drawn through P-O will be perpendicular to A-B.

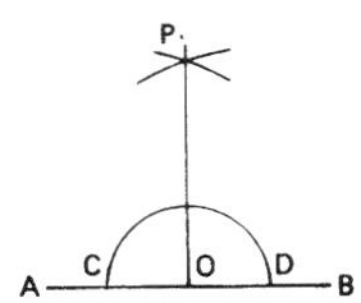

Fig. 4.21. *Constructing a perpendicular to a given point in a line.*

Angles

(*See* Fig. 4.22.) An angle is formed by the intersection of two straight lines, the magnitude of the angle being measured in degrees. Figure 4.22*(a)*, *(b)* and *(c)* shows *acute* (less than 90°), *right* (90°), and *obtuse* (greater than 90°) angles respectively.

Fig. 4.22. *Angles.*

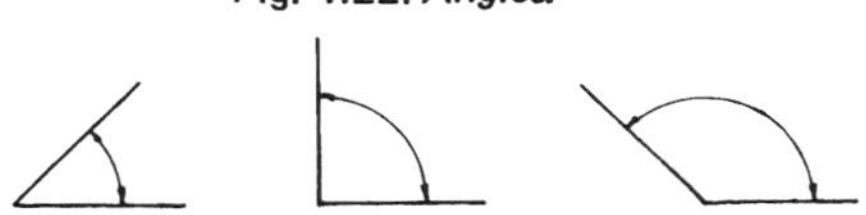

(a) Acute angle. *(b) Right angle.* *(c) Obtuse angle.*

To reproduce a given angle

(*See* Fig. 4.23.) Let A-B-C be the given angle, and D-E be the base line of the required angle. With B as centre, draw arc P-Q. With same radius and E as centre, draw arc S-T. With P as centre and radius P-Q, draw an arc through the intersection of B-C and P-Q. With same radius and centre T, draw arc to intersect S-T at X. A line drawn through E and X will reproduce angle A-B-C.

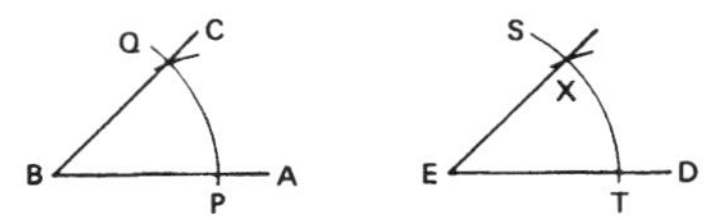

Fig. 4.23. *Reproducing a given angle.*

To bisect an angle

(*See* Fig. 4.24.) Let A-B-C be the given angle. With centre B, draw an arc to cut A-B and B-C at D and E. With D and E as centres, draw arcs to intersect at F. A line drawn through B-F will bisect angle A-B-C.

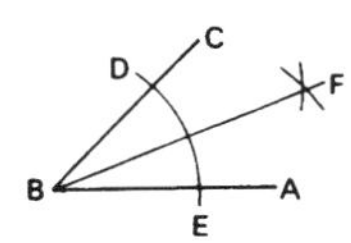

Fig. 4.24. *Bisecting an angle.*

Parts of a circle

(*See* Fig. 4.25.) The various terms relating to parts of a circle are illustrated in this figure.

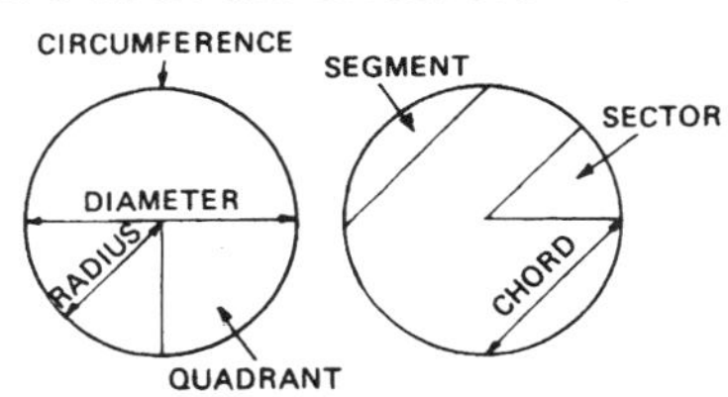

Fig. 4.25. *Parts of a circle.*

To find the centre of a circle

(*See* Fig. 4.26.) Draw any two chords A-B and B-C as shown. Bisect the two chords and produce the bisectors to intersect at O – the centre of the circle.

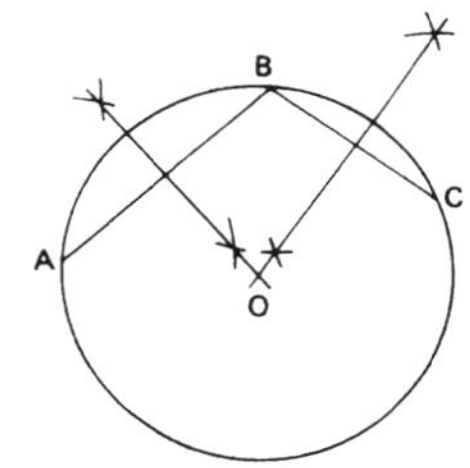

Fig. 4.26. *Finding the centre of a circle.*

To construct a regular hexagon of given side

(*See* Fig. 4.27.) Let A-B be the length of the side. With radius A-B and centres A and B respectively, draw arcs A-C and B-D to intersect at O. With O as centre and radius A-O describe a circle around point O and passing through A and B. Still using radius A-B, step off around the circumference to complete the hexagon.

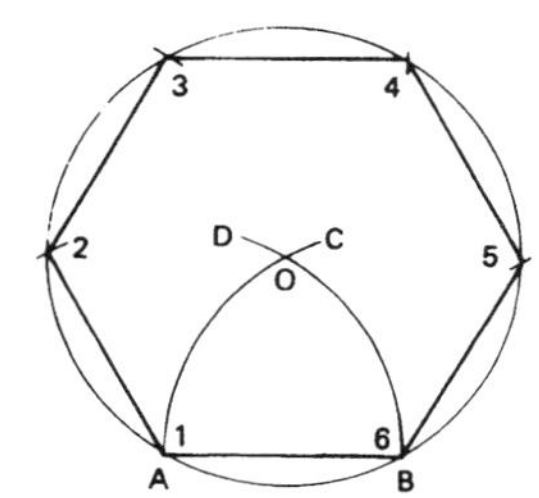

Fig. 4.27. *Constructing a regular hexagon.*

To construct a regular octagon within a square

(*See* Fig. 4.28.) Let A-B-C-D be the given square. Draw diagonals A-C and B-D to intersect at O. With radius A-O and centre A draw the quadrant 1-6. Repeat from centres B, C and D to obtain points 3-8, 2-5 and 4-7, to complete the octagon.

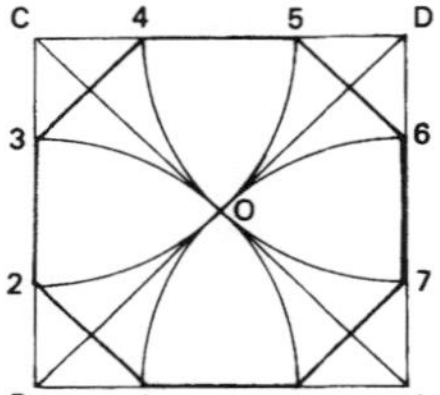

Fig. 4.28. *Constructing an octagon within a square.*

To construct any regular polygon of given side

(*See* Fig. 4.29.) Let A-B be the given side. Produce A-B as shown. With B as centre, describe a semi-circle of radius B-A. Divide (by trial and error) the semi-circle into the same number of parts as the required polygon has sides (say seven). Draw a line from B through point 2 (point 2 is always used, irrespective of the number of sides involved), making B-C equal to A-B. Bisect A-B and B-C and produce the perpendicular bisectors to intersect at O. With O as centre and radius O-C, describe a circle passing through points A, B and C. With radius A-B step off the remaining points D, E, F and G and join up to complete the polygon — in this case a heptagon.

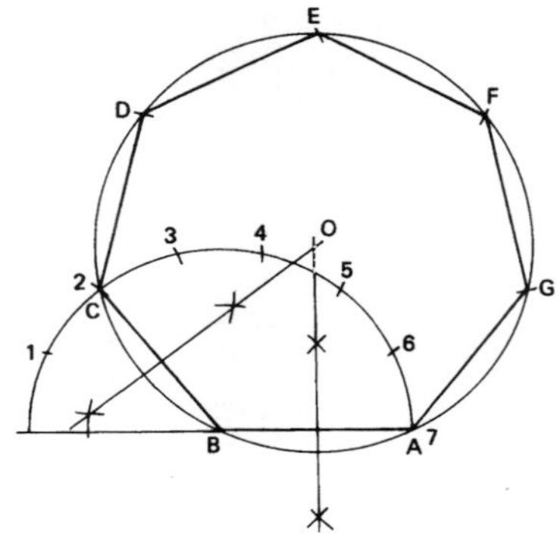

Fig. 4.29. *Constructing any regular polygon of given side.*

To draw a tangent to a circle through a given point in its circumference

(*See* Fig. 4.30.) Let P be the given point. Draw a line from centre O through P and produce it beyond P as shown. Since line O-P passes through the centre of the circle it is *normal* to the circumference. With P as centre describe a circle of suitable diameter to intersect the normal at A and B. Bisect A-B (as in Fig. 4.19), the bisector being a tangent to the circle at the given point.

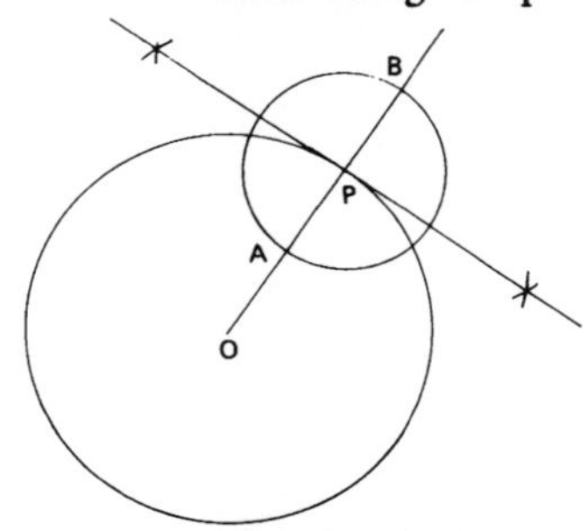

Fig. 4.30. *Drawing a tangent through a given point on the circumference of a circle.*

To draw a tangent through a given point in an arc

(*See* Fig. 4.31.) Let P be the given point on the arc. Describe a suitably sized circle around P to cut the circumference at A and B, and join A to B to form a chord. Bisect A-B to produce the normal C-D. When produced, this line would pass through the centre of the circle of which this arc is a part. Bisect C-D, the bisector E-F being the required tangent.

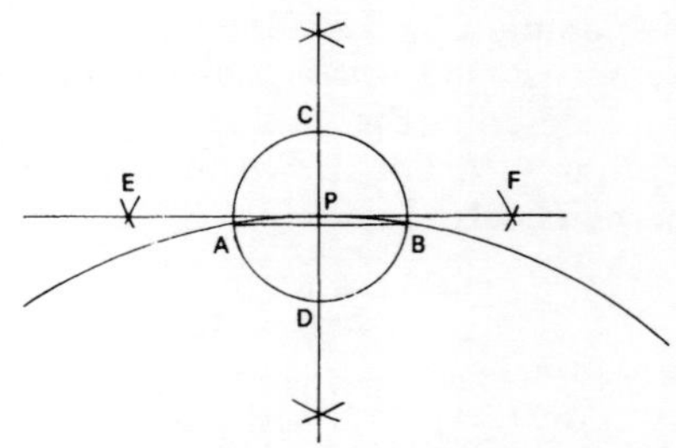

Fig. 4.31. *Drawing a tangent through a given point in an arc.*

To draw an ellipse — Method 1

(*See* Fig. 4.32(*a*).) Let A-B and C-D be the major and minor axes respectively. With O as centre and radius O-C describe a circle passing through C and D. With O as centre and radius O-A describe a circle passing through A and B. Using a 30°/60° set square, trisect each quadrant to plot points 1 to 12 and 1′ to 12′ on the outer and inner circles respectively. Lines drawn horizontally from points on the inner circle to intersect lines drawn vertically from points on the outer circle will give points 1^2 to 12^2. A fair curve drawn through these points will give the required ellipse.

Fig. 4.32. *Construction of ellipses.*

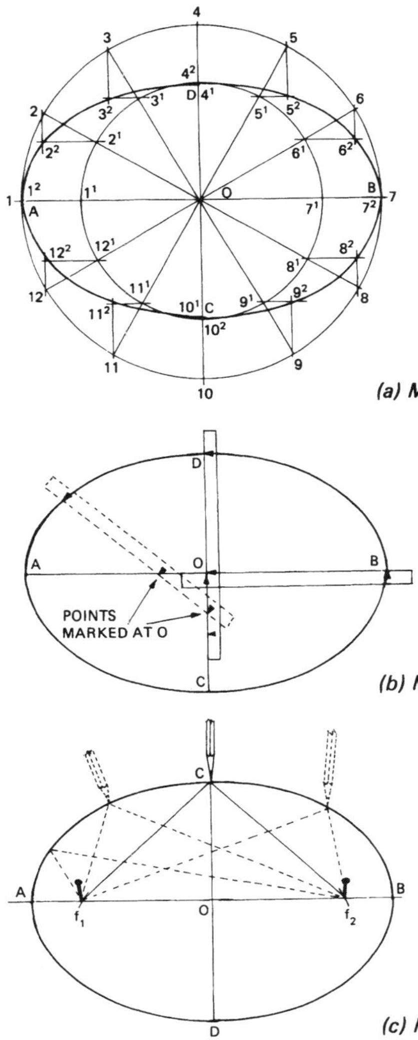

(a) Method 1.

(b) Method 2.

(c) Method 3.

To draw an ellipse – Method 2

(*See* Fig. 4.32(*b*).) This method involves the use of a "trammel" – a straight strip of paper serving well enough to draw small ellipses – and is probably the quickest and simplest method of constructing an ellipse. Let A-B and C-D be the major and minor axes respectively. Lay the strip of paper along the semi-major axis and make marks on its edge at points O and B as shown. Turn the trammel upright, placing the point *which was at B* at D. Make a further mark at O. We now have three marks on the trammel. The trammel may now be moved around either clockwise or anti-clockwise, keeping the points on the trammel *which were marked at O* on the two axes, and using the third point to mark off the outline of the ellipse. A fair curve through the points plotted will give the required ellipse.

NOTE: The number of points plotted in this manner can be varied to suit the size of the ellipse and/or the skill of the draughtsman.

To draw an ellipse – Method 3

(*See* Fig. 4.32(*c*).) This method is very useful in the workshop for drawing large ellipses and involves the use of string and pins, the outline of the ellipse being plotted by *loci* (the path traced by a moving point, the position of which is constantly fixed in relation to two other points). Let A-B and C-D be the major and minor axes respectively. With radius O-B (the semi-major axis) and centre C or D, strike arcs to intersect the major axis at f_1 and f_2, the two focal points. Drive small pins into the board at f_1 *and* f_2 and C. Pass a string around the three pins and knot tightly, carefully avoiding too much or too little tension in the string. Remove the pin at C and replace with a sharp pencil point as shown. Move the pencil point clockwise or anti-clockwise keeping a constant tension on the string to draw the ellipse.

FURTHER READING

British Standards and Codes of Practice

BS 1192:1969	Building drawing practice
BS 3763:1976	The International System of Units (SI)
BS 4471:	Specification for dimensions of softwood
	Part 1: 1978 Sizes of sawn and planed timber
BS 5450:1977	Sizes of hardwoods and methods of measurement

SELF-TESTING QUESTIONS

All the information required to answer the following questions is contained within this chapter. Attempt each section *as fully or as briefly* as the question demands, and then check your answers against the information given in the chapter.

1. (*a*) State six reasons for the use of a setting out rod.

(*b*) State six of the requirements of "marking out".

2. Prepare a setting out rod for the two light casement window shown in Fig. 4.3. The rod should show:

(*a*) overall frame and sash sizes;
(*b*) sections through all the members;
(*c*) positions of all the mortices.

(This rod should preferably be made full size, but may be drawn to scale of 1:2 – half full size – if drawing paper is used.)

3. Prepare a fully detailed cutting list from the setting out rod prepared in answer to 2. (Check your *finished* list against that given in Table II, making due allowance for the fact that the example given in Table II is for *four* such windows.)

4. Calculate the cost of the timber for the window in 2 and 3, given that the price per linear metre of the members involved is as follows:

Member	Price per metre
Head and jambs	75p
Mullion	85p
Cill	£1.40
Stiles and top rails	40p
Bottom rails	50p

5. Calculate the cost of the floor boards required to cover a floor measuring 6.25 m long by 4.200 m wide, given that the floor boards have a covering width of 92 mm and are charged at £5.50 per m^2. Allow 6 per cent extra for cutting and waste.

6. Calculate the number of linear metres of timber in the following sectional sizes that are contained in one cubic metre (1 m^3).

(*a*) 50 mm × 200 mm;
(*b*) 25 mm × 150 mm;
(*c*) 75 mm × 225 mm;
(*d*) 19 mm × 175 mm.

7. (*a*) Draw to a scale of 1:5 the plan, front and end elevations of the mitre block shown in Chapter 1, Fig. 1.42. The dimensions of the mitre block are as follows:

Base: 400 mm × 125 mm × 25 mm;
Guide block: 400 mm × 50 mm × 50 mm;
Saw cuts start 60 mm from each end.

(*b*) Draw rectangles 50 mm × 25 mm and cross-hatch them in accordance with BS 1192 to indicate the following materials: (*i*) brick, (*ii*) concrete, (*iii*) sawn (unwrot) timber; (*iv*) planed (wrot) timber, (*v*) hardcore, (*vi*) lightweight building blocks, (*vii*) stone.

8. Show by means of plane geometry how to:

(*a*) reproduce a given angle;
(*b*) bisect a given angle;
(*c*) find the centre of a circle;
(*d*) construct a regular hexagon of 50 mm side;
(*e*) construct a regular pentagon (five sided figure) of 50 mm side;
(*f*) draw a tangent through a given point in an arc;
(*g*) construct an ellipse (any method) having major and minor axes of 150 mm and 100 mm respectively.

5. Doors, Frames and Linings

After completing this chapter the student should be able to:

1. Describe the construction of a ledged and braced door.
2. Describe the construction of a panelled door.
3. List the sequence of operations in the making of a panelled door.
4. Draw sections and construction details of door frames and linings.
5. List the sequence of operations in hanging a door.
6. Name, sketch and specify suitable ironmongery and furniture for specific types of doors.
7. State methods used to protect joinery items from damage and deterioration.

LEDGED AND BRACED DOORS

Ledged and braced doors are among the simplest items of joinery that the woodworker can be called upon to make. There are, nevertheless, certain principles involved in the construction of these doors to which it is necessary to adhere in order to produce a door which performs satisfactorily in service. The main points which must be considered before work can be commenced on the door and its frame are:

(a) Will the door be exposed to the weather?

(b) On which edge will it be hung?

(c) Which side of the door will be flush with the face of the door frame?

These points vitally affect the construction of the door and its frame, as will be seen when the general design and construction are studied.

General construction and design

Figure 5.1 shows the front and rear elevations of a typical ledged and braced door, and it will be immediately apparent that the door comprises of match boards, cleated into a door sized panel by means of three ledges, fastened to the boards by nails and screws.

Particular note should be made of the two diagonal braces which are vital if the door is not to drop out of square. The braces should always be fitted with the lower ends to the hinged edge of the door

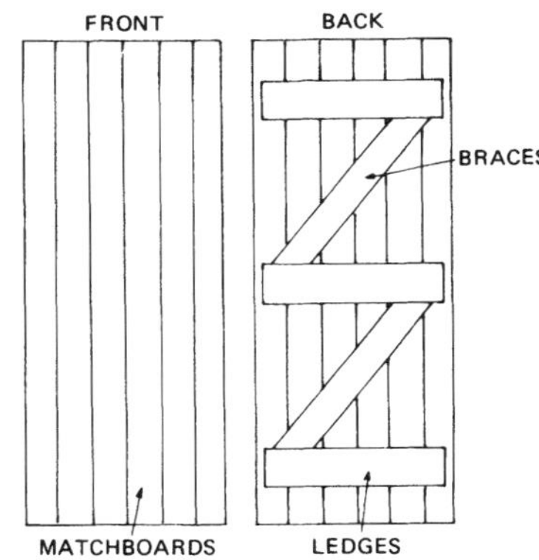

Fig. 5.1. *Ledged and braced door.*

so that they are in compression when resisting the tendency of the door to drop. The braces are fastened to the door by nails driven through the face of the match boards.

Figure 5.2 shows two methods of fitting the braces to the ledge, the first method being very common whilst the second is to be preferred in good class work.

Figure 5.3 is a horizontal section taken through part of one of the ledges and shows the two countersunk screws at the end of the ledge, the nails holding the board to the ledge (these are sometimes

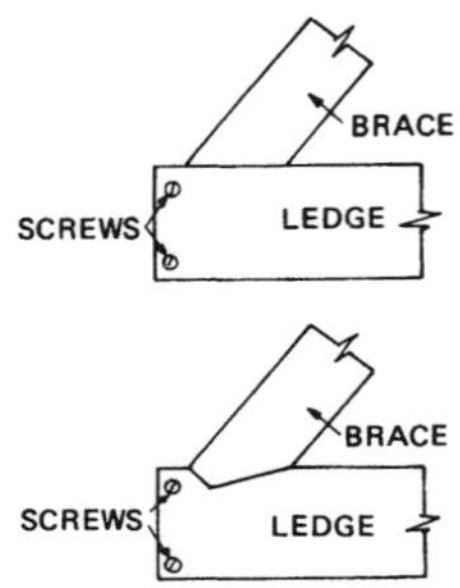

Fig. 5.2. *Methods of fitting braces.*

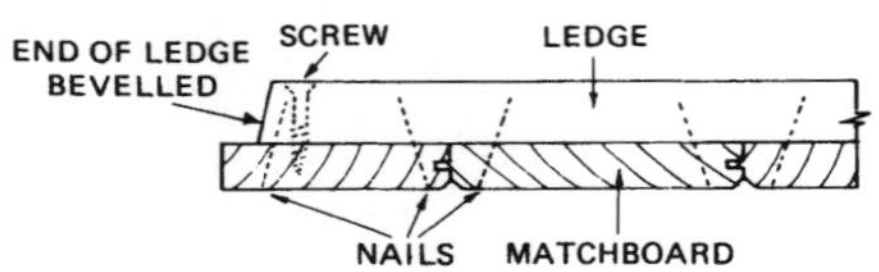

Fig. 5.3. *Section through edge of ledged and braced door.*

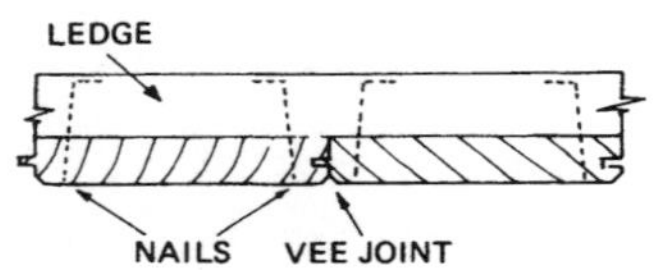

Fig. 5.4. *Boards held to ledges with clenched nails.*

driven right through and clenched – *see* Fig. 5.4), and the shape of the tongued, grooved and veed match boards. The vee joints are used to emphasise the joints between the boards which would otherwise, sooner or later, appear as an objectional crack. (It has long been a joiner's maxim that a joint which cannot be hidden should be emphasised.) Note also how the end of the ledge is finished with a bevel for the sake of neatness, and that it is set back 12 mm or so from the edge of the door. This latter

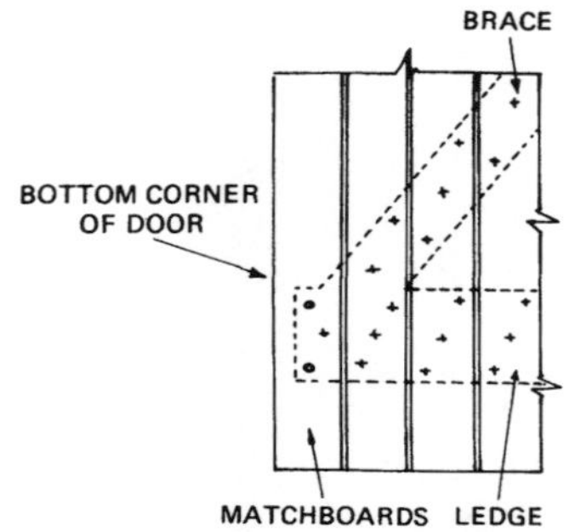

Fig. 5.5. *Nailing pattern for matchboards.*

feature allows for the back side of the door to fit closely to a door stop as shown in Fig. 5.7(*b*).

Figure 5.5 shows the nailing pattern for the match boards and braces. Note that nails which are not clenched should be dovetailed (driven in at a slight angle) for greater holding power. Cut clasp nails are ideally suited for this purpose.

Construction and assembly

First, select the match boards, trimming the tongues and grooves off the two outside boards to obtain the correct overall width of the door. Next prepare the timber for the ledges, bevelling the edges as described previously. A pair of holes should be drilled at the end of each ledge and countersunk to receive 38 mm × 10 screws. The material for the braces can now be prepared, the edges bevelled to match the ledges, and an allowance of 25 mm or so made in the length for fitting later.

Having trimmed the match boards to length, the tongues and grooves should be painted with a good quality primer so no unpainted timber is exposed should shrinkage occur and to prevent absorption of moisture should water penetrate the joints. On really good work the backs of the braces and ledges would be primed as might also that part of the match boarding covered by them.

The boards should now be laid face down on bearers laid across the bench – one bearer to each ledge – and held together with a pair of sash cramps. It is now possible to line up the ends of the match boards to make the door exactly square.

Mark the position of the ledges – one about 150 mm from each end and the third in the centre. Place in position and insert the screws, making sure the ends of the ledges leave an equal amount of match board exposed on either side. Turn the door over carefully so as not to allow the unfixed central boards to collapse, and nail through the face of the boards into the ledges. Punch the nails below the surface.

Fitting the braces

Turn the door so the ledges are again uppermost and decide on the direction in which the braces must slope. Lay the braces in position on top of the ledges and mark for length, sighting along the edge of the ledge as shown in Fig. 5.6. Cut the braces to length, trimming the ends with a plane if necessary until a good tight fit is obtained and fix into position by skew nailing into the ledges. Turn the door over again and nail through the face of the match boards into the braces (the position of the braces can be

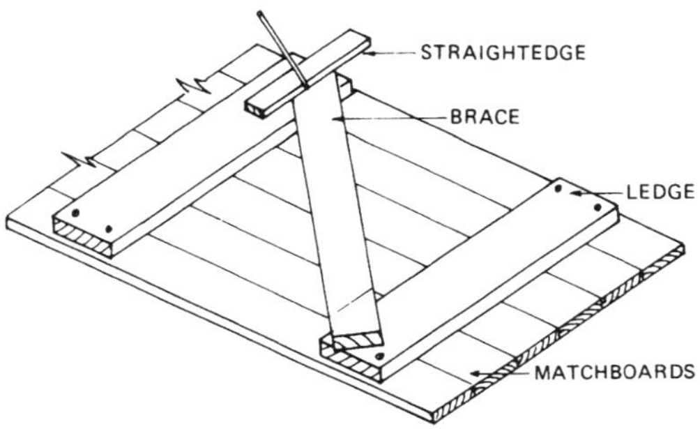

Fig. 5.6. *Marking the braces for the door.*

transferred onto the face of the door by driving small nails or pins between the match boards alongside the braces).

The door is now ready for cleaning up and priming.

FRAMES FOR LEDGED AND BRACED DOORS

These comprise a pair of jambs and a head which are fitted together with mortice and tenon joints, these being secured with wedges or by draw boring as described in Chapter 3. The joints may be further strengthened by the use of a good water-proof adhesive or priming paint.

Figure 5.7*(a)* and *(b)* shows sections through the head/jamb of typical door frames suitable for use with ledged and braced doors, and indicate a solid rebated frame and a frame with a planted (nailed on) door stop respectively. Note that the depth of the rebate in the solid rebated frame is governed by the position of the door within the frame (*see* Fig. 5.7*(a)*).

Fig. 5.7. *Door frames.*

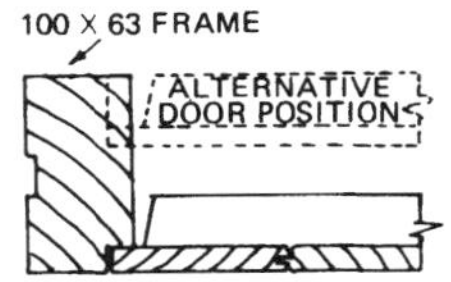

(a) Solid rebated frame showing alternative door position.

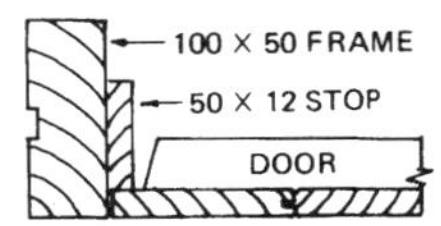

(b) Door, frame and planted stop.

Bracing the frame

To hold the door frame square and parallel during storage, transit and fixing, the frame must be braced as shown in Fig. 5.8. Bracing is done with scrap timber, rippings, etc., which are nailed to the face or back of the frame so forming a triangulated structure able to withstand the tendency to "rack" out of square. Braces are removed after the frame has been fixed. Horns are left on the frame for building into the brickwork whenever possible. Like the door, the frame should be primed before it leaves the workshop.

Hanging the door

The first stage in hanging the door is to stand it in position against the frame to check for size. The four edges of the door can then be "shot" with a jack plane until the door fits the opening with a 2-3 mm joint at the top and each side and about 4 mm at the bottom for ground clearance. Remove all sharp arrises (sharp corners) with a stroke or two of the plane, and the door is ready to hang.

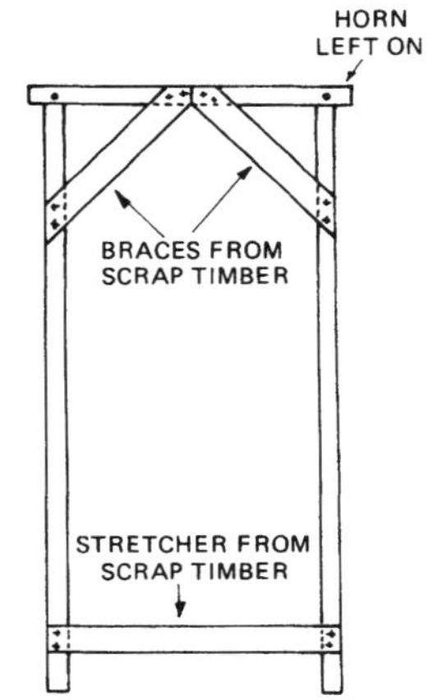

Fig. 5.8. *Assembled frame showing bracing.*

Tee hinges are a good choice for use on a ledged and braced door, and these are first screwed to the face of the door with countersunk or round headed screws about 25 mm long. The hinges should be positioned over the ledges as shown in Fig. 5.9, two being sufficient in most cases, but three for better quality work.

Next the door should be wedged into position so that the joints around the edge are as required on the finished job and the flap of the hinge screwed to the frame. If the door is an external one, then it becomes important to paint the bottom edge before the door is hung. The wedges can now be removed to free the door, which should

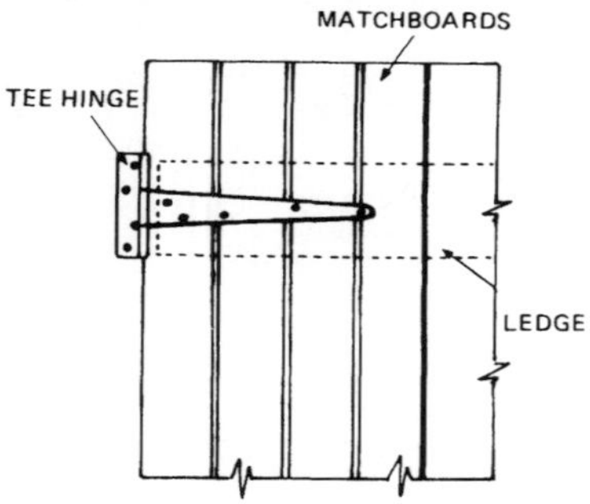

Fig. 5.9. *Part of door showing position of tee hinges.*

be opened and closed once or twice to check for proper action, and all is ready for fitting the furniture – the lock or latch.

Figure 5.10*(a)* shows a "thumb" or "Norfolk" latch which is commonly used on this type of door. It is very simple to fit, requiring no more than a small slot cut into the door to enable the lever to pass through. The latch, guide and keeper are fastened to door and frame with round-head screws.

Figure 5.10*(b)* shows a rim lock, used where some degree of security is required, which is also very simple to fit. The rim lock shown would normally be used as an alternative to the thumb latch and is also available in the form of a rim "dead" lock as shown in Fig. 5.10*(c)* which can be used in addition to the thumb latch. A dead lock is furnished with keys only – no provision being made for handles. Planted door stops are fitted and nailed to the frame as the final operation, the lock/latch holding the door closed whilst this is done.

Fig. 5.10. *Latches and locks.*

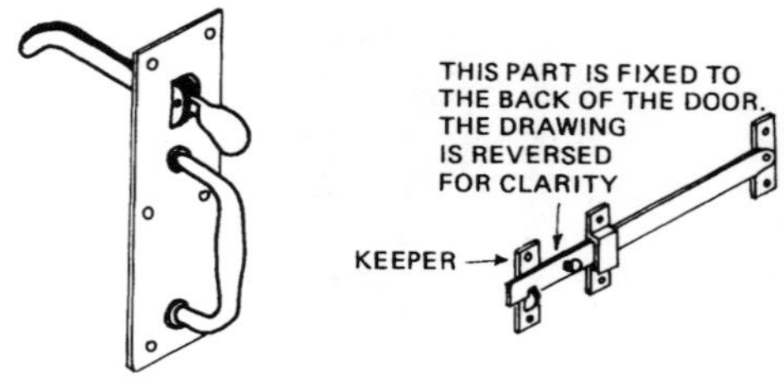

(a) Thumb or Norfolk latch.

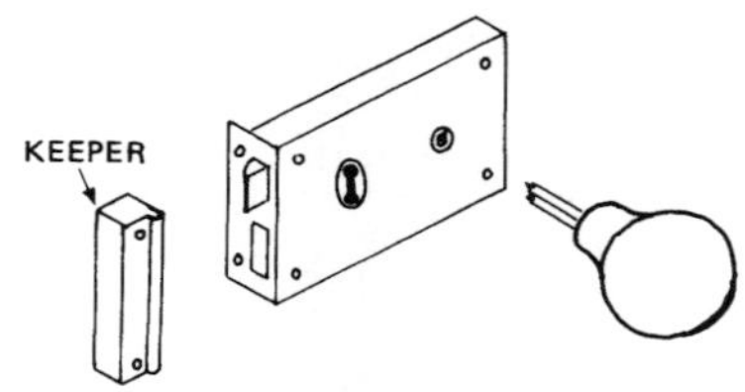

(b) Rim lock and keeper with knob handles.

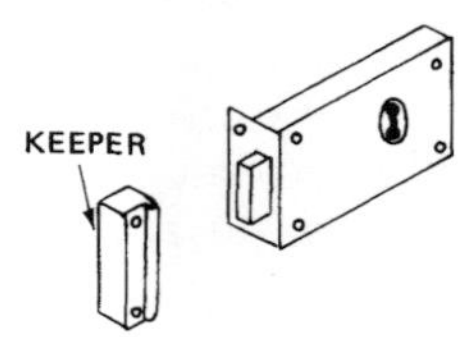

(c) Rim dead lock (no latch).

PANELLED DOORS

Panelled doors are made for both interior and exterior use and are generally described by the number of panels they contain. It should, however, be appreciated that there are many possible variations in the design of a panelled door, and therefore for constructional purposes, a much fuller and more detailed description is required to avoid any misunderstanding. Indeed, a detailed drawing or accurate sketch plus a good specification is highly desirable to ensure the finished door is exactly as required.

Doors for both interior and exterior use are basically similar in construction but differ somewhat in the type of adhesive and the moisture content of the timber which is used – and often in their overall size, external doors often being rather wider and thicker. Extra care should also be given to the design of an exterior type door to ensure there is no penetration of rainwater around or into the door, and that water effectively runs off the exposed face – in short to ensure the door is fully and properly "weathered".

British Standard specifications

Specifications for the construction, design and sizes of standard doors are given in BS 459, Part 1 and Part 4, these specifications relating to panelled and match boarded doors respectively.

Other specifications relevant to the construction of these doors include the following:

- BS 745 – Animal glue for wood
- BS 1444 – Cold-setting casein glue for wood
- BS 1204 – Synthetic resin adhesives for wood
- BS 1186 – Quality of timber and workmanship in joinery

Standard doors made in conformity with BS 459 are available in both metric and imperial sizes as follows.

Metric

826 × 2040 × 40 or 44	Interior doors –
726 × 2040 × 40 or 44	40 mm thick
626 × 2040 × 40 or 44	Exterior doors –
526 × 2040 × 40 or 44	44 mm thick

Imperial

840 × 1980 × 35 or 44
(2′ 9″ × 6′ 6″ × 1 3/8″ or 1 3/4″)
760 × 1980 × 35 or 44
(2′ 6″ × 6′ 6″ × 1 3/8″ or 1 3/4″)
680 × 1980 × 35
(2′ 3″ × 6′ 6″ × 1 3/8″)

Methods of construction

Panelled doors are constructed in one of two ways:

(*a*) with mortice and tenon joints;
(*b*) with dowelled joints (*see* Chapter 3).

Provided the work is properly carried out there is little to choose between the two methods, but generally speaking dowelling is a method best suited for mass production purposes owing to the sophisticated equipment required, whilst the mortice and tenon joint is eminently suitable for the special or "one off" type of job and for the small to average size joinery works.

Design of panelled doors

Figure 5.11*(a)-(e)* shows the elevations of single, two, three, four and six panel doors respectively, these being typical – but by no means the only – designs for doors with similar numbers of panels. The general construction of all these doors is basically the same, varying only in accordance with the number of members involved (and thus the number of joints), and by the possible treatments given to the panels and the inside edges of the door framework. All the doors illustrated are framed up with mortice and tenon joints, and all have panels either of plywood or solid timber which are fitted into grooves ploughed into the edges of the framework. Figure 5.12 is a more detailed illustration of a six panelled door, the various members all being named and the proportions of the tenons indicated by dotted lines.

Fig. 5.11. *Types of panelled door.*

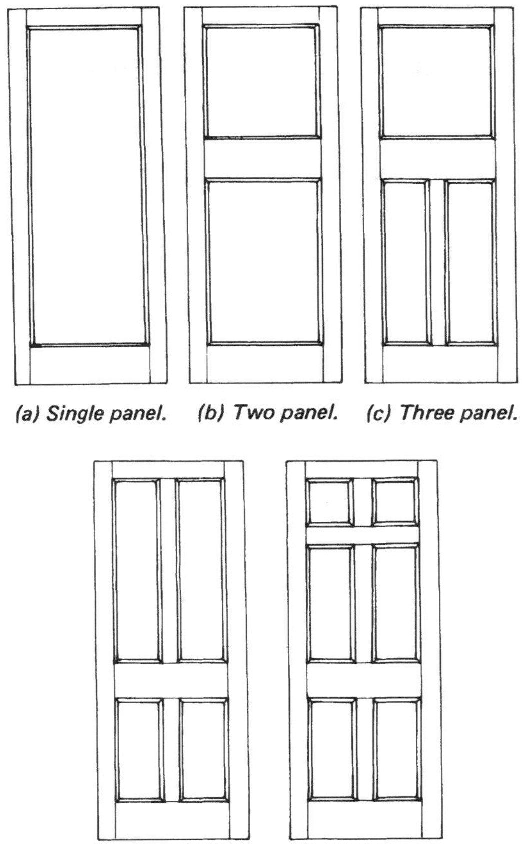

(a) Single panel. *(b) Two panel.* *(c) Three panel.* *(d) Four panel.* *(e) Six panel.*

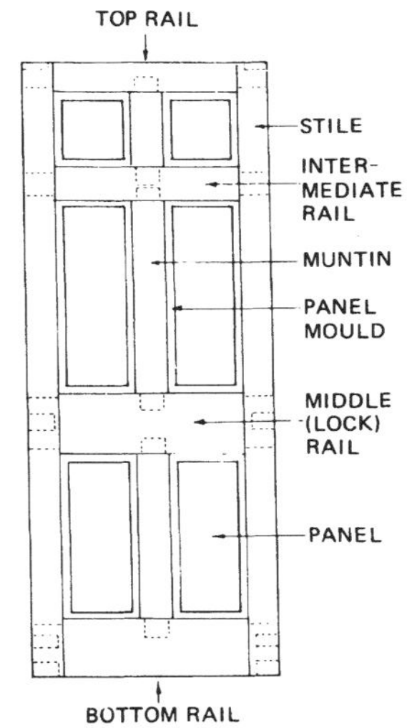

Fig. 5.12. *Parts of a panelled door.*

Panel treatments

Figure 5.13*(a)* shows a very simple type of treatment, generally referred to as "square edged" or "square sunk" – the panel being sunk below the surface of the framework. Figure 5.13*(b)* shows a similar treatment, relieved with "planted" panel moulds used to improve the appearance of the door. Figure 5.13*(c)* shows a section with "stuck mouldings", the term "stuck" being the woodworker's expression for a moulding or rebate which is worked on the solid timber.

Fig. 5.13. *Types of panel treatment.*

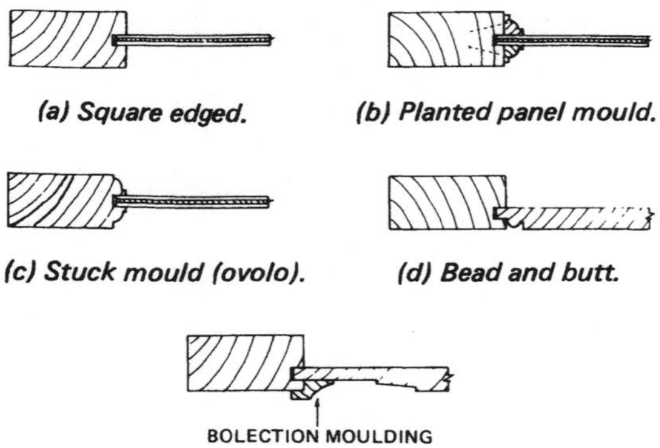

(a) Square edged. *(b) Planted panel mould.*

(c) Stuck mould (ovolo). *(d) Bead and butt.*

(e) Raised and fielded.

The section has the same appearance as Fig. 5.13*(b)* but is obviously better construction and is to be preferred in most cases. Figure 5.13*(d)* shows the type of panel known as "bead and butt". This panel, which is suitable for exterior doors, is made from solid timber about 22 mm thick, and has a bead worked on the vertical edges, the horizontal edges being "butted" to the rails on the face and tongued into the grooves. Figure 5.13*(e)* shows a superior type of panel treatment, the solid panel being known as "raised and fielded". Note the "bolection mould" often used in conjunction with this type of panel in good class work.

Construction details

Figure 5.14 shows an exploded view of the joints between the various members of a four panel door, in this case with square edges. One or two points here are worthy of mention.

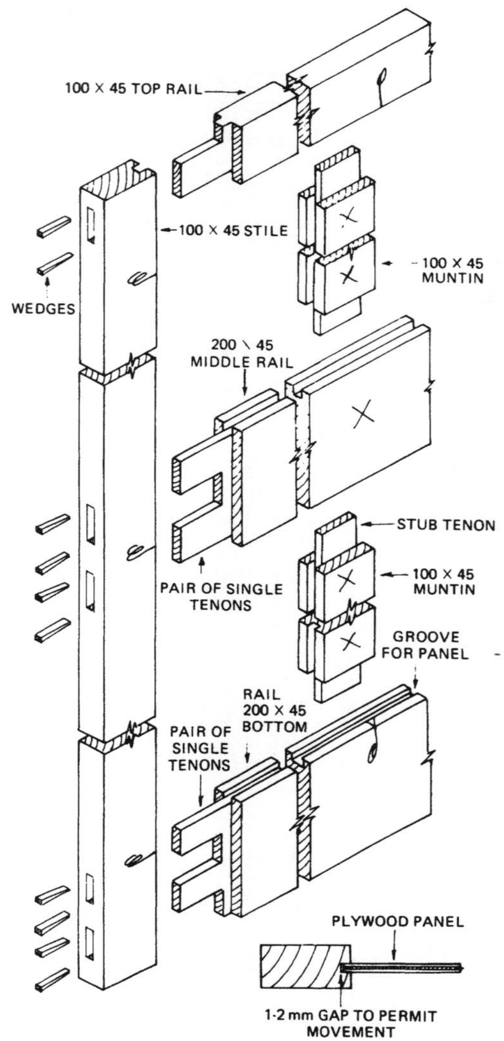

Fig. 5.14. *Panelled door showing details of joints.*

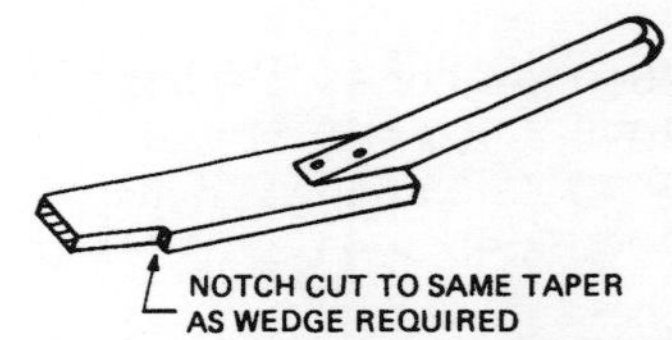

Fig. 5.15. *Jig for cutting wedges on a circular saw.*

Note first the proportion of the tenons and the use of haunches to obtain maximum strength (*see* Chapter 3).

Note secondly the wedges used to secure the mortice and tenon joints. These are normally cut from waste timber (often from that part of the tenon which is cut away to form the haunch) but where several doors are under construction, are cut by machine using a jig as shown in Fig. 5.15. The wedges should be long and tapering – long enough to be driven three quarters of the way into the stile, about 9 mm wide at the thick end tapering to about 2 mm at the thin end. They must, of course, be exactly the same thickness as the tenon itself.

Note thirdly the panels. These must be free to move fractionally within their grooves as may happen due to absorption or loss of moisture (*see* Chapter 2), and therefore must in no way be fixed, either by glue, nails or screws. Indeed, if panel moulds are fixed care must be taken to ensure that the nails *do not* enter the panel. To allow for a possible slight swelling, a panel should not be fitted tightly to the bottom of the groove, but should have a small gap of about 2 mm all round.

Making a four panel door

For the purpose of this exercise, assume that the door is to be made by hand (machine production is discussed in Chapter 11), has square edge framing, plywood panels and planted panel moulds.

The sequence of operations involved will be as follows.

(a) Prepare setting out rod as described in Chapter 4.

(b) Prepare cutting list, also described in Chapter 4.

(c) Select the materials for the door taking care to choose the straightest possible timber for the stiles. Select rift sawn timber as far as is possible.

(d) Cross cut to length and rip with width.*

(e) Plane up face sides accurately; apply face marks.*

(f) Plane up face edges, squaring exactly to the face sides. Apply marks.*

(g) Gauge and plane to width.*

(h) Gauge and plane to thickness.*

(i) Mark out the stiles as shown in Fig. 5.16*(a)*.

(j) Mark out the rails as shown in Fig. 5.16*(b)*.

(k) Mark out the muntins as shown in Fig. 5.16*(c)*.

(l) Chop mortices in stiles and rails – through mortices must be chopped from both edges. Cut away space for wedges on outside of stiles.

(m) Chop out for haunches on face edges of stiles (minimum depth = depth of groove).

(n) Saw cheeks of tenons.

(o) Plough the grooves for the panels, taking care that the fence of the plough runs along the face side of all the members.

(p) Saw tenon shoulders. Remove feathers left alongside plough groove on the edges of tenons.

Fig. 5.16. *Marking out for a four panel door.*

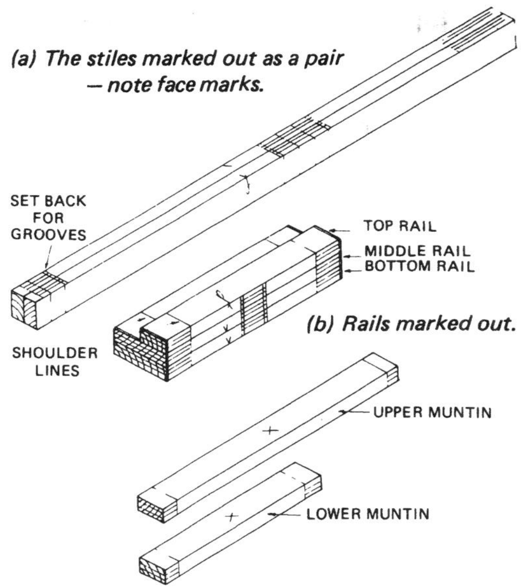

(q) Cut the tenons to form haunches.

(r) Cut panels and trim to size.

(s) Assemble door and examine joints for proper fit.

(t) Take door apart and clean up all inside edges.

(u) Place bearers on bench, loosely assemble door, apply glue to exposed part of tenons (40 mm or so from shoulders), knock up tightly and check for squareness.

(v) Cramp up door as shown in Fig. 5.17 and drive in wedges (wedges on outside of tenons first).

(w) Remove cramps, set door aside to dry – door *must* be kept flat and true.

(x) Clean off door on both sides, glass papering diagonally across the grain for painted softwood and with the grain for polished hardwood. Cut and fix panel moulds.

(y) Set door aside for priming or polishing.

NOTE: Horns are not generally sawn off until the carpenter hangs the door, since they serve to protect the ends of the stiles during transit and storage.

*These operations would normally be carried out by machine, even where a door is termed *hand made.*

FRAMES AND LININGS FOR PANELLED DOORS

Frames suitable for panelled doors are basically of two types:

(a) those for external doors – usually referred to as "frames";

(b) those for internal use (generally much lighter in construction – referred to as door "linings" or "casings".

Exterior frames

These are generally solid rebated and moulded, framed up with mortice and tenon joints, and rebated to allow the door to open inwards. (Door frames for public buildings are normally rebated for the door to open outwards so that in the case of panic due to fire the doors would burst open rather than become jammed shut.)

Figure 5.18 shows a section through the jamb of an external door frame which is described as "solid rebated and ovolo moulded". The ovolo mould on this type of frame would

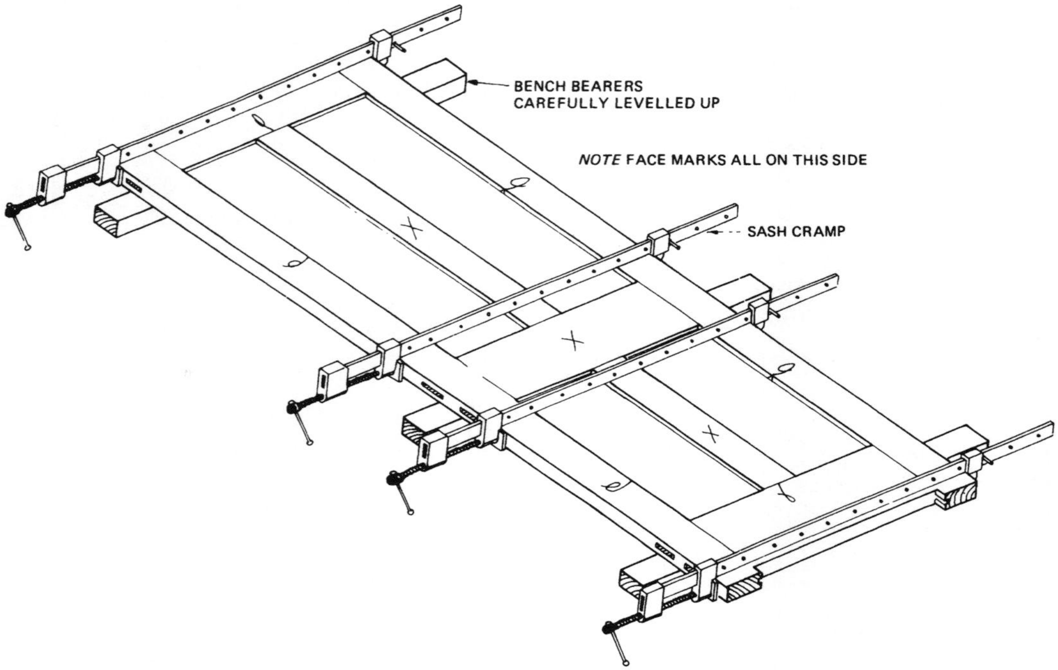

Fig. 5.17. *Cramping up a four panel door.*

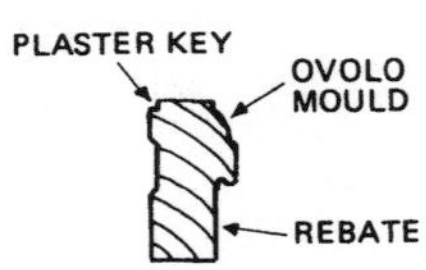

Fig. 5.18. *Section through jamb of a solid rebated and ovolo moulded frame.*

be scribed at the junction of head and jamb as shown in Fig. 5.19.

Figure 5.20(*a*) shows a door frame section utilising a bevel in place of an ovolo mould – again the junction would be formed by scribing. Figure 5.20(*b*) shows a further example of a solid rebated frame, this time the edge being relieved with a "pencil round" moulding. This type of moulding is applied after the frame has been assembled and forms a "mason's mitre" (*see* Chapter 3) at the intersection.

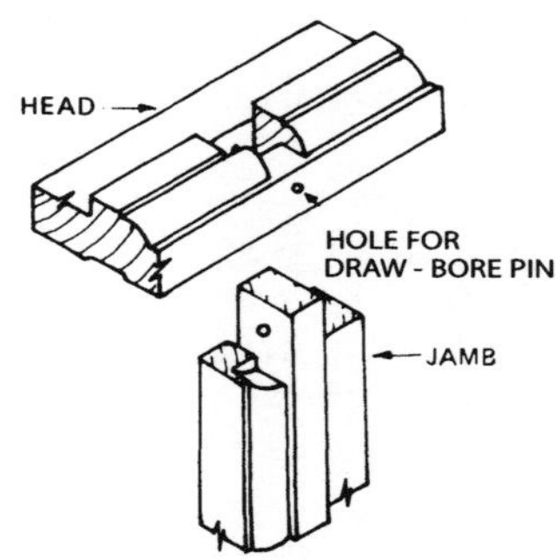

Fig. 5.19. *Joint at head of door frame.*

Fig. 5.20. *Solid rebated door frame.*

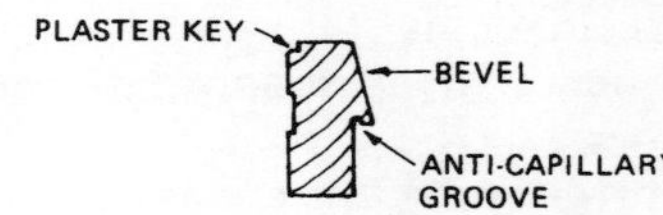

(a) With bevel.

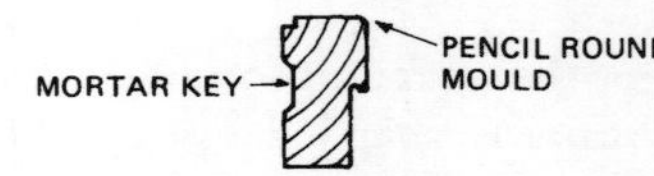

(b) With a pencil round moulding.

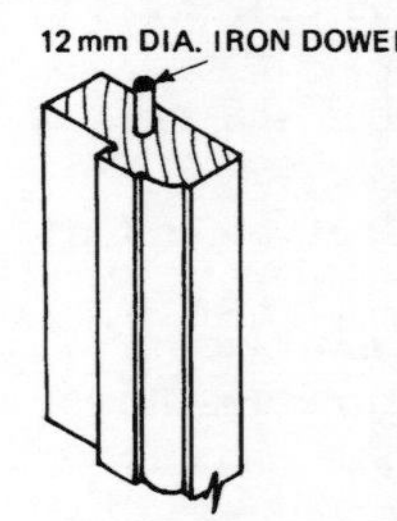

Fig. 5.21. *Bottom of jamb fitted with iron dowel.*

Door frames of the foregoing type are assembled and braced in the joiner's workshop, the horns being left on for building into the walls. Where a concrete step is to be cast under the bottom of the jambs, iron dowels are often fitted to secure them firmly to the step as shown in Fig. 5.21. Points worthy of consideration in the design of external door frames and shown in Figs. 5.18 and 20*(a)* and *(b)* include the following:

(a) the anti-capillary groove to prevent ingress of rain water by surface tension – this forms a gap which is too wide for the surface tension of the water to bridge;

(b) the mortar key or "frog" recessed into the back of the frame and which becomes partly filled with mortar as the frame is built in;

(c) the plaster key, not always included in the design of door frames, but nevertheless serving a useful purpose in forming a good joint between frame and plaster in the reveal.

Threshold

A hardwood threshold or cill is frequently used to prevent ingress of water under the bottom of the door. Used in conjunction with a weather board, tongued into the face of the door and a galvanised iron water bar (GIWB), thresholds form a very satisfactory weather seal to domestic external door frames. Details of such a threshold are given in Fig. 5.22.

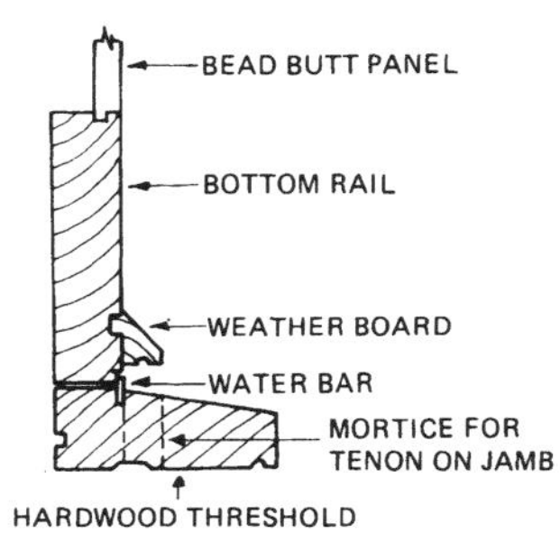

Fig. 5.22. *Detail of threshold under bottom edge of an external door.*

Fig. 5.23. *Plain door linings.*

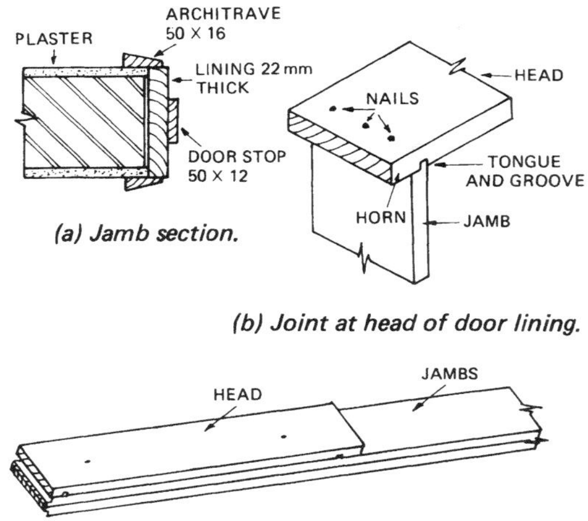

(a) Jamb section.

(b) Joint at head of door lining.

(c) Linings nailed together for delivery as a "set".

Door linings

These are almost invariably used for doorways in internal partition walls and may have planted door stops or be solid rebated, the former being the more common.

Figure 5.23*(a)* shows the section through a plain door lining with a planted stop. The lining is 22 mm thick and extends the full thickness of the wall – plaster included. Door linings of this type are made up in the joiner's shop, the joint between head and jamb being a tongue and groove as shown in Fig. 5.23*(b)*. Although such door linings are generally assembled, braced and primed before leaving the workshop, they are occasionally despatched to the site in the form of "sets", as shown in Fig. 5.23*(c)*, for assembly by the site carpenter. This practice makes for easier stacking and storage, both in workshop and on site. Transport is also simplified but the fact remains that assembly on site is rarely as efficient and accurate as when done in the workshop. Softwood linings should always be primed before fixing to prevent undue absorption of water when the building is plastered.

Fig. 5.24. *Solid rebated door lining.*

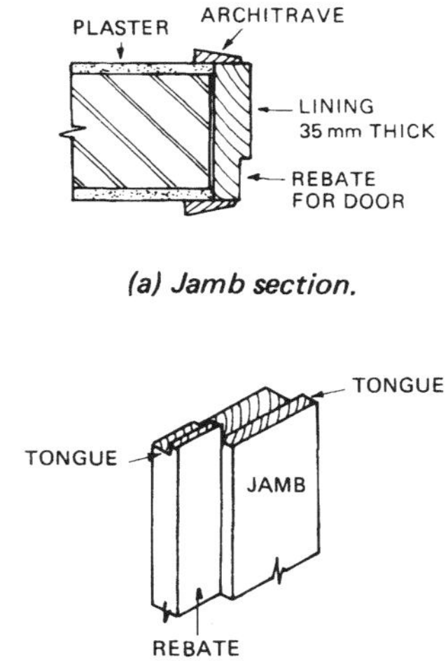

(a) Jamb section.

(b) Joint at head of rebated lining.

Figure 5.24*(a)* shows a solid rebated door lining, again extending the full thickness of the wall and generally 35 mm thick. Solid rebated linings are obviously a superior type of construction and to be preferred on good class work. The joint between head and jamb may be a form of tongue and groove as shown in Fig. 5.24*(b)*, or a mortice and tenon, this depending largely on the thickness of the lining.

Hanging the door

Hinges

Panelled doors, either interior or exterior, are hung to the frame on "butt" hinges, two hinges being usual on interior doors and three on heavier exterior types. Butt hinges, shown in Fig. 5.25 are available in a variety

Fig. 5.25. *Butt hinge.*

of sizes from 25 mm to 150 mm. 100 mm hinges are suitable for exterior doors whilst 75 mm or 87 mm are more usual for internal doors. Butt hinges are made of steel, cast iron, plastic and bronze, the brass and bronze variety being very expensive and used mainly on large, heavy, hardwood doors. Pressed steel hinges are relatively cheap and are commonly used on both internal and external doors.

Fitting the door

First, determine from the plans on which side of the frame the door is to be hung, and whether the door is to open inwards or outwards — on solid rebated frames the latter will be self evident. Next, check the door for overall size, support it on a pair of sawing stools and carefully remove the horns. Hold the saw at a low angle to prevent the grain shattering on the underside. Alternatively, cut from both sides.

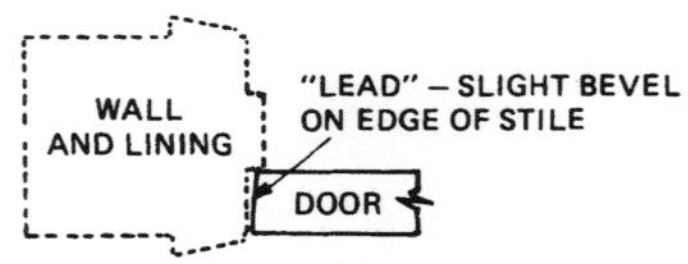

Fig. 5.26. *Sketch showing "lead" on closing edge of door.*

Having removed the horns, shoot the edges and ends of the door, removing as little material as is possible until the door will just fit inside the frame. The door has now to be fitted accurately to the frame and so it is best to work in a proper sequence as follows:

(*a*) shoot hanging stile to fit jamb;

(*b*) shoot top edge of door to fit head;

(*c*) shoot bottom edge to fit floor, allowing about 4 mm ground clearance;

(*d*) shoot closing stile until joint is parallel and twice that required on the finished door (when testing the bottom of the door and the closing stile for correct joint the door should be wedged tightly against the head and hanging jamb);

(*e*) plane a slight bevel or lead on the closing stile of the door to give the necessary clearance when opening and closing as shown in Fig. 5.26, and the door is ready to have the hinges fitted.

Fitting the hinges

Position the door in its frame and insert small wedges in the closing joint to force the hanging stile tightly against the jamb. Insert another small wedge under the bottom edge of the door to give the correct clearance top and bottom. Using a sharp pencil or marking knife, scribe a short mark across the edge of the door and frame to indicate the positions of the top edge of the top hinge

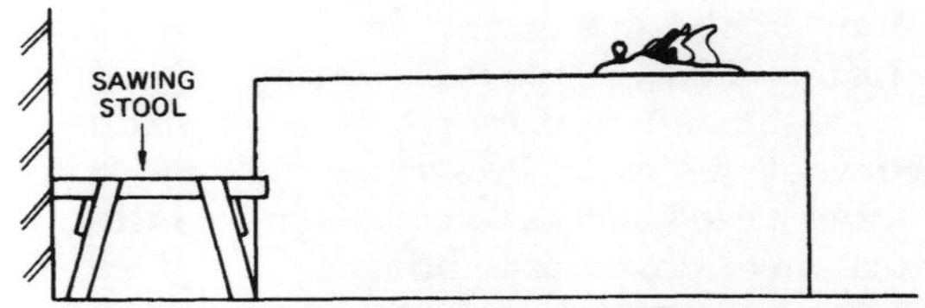

Fig. 5.27. *Holding a door whilst "shooting-in" and fitting hinges.*

Fig. 5.28. *Setting gauges to hinge.*

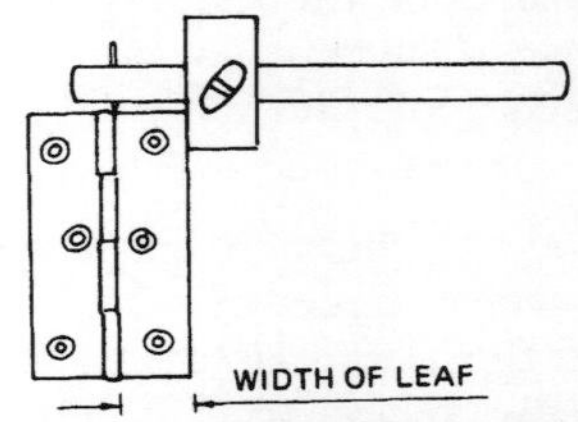

(a) Setting gauge to width of leaf.

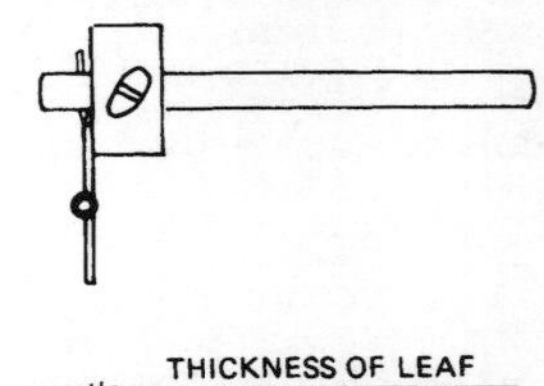

(b) Setting a gauge to thickness of leaf.

— about 150 mm from the top of the door — and the bottom edge of the bottom hinge — about 200 mm from the floor. Remove the door from the frame and stand it on edge on the floor, supporting it with the jaw of the sawing stool as shown in Fig. 5.27.

Using a hinge as a guide, mark the length of the recess on both door edge and door jamb. Set a marking gauge to the width of the hinge leaf as shown in Fig. 5.28(*a*) and

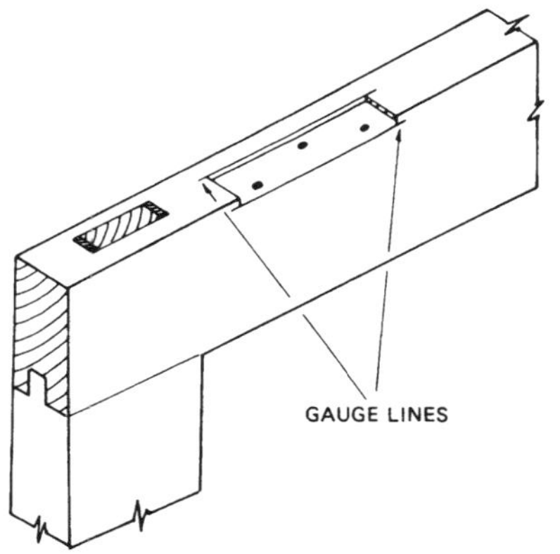

Fig. 5.29. *Recess chopped out for butt hinge.*

gauge the width of the recesses on door and frame. Repeat this process with the gauge set to the leaf thickness as shown in Fig. 5.28*(b)*.

The marked recess can now be cut out as shown in Fig. 5.29 using a sharp 25 mm chisel and a mallet.

Again using a hinge as a guide, make pilot holes for the screws with a bradawl. Screw the hinges to the door. Stand the door against the jamb in a 90° open position and insert one screw into each of the remaining leaves, top hinge first. The door can now be tested for proper action when opening and closing, and, if all is satisfactory, the remaining screws can be inserted. If the door has insufficient clearance on the closing stile, the hinges must be sunk a shade further into the door jamb, whilst if the door appears to be "hinge-bound" (has a tendency to spring open due to too tight a joint on the hanging side) there is no alternative but to insert a thin cardboard packing behind the leaf which is sunk into the door jamb to increase the joint. Before doing so, however, make sure that the trouble is not due to the screw heads protruding from their countersinking.

> NOTE: The correct and proper hanging of a door is not the easiest of tasks, but a little practice soon gives the necessary experience and skill enabling this to be carried out without difficulty.

Door furniture

The type of furniture required for a door depends to a large extent upon the degree of security required. Internal doors rarely require other than a latch, fitted with suitable handles, whilst exterior doors require a lock as well as a latch, often incorporated in the same piece of ironmongery. Figure 5.30*(a)* shows a typical mortice latch together with a suitable lever handle. Mortice latches of this type are used on internal doors. Figure 5.30*(b)* shows an upright mortice lock and its lever handle. This type of lock incorporates a latch as well as a lock and is thus suitable for both internal and external doors. Upright mortice locks of this type can be fitted into doors with relatively narrow stiles since the lock is only about 75 mm deep. Figure 5.30*(c)* shows a horizontal type of mortice lock, also incorporating a latch, which is suitable for panelled doors having a lock rail since the length of the lock is too great for the average door stile. Used mainly for exterior doors, this type of lock is normally furnished with door knobs rather than lever handles.

Figure 5.30*(d)* shows a very useful type of lock known as a "cylinder night latch". This

Fig. 5.30. *Door furniture.*

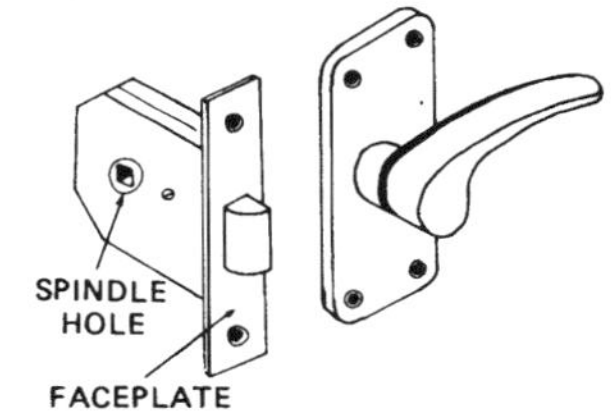

(a) Mortice latch and lever handle.

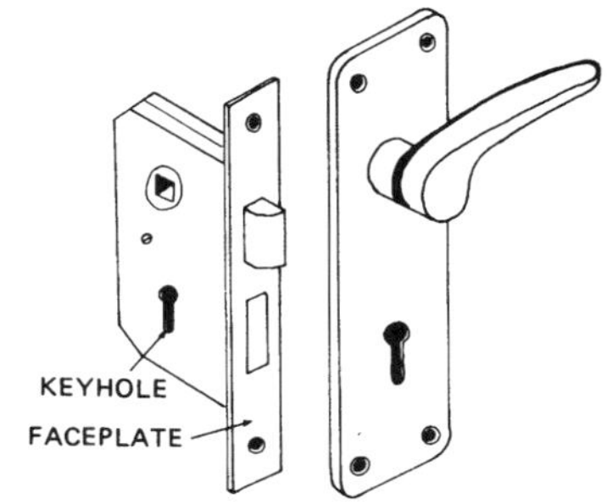

(b) Upright mortice lock and lever handle.

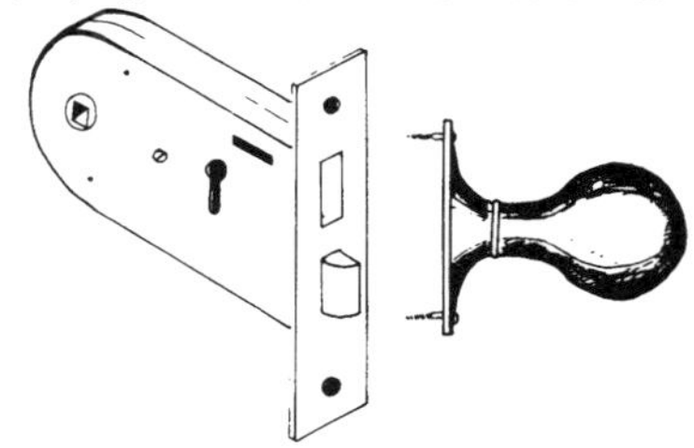

(c) Horizontal mortice lock and knob handle.

(d) Cylinder night latch.

type of lock is used principally as a second lock for added security to external (and sometimes internal) doors. The fitting of locks and door furniture is discussed in Chapter 10.

STORAGE AND PROTECTION OF DOORS AND FRAMES

Since doors and frames are expensive items of joinery, it becomes important to ensure they are not damaged or allowed to deteriorate during the construction of a building or whilst alteration or repair work is in progress.

Fortunately the protection of joinery is largely a matter of commonsense. The main points for consideration are outlined below.

(a) Doors should be primed before delivery to site and horns left on.

(b) Doors must be stored under cover and stacked flat to prevent twisting.

(c) Delivery to site should be at the anticipated time of use – not too early.

(d) Door frames and linings should be primed before delivery. Assembled frames should be securely braced and stored flat to prevent twisting.

(e) Joinery of any description should be handled carefully, especially during loading and unloading.

(f) Fixed frames can be protected from accidental damage by nailing scrap timber to the lower part of the jambs – essential when heavy or awkward materials are being moved through the building.

(g) Doors should be fastened or locked as soon as they have been hung to keep out intruders and to prevent wind damage.

(h) Joinery should not be delivered to or kept on site unless it can be stored in a secure place under lock and key and under the supervision of a responsible person.

FURTHER READING

British Standards and Codes of Practice

CP 151	Doors and windows including frames and linings Part 1:1957 Wooden doors
BS 459	Specification for wooden doors Part 1:1954 Panelled and glazed wood doors Part 2:1965 Match boarded doors
BS 565:1972	Glossary of terms relating to timber and woodwork
BS 584:1967	Specification for wood trim (softwood)
BS 1186	Specification for quality of timber and workmanship in joinery Part 1:1979 Quality of timber Part 2:1979 Quality of workmanship
BS 1567:1953	Wood door frames and linings

Building Research Establishment Digests

No. 73	Prevention of decay in external joinery
No. 175	Choice of glues for wood
No. 201	Wood preservatives – application methods

SELF-TESTING QUESTIONS

All the information required to answer the following questions is contained within this chapter. Attempt each section *as fully or as briefly* as the question demands, and then check your answers against the information given in the chapter.

1. *(a)* Draw a full size horizontal section through the edge (150 mm) of a ledged and braced door to show:

(i) how the *end* of a ledge is fastened to the matchboard;

(ii) how the matchboards are fastened to the ledges;

(iii) the joint between two adjacent match boards.

(b) *(i)* State the treatment to be given to the edges of the match boards of a ledged and braced door *before* assembly.

(ii) Explain the relationship between the braces and the hinges.

(iii) Sketch and name the type of hinge used to hang the door.

2. *(a)* Name the two basic joints which are used in the construction of panelled doors.

(b) State which of the two joints named in *(a)* is most commonly used in mass production.

(c) Sketch details to show the forms of the following types of panel treatment:

(i) square edged (square sunk);

(ii) plywood panel with "planted" panel moulds;

(iii) plywood panel with "stuck" moulds;

(iv) raised, fielded and bolection moulded.

3. Sketch details to show the general shape and proportions of the tenons used in

panelled door construction in the following instances:

(a) top rail to stile;
(b) wide (200 mm) lock rail to stile;
(c) wide (200 mm) bottom rail to stile.

4. List the sequence of operations involved in the following *stages* in the construction of a four panelled door:

(a) from receiving details of the door to the completion of marking out;

(b) from completion of all the joints to the point at which the door is ready for delivery.

5. *(a)* Sketch a section through the head (or jamb) of a solid rebated and ovolo moulded door frame, indicating on your sketch the following details:

(i) mortar key or "frog",
(ii) plaster key,
(iii) anti-capillary groove,

stating the function of each.

(b) Sketch a detail to show the sectional shape of a hardwood threshold (or cill) at the bottom of an inward opening external door frame. Include in your sketch the following details: *(i)* a water bar, and *(ii)* the bottom of the door, showing a weather board.

6. *(a)* Sketch sections through the jambs of:

(i) a plain door lining with planted stops;
(ii) a solid rebated door lining.

(b) Sketch the type of joint normally used between the head and the jamb of a door lining.

7. *(a)* List the sequence of operations involved in hanging a door to an internal door lining.

(b) Name and describe the hinges and ironmongery that would normally be required to hang and furnish:

(i) an interior door (lounge to hall);
(ii) an external door (front door) to a domestic dwelling.

8. Outline briefly the main points which must be considered in order to prevent damage to doors, frames and linings, from the time they are completed by the joiner until the building is occupied.

6. Traditional Casement Windows

After completing this chapter the student should be able to:

1. State the functions of a casement window.
2. Name the component parts of a casement window.
3. Sketch sections through the members of a casement window.
4. Sketch the joints used in casement window construction.
5. Prepare a setting out rod from given data.
6. State the sequence of operations involved in making a casement window.
7. Select and name suitable ironmongery.
8. Describe the fixing procedure for casement frames.

FUNCTIONS OF A CASEMENT WINDOW

The primary function of a casement window is, of course, to admit light into a room or building. This, however, is not necessarily its sole function. Most casement windows are fitted with opening "lights" or sashes to fulfil the secondary function of providing a means of ventilation.

In addition, most windows are designed to allow the occupants of the building or room to see out -- to afford a view of what is going on outside. Other less fundamental but nevertheless important functions include:

(a) retention of heat within the building;

(b) reduction of sound from exterior sources;

(c) prevention of ingress of water and draughts;

(d) aesthetic considerations — windows assist materially in enhancing the design of a building.

Types of wooden casement window

There are basically two types of wooden casement window.

(a) Traditional type casements. These are described in this chapter and are to be found in great quantity in older type buildings. They are occasionally still used in newer, more modern buildings.

(b) Stormproof casements. These are beyond the scope of this volume (but see *Carpentry and Joinery Book 2* by the same author).

Both types should be constructed in conformity with BS 644.

DESIGN OF A TRADITIONAL CASEMENT WINDOW

There are two main elements to a casement window:

(a) the frame, nowadays made so its overall height and width conform to the metric modules specified in BS 644, or made as a replacement component in an existing building;

(b) the lights or sashes, made to fit the openings in the frame. The lights may be fixed into the frame solely for the purpose of admitting light, or hung on butt hinges to allow for ventilation also. In either case, sashes are essential in traditional type casement windows as the rebates in the frames are too big to allow for direct glazing.

Casement windows are described by the number of lights they contain, e.g. single light, two light, four light, etc.

Components of a traditional casement frame are named and illustrated in Fig. 6.1 which shows the front elevation of a typical

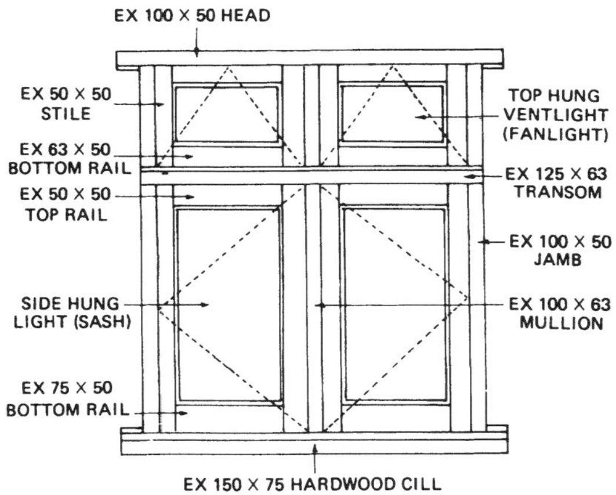

Fig. 6.1. *Elevation of a four light traditional casement frame showing members.*

four light window. As will be seen, the design of the frame closely follows that of the solid rebated and moulded door frame described in Chapter 5, the top horizontal member being referred to as the head, the vertical side members as jambs and the lower horizontal member as the cill. In addition, there is an intermediate vertical member (in two sections) known as a mullion and also an intermediate horizontal member known as a transom.

The sashes comprise of stiles, top and bottom rails, and are occasionally sub-divided into smaller panes by members known as glazing bars.

All the joints, both in the frame, and the sashes are mortices and tenons, secured with wedges and a good waterproof adhesive.

The diagonal dotted lines on the elevation indicate that the sashes are hung to open (the "vee" pointing towards the hinged side). The lower two sashes are generally referred to as "side hung" and the upper two as "top hung". Top hung sashes fitted above the transom – as shown in the elevation – are usually known as "fanlights" or "ventlights".

Casement windows may have sashes which open either inwards or outwards, the latter type being by far the more common. Inward opening casements are less easy to weatherproof and rather more inconvenient in use. However, their constructional features are beyond the scope of this volume. (See *Book 2* for a discussion of constructional features.)

CASEMENT WINDOW DETAILS

The head of a typical four light casement is shown in section in Fig. 6.2. Points worthy of note are as follows.

(a) The weather strip above the sash ploughed into the face of the head – an essential feature on an exposed window of this type.

(b) The anti-capillary grooves in both head and top rail. These should be at least 6 mm wide in order to perform their function satisfactorily.

(c) The moulding, in this case an ovolo, is so proportioned that the quirk on the face is level with the line of the rebate in the frame. This makes for easier setting out, since it brings the shoulders on the tenons into line.

(d) Rebates in both frame and sash are on the outside (the weather side) of the window, this being the easiest and most satisfactory way in which to exclude rain water. The frame is rebated 12 mm deep, the width of the rebate being determined by the thickness of the sash. It is good practice to make this rebate 2-3 mm wider than the thickness of the finished sash, this giving better appearance and weather protection. The rebate in the sash (the glazing rebate) is normally about 8 mm deep by 15-16 mm wide, this being quite sufficient for normal window glass, bedded and face pointed with putty.

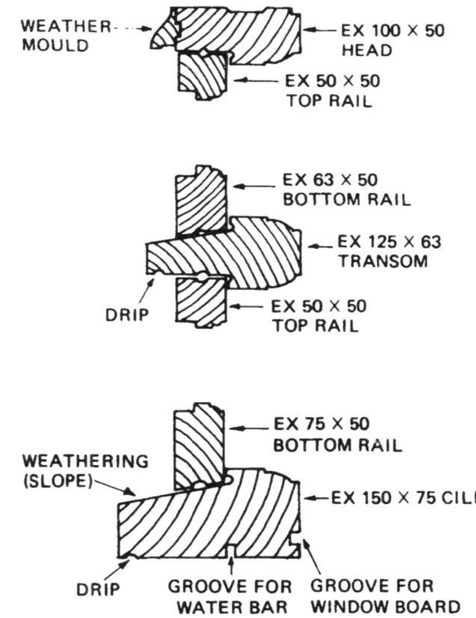

Fig. 6.2. *Vertical section through a four light casement frame.*

The jambs

Shown in Fig. 6.3, these have a section identical with that of the head with the exception of the weather strip which would serve no useful purpose in a vertical position. Note the section of the stile, this also being the same as that of the top rail.

Fig. 6.3.

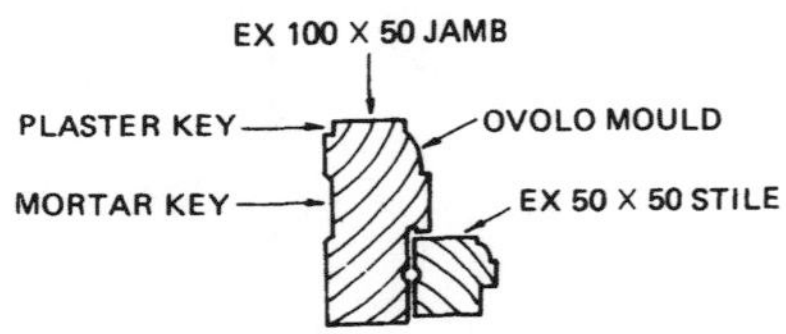

(a) Section through jamb and stile.

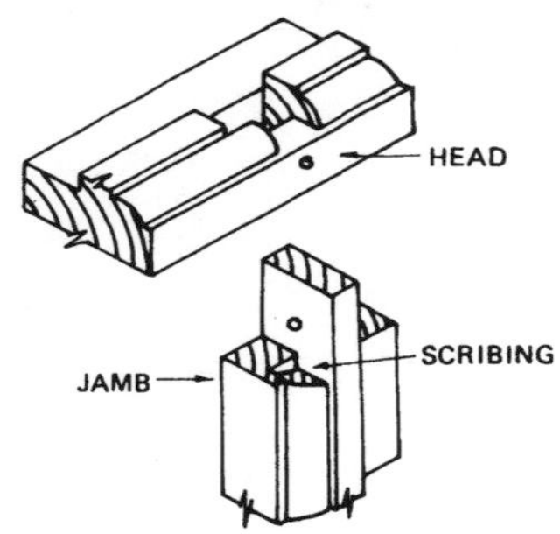

(b) Joint between head and jamb (hand scribed).

The cill and bottom rail

Sections of these are also clearly shown in Fig. 6.2, the following points being worthy of note.

(a) The cill is best made of a suitable hardwood such as oak, iroko, etc. (*see* Chapter 2) and is rebated and weathered to ensure the exclusion of rainwater and to prevent any tendency for water to "lay on" and possibly give rise to decay.

(b) The *drip* or throating under the front edge of the cill is also an essential feature to prevent the tailing back of water under the cill.

(c) The groove ploughed into the inside face of the cill is to receive the back edge of a window board during first fixing (*see* Chapter 10).

(d) The groove ploughed into the centre portion of the underside of the cill is to allow for a water bar to be fitted further reducing the possibility of water penetration.

(e) The bottom rail has the same moulding and glazing rebate as the stiles and top rails but it will be seen that the bottom rail is wider than these in order to give extra strength and better appearance. The bottom edge of the rail is bevelled to fit the weathering on the cill.

The transom

(*See* Fig. 6.2.) This really serves a dual purpose, acting as a cill to the top sash (the fanlight) and as a head to the bottom sash. The transom usually projects about 25 mm in front of the jambs, and has a double throating on its under surface – again to check the entrance of water by capillary action. The transom spans the complete width of the window, being housed and tenoned into the jambs on either side.

Mullions

(*See* Fig. 6.1.) These serve the purpose of central jambs and have mouldings and rebates on both sides. Mullions should be at least 63 mm thick to allow for this *double sticking* whilst their width is the same as for the jambs and head.

Where a transom is present as in Fig. 6.1, this cuts across the mullion which must therefore be in two pieces, these being stub tenoned into the transom from either side. Mullions are also tenoned into into the head and cill, the tenon being either a stub or a "through" type.

Nominal sizes of members

Reference to Fig. 6.2 will show that the sectional sizes of the members are indicated as *ex* 50 X 100 head, *ex* 75 X 150 cill, etc. These stated sizes represent the sawn sizes – the original dimensions of the timber from which each member was produced and therefore the size of timber which the customer must pay for. A moment's consideration, however, should make it amply clear that when any piece of timber is planed or *wrot* it becomes reduced in size. A piece of 63 X 100 for instance which has been planed up on four faces (p.a.r.) ready for setting out may well have a finished size of 58 X 95 – a reduction of 2½ mm on each face. Thus we have two sizes to contend with for joinery timbers – the original sawn or "nominal" size, and the "planed" or "finished" size. Most timber components in joinery are given in nominal sizes unless it is required to finish a particular member to a specific dimension, when its "finished" size may be stated.

As far as the joiner is concerned, it makes little difference whether the finished size of a member be 97 X 60 or 95 X 58, *provided that he knows* the exact dimensions he is dealing with at the setting out stage of the job. British Standard 4471 lists the maximum permissible reduction in size due to machining for most items of work with which the joiner may be concerned.

Constructional details

The frame

Figure 6.3*(b)* shows the joint between the head

Fig. 6.4. *Casement frame joints.*

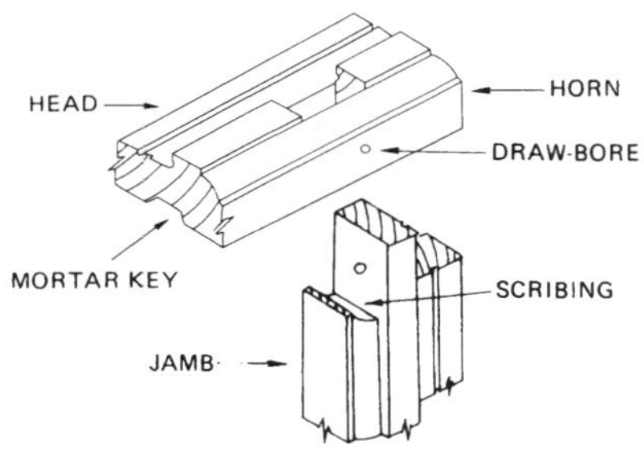

(a) Joint between head and jamb (head reversed for clarity).

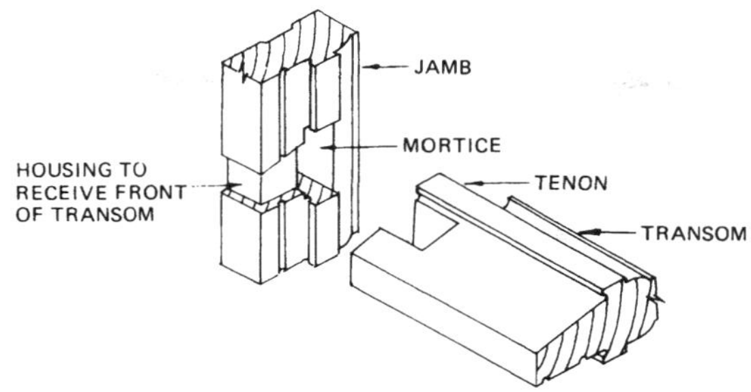

(b) Joint between jamb and transom.

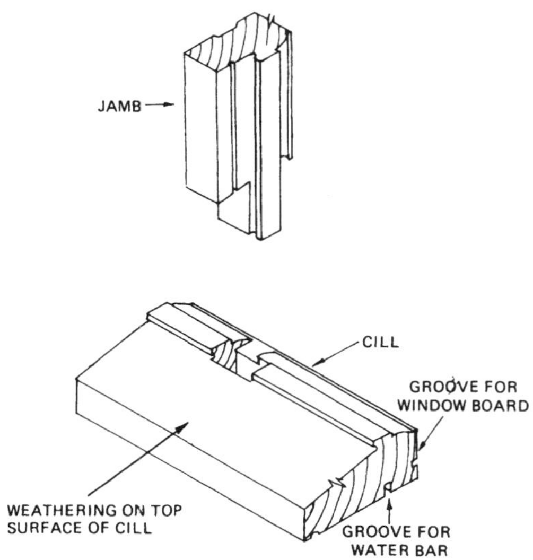

(c) Joint between jamb and cill.

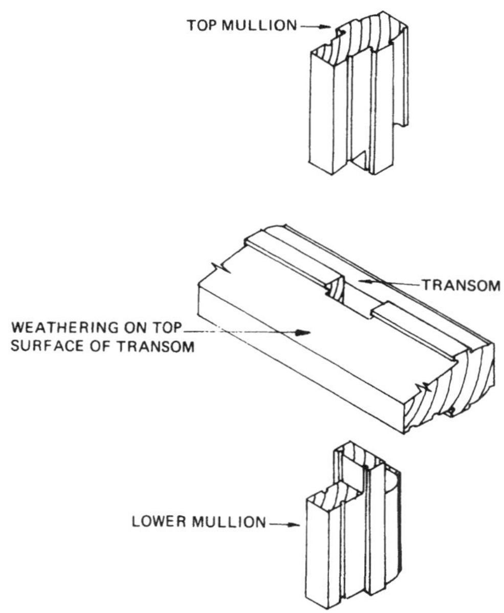

(d) Joints between mullions and transom.

and jamb of a traditional casement frame. Note that the ovolo mould on the jamb is scribed to fit over that on the head. Where the moulding is scribed by hand, only part of the section is cut out of the jamb shoulder, the remainder of the moulding being cut away from the head. When similar tenons are cut and scribed by machine, the scribing is carried right across the shoulder as shown in Fig. 6.4*(a)*, this being more convenient in machine production.

Figure 6.4*(b)* shows the joint between the jamb and the transom. Note the outside weathered edge of the transom which extends beyond the front of the jamb and is housed into it on two faces. This is good practice and gives added resistance to weather penetration.

Figure 6.4*(c)* shows the joint between the jamb and the cill, the outside shoulder of the jamb being cut to fit the weathering on the cill.

Figure 6.4*(d)* shows the joints between the mullions and the transom, the stub tenons on the mullions entering the transom half way (or just less) from each side.

The sashes

Figure 6.5 shows alternative joints between the top rail and the stile of the sashes. In each case, the ovolo mould is scribed at the intersection, the scribing being carried either partly or wholly across the rail shoulder. Figure 6.5*(a)* shows a "haunched" joint whilst Fig. 6.5*(b)* shows a "franking". There is little to choose between the two methods, some craftsmen preferring the one and some the other. The "franked" joint is probably a little easier and quicker to execute, whilst the

Fig. 6.5. *Sash joints.*

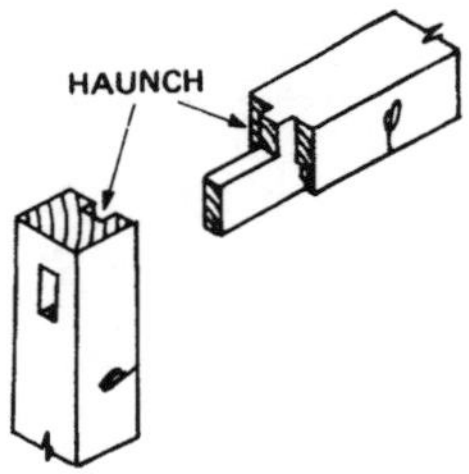

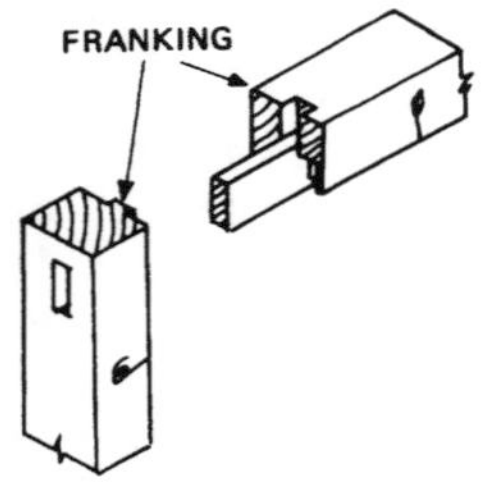

(a) Haunched tenon. *(b) Franked tenon.*

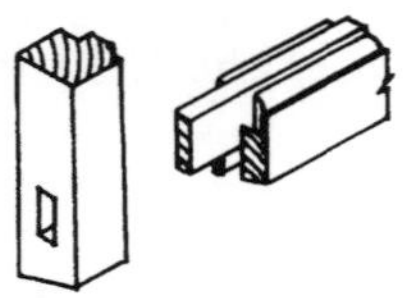

(c) Joint between bottom rail and stile.

haunched joint would appear to be rather stronger.

The joint between the bottom rail and the stile is shown in Fig. 6.5*(c)* and is similar to that used on the top rail.

Glazing bars

Where glazing bars are incorporated in the design of the sash, the joints are as shown in Fig. 6.6*(a)*. Scribings are carried right across the shoulder, vertical glazing bars being stub tenoned into the horizontal bars and also into the top and bottom rails. Horizontal bars are usually stub tenoned into the stiles and extend the full width of the sash, bisecting the vertical bars. On large sashes, where three or more horizontal glazing bars are included, it would be good practice to run at least one tenon right through the stiles, for

Fig. 6.6. *Glazing bars.*

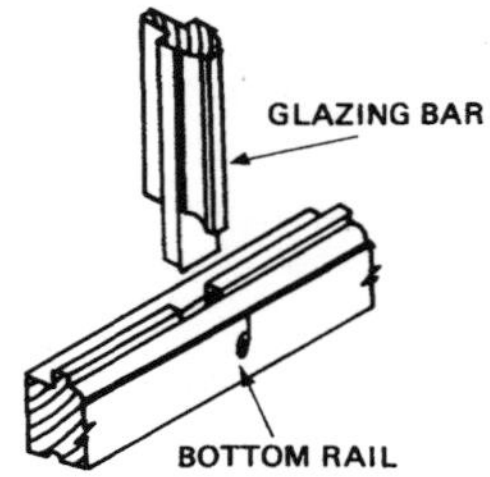

(a) Joint between glazing bar and bottom rail.

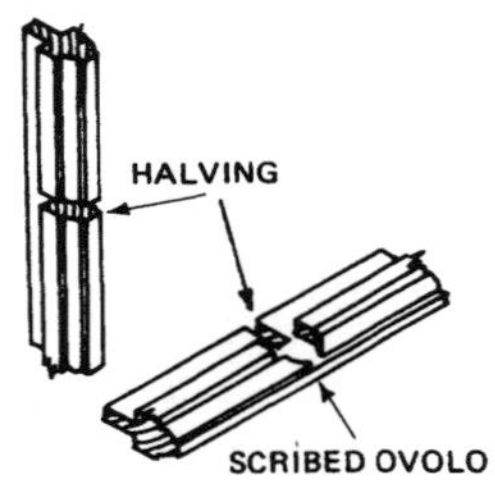

(b) Halved and scribed joint for intersection of glazing bars.

wedging up, thus forming an effective "tie". Figure 6.6*(b)* shows an alternative joint for use at the intersection of glazing bars. This joint, known as a "scribed halving", is considerably more time consuming than the stub tenon, but is useful on large sashes where a continuous vertical tie is required between top and bottom rails.

SETTING OUT A CASEMENT WINDOW

The marking out of a casement frame and sashes is fully illustrated in Fig. 6.7.

Setting out rod

Unless the casement frame is a very simple one, it would be good policy to prepare first a setting out rod. This is invaluable as it gives accurate overall sizes of the frame and sashes, the position and sections of the various members, and also the position and dimensions of the mortice and tenon joints. A typical setting out rod for a four light casement is shown in Fig. 6.7*(c)*.

Face and face edge marks

A brief glance at a wooden casement frame is sufficient to see that one side of the frame is flush (the side having the mouldings) and the other is not, the transom and cill projecting beyond the outside face. Since it is good policy to do the marking out – the squaring and gauging – from one side only, then obviously it is the inside of the frame – the moulded side – which will be the "face". The face edge marks should be applied to the inside faces of the material, i.e. the sides of the material which are to be both moulded and rebated. Transoms and mullions thus have *two* face edges and should be marked accordingly.

NOTE: It is important to mark the transom to indicate which side is to be the top since this member has different "stickings" top and bottom and is not therefore reversible.

On the sashes, it is the moulded side again which is the "face", a precaution which ensures that this side finishes flush making the scribing of the moulding positive. Any slight variation in the thickness of the material

Fig. 6.7. *Marking out for a casement frame and sashes using a setting out rod.*

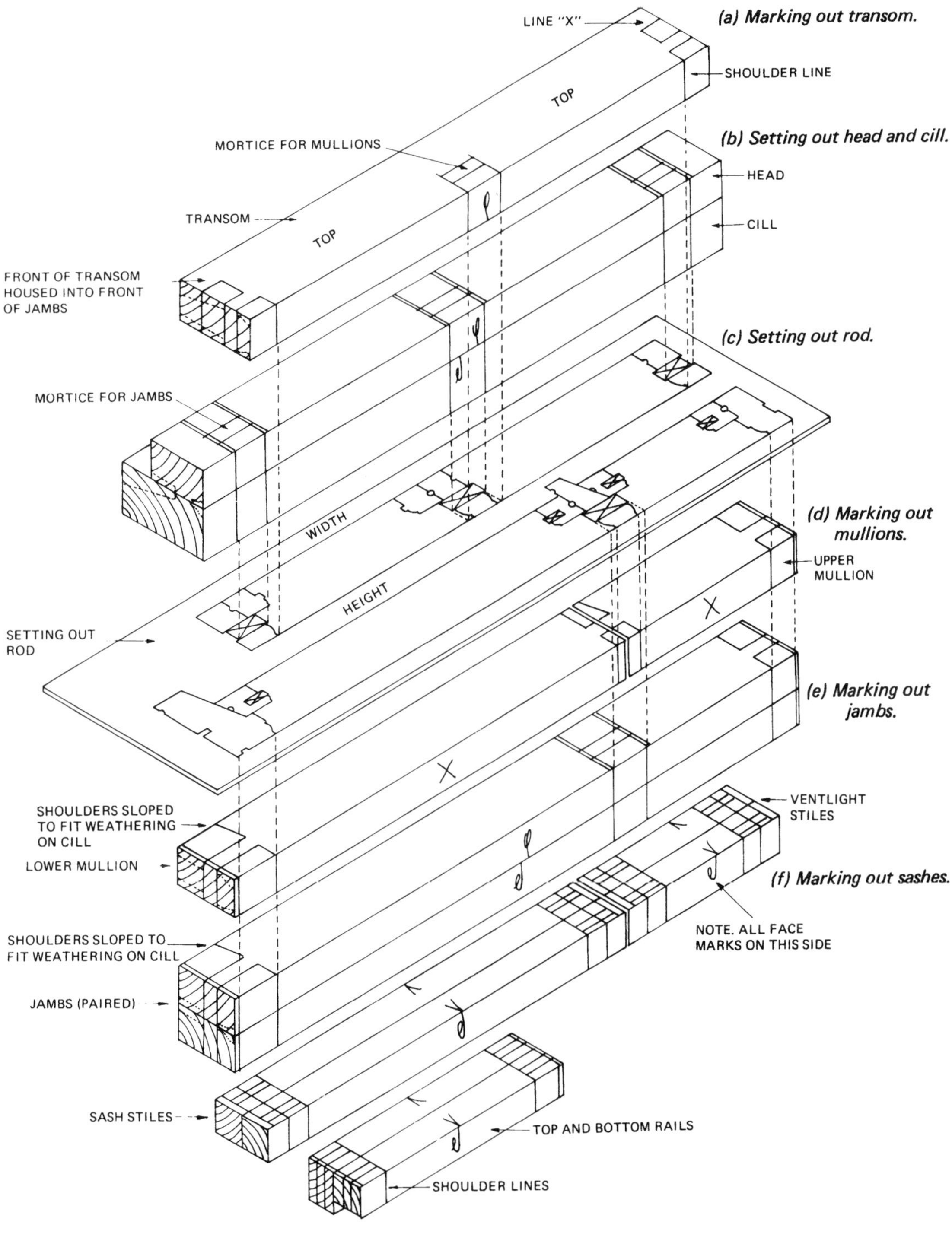

therefore shows up on the rebated side and is easily cleaned off.

This principle applies to all casement frames, irrespective of the type of moulding, or indeed whether a moulding is to be applied or not.

Setting out procedure

First set out the head and cill as shown in Fig. 6.7*(b)*, taking the position of the mortices directly from the setting out rod and taking care to gauge from the face side. Note the "wedge spaces" which are marked on the outside face (the face which will be in contact with the brickwork).

Next, mark out the transom as in Fig. 6.7*(a)* again directly from the rod to establish the exact position of the mortice for the mullions (no wedges here) and the shoulder lines for the jambs. Cut the shoulder lines with a marking knife to ensure a clean accurate edge to the shoulder. The shoulders on the transom are marked on the face side, and extend across the upper and lower surfaces only as far as the gauge lines. Set a marking gauge to the width of the jambs – less 8 mm – and gauge the line marked X. This shows the depth to which the transom is housed across the side of the jamb. Mark the intermediate shoulder line on the upper and lower surfaces of the transom, allowing for this to be housed 8 mm into the face of the jamb rebate.

Now mark out the two jambs as shown in Fig. 6.7*(e)*, again marking the shoulders with a marking knife. Note that the outside shoulder must be bevelled to fit the weathering on the cill. Gauge the mortices and tenons from the face side.

Finally, mark out the mullions as shown in Fig. 6.7*(d)* again making sure to gauge from the face side and using the marking knife for the shoulders.

The sashes are marked out as shown in Fig. 6.7*(f)*, this being a relatively simple operation. Once again the position of the mortices and the tenon shoulders are taken directly from the setting out rod, mortices being marked with a pencil and tenon shoulders with a marking knife. All the gauging must be done from the face side.

> NOTE: The thickness of the sash tenons is governed largely by the thickness of the sash itself and should approximate to one third of this, the flat section between moulding and rebate being deliberately proportioned to suit the mortice chisel (*see* Fig. 6.8).

SEQUENCE OF CONSTRUCTION

The following sequence of operations in the construction and assembly of a typical four light frame assumes the use of woodworking machinery in the initial preparation of material and for the rebating and moulding of the members. Other than this, hand methods are described.

(a) Prepare a setting out rod for frame and sashes.
(b) Prepare a cutting list for all members.
(c) Select timber for job, cross cut to length and saw to nominal sizes.
(d) Plane face sides of timber and apply face marks.
(e) Plane face edges of timber and apply face edge marks.

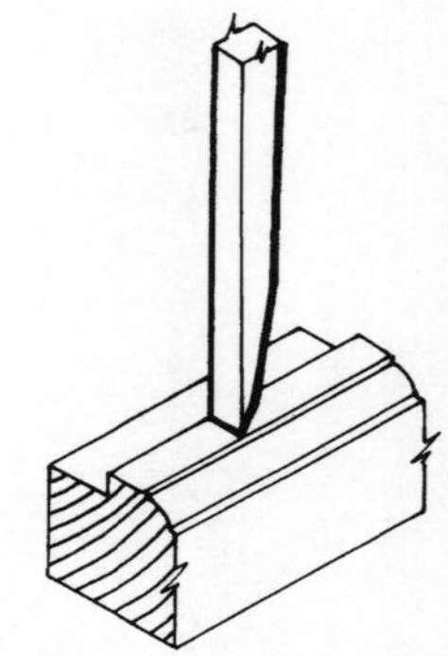

Fig. 6.8. *Mortice chisel to suit sash stuff.*

(f) Plane timber to width and thickness.
(g) Set out *all* members.
(h) Chop all mortices.
(i) Saw tenon cheeks.
(j) Cut rebates and weatherings on cill, transom, head, mullions and jambs.
(k) Apply mouldings to face and face edge corners. Rebate sashes.
(l) Cut grooves, throatings and drips.
(m) Cut tenon shoulders.
(n) Scribe mouldings – *see* Fig. 6.9.
(o) House transom into jambs.
(p) Cut and fit haunches/frankings on sashes.
(q) Assemble frame and sashes; examine for proper fit.
(r) Dismantle, and clean up *all* inside edges, mouldings, etc., and also the *outside face* of the jambs. This cannot be done easily after assembly because of the projection of the transom and cill.
(s) Glue, cramp and wedge up frame and sashes, first checking for squareness as shown in Fig. 6.10.

Fig. 6.9. *Scribing mouldings.*

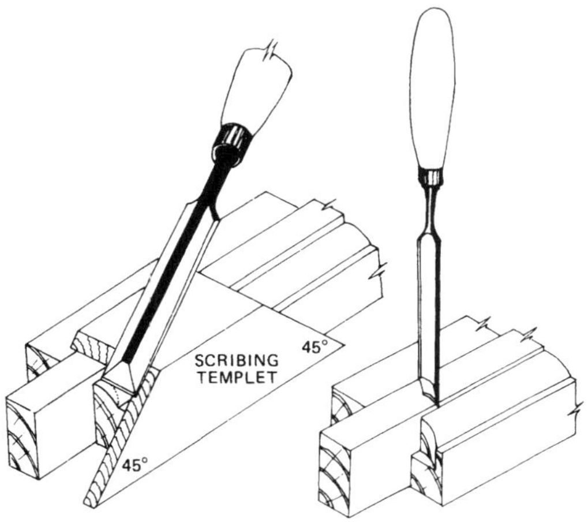

(a) Use of scribing templet to mark scribings.

(b) Use of inside ground gouge to scribe moulding.

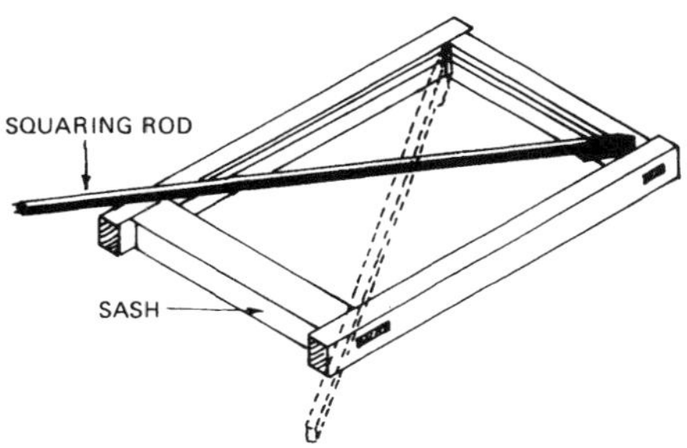

Fig. 6.10. *Use of a squaring rod (the diagonals must be equal in length).*

(*t*) Clean off both sides of frame and sashes, remove sharp arrises and brace frame.

(*u*) Fit sashes into frame, making a tight joint for fixed sashes and allowing 2-3 mm all round on those which are hinged.

(*v*) Prime window, fit ironmongery, and ensure window is stored under cover and kept flat and true prior to delivery.

Fixed sashes

Fixed sashes or "dead lights" are sashes which are fitted into the casement frame but do not open. These should be fitted tightly into the rebate and nailed as shown in Fig. 6.11.

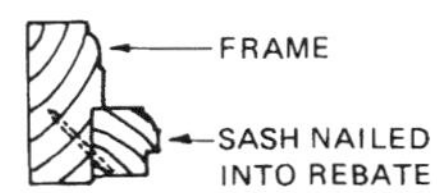

Fig. 6.11. *Fixing a "fixed" sash.*

The edges of the dead light should be painted before fixing.

Storage and protection of casement windows

Wooden casement windows, though perhaps less liable to site damage than door frames, require exactly the same care in storage and handling (*see* Chapter 5). It is, however, worthwhile protecting the cill by covering the weathered part with a length of scrap timber so as to avoid possible damage by cement droppings, pieces of falling brick, etc. Such an elementary precaution is also necessary where materials such as floor boards may be passed into the building through an open window. Casement windows, whether glazed or not, should be kept fastened to prevent wind damage.

IRONMONGERY FOR CASEMENT WINDOWS

Casement fasteners

These are used to fasten side hung, outward opening lights and are illustrated in Fig. 6.12(*a*). They are available in various finishes

Fig. 6.12. *Window fasteners and stays.*

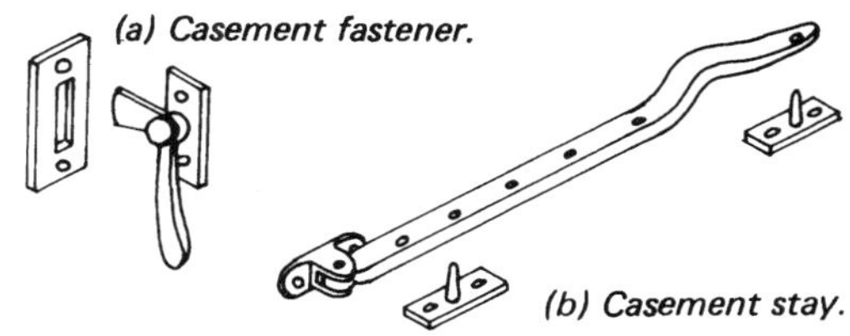

such as enamel, black Japan, stainless steel, brass and BMA.

Casement stays

Shown in Fig. 6.12(*b*) these are used to hold side hung lights in an open position and to prevent wind damage. The series of holes in the stay allow the sash to be held either fully or partly open.

Fanlight stays

These are similar in appearance to casement stays but require a pair of pins to enable the fanlight to be locked. Both fanlight and casement stays are available in the same finishes as casement fasteners.

FIXING WOOD CASEMENT WINDOWS

New openings

Casement window frames are usually "built

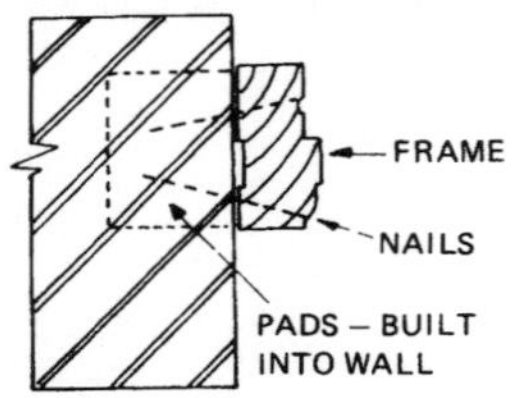

Fig. 6.13. *Fixing a casement frame.*

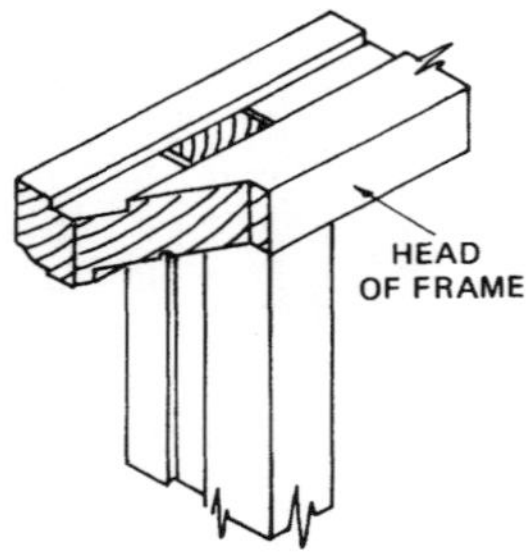

Fig. 6.14. *Horn cut back for building into wall.*

in" by the bricklayer during the construction of the walls, the horns on the head and cill being built into the wall, firmly securing the frame at the top and bottom. All that is required by the carpenter, therefore, in this instance is the fixing of the jambs to the completed wall. This is done by nailing through the jambs into "pads" which the bricklayer will have previously set into the brickwork for this purpose as shown in Fig. 6.13.

Where the casement frame is to be fixed flush or close to the face of the outer skin of brickwork, the bricklayer will generally require the horns on the head to be cut back as shown in Fig. 6.14 so they do not show on the face of the wall.

Existing openings

If the window is a replacement one, or where the window opening is built prior to positioning the frame, it is necessary for the carpenter to "plug" the wall (as will be described in Chapter 10) to obtain a fixing. In this case it may sometimes be necessary to remove the horns from the frame in order to avoid cutting away the brickwork, and if such is the case it is obviously good policy to draw bore the frame joints to avoid undue weakening of the joint which becomes, in effect, an open bridle (*see* Chapter 3).

FURTHER READING

British Standards and Codes of Practice

CP 153	Windows and rooflights Part 2:1970 Durability and maintenance
BS 565:1972	Glossary of terms relating to timber and woodwork
BS 644	Part 1:1951 Wood casement windows
BS 1186	Quality of timber and workmanship in joinery Part 1:1971 Quality of timber Part 2:1971 Quality of workmanship

Building Research Establishment Digests

No. 73	Prevention of decay in external joinery
No. 140	Double glazing and double windows
No, 175	Choice of glues for wood

SELF-TESTING QUESTIONS

All the information required to answer the following questions is contained within this chapter. Attempt each section *as fully or as briefly* as the question demands, and then check your answers against the information given in the chapter.

1. *(a)* State four main functions of a casement window.

(b) Name the two main elements of a traditional type casement window.

(c) Show by means of a sketch how top and side hung sashes are denoted on a drawing.

2. *(a)* Sketch the elevation of a typical four light casement window and name and dimension all the members.

(b) State suitable dimensions for:

(i) the rebates in the sashes (for the glass);

(ii) the rebates in the frame (for the sashes).

3. *(a)* Draw half full size (scale 1:2) sections through the following members of a casement window: *(i)* head and top rail; *(ii)* jamb and stile; *(iii)* cill and bottom rail; *(iv)* mullion and stiles; *(v)* transom, top and bottom rails.

(b) Indicate on your drawings for *(a)* the features which are designed to:

(i) prevent water "laying on" the cill;

(ii) prevent water entering by capillary action;

(iii) prevent water tailing back under the cill and transom;

(iv) facilitate the fitting of a water bar;

(v) ensure a good joint with a window board.

4. *(a)* Name the type of joint used in the

construction of traditional casement windows.

(b) Name a suitable adhesive for window assembly.

(c) State why "hardwood" is recommended for a cill.

(d) Explain clearly what is meant by *(i)* nominal size, and *(ii)* finished size, of joinery timbers.

5. *(a)* Sketch joints between the head and jamb of a solid rebated and ovolo moulded casement to show:

(i) hand scribing of the moulding;

(ii) machine scribing of the moulding.

(b) Sketch joints between the top rail and stile of a sash to show *(i)* a haunched joint, and *(ii)* a franked joint.

6. *(a)* List the sequence of operations involved in the construction – from the completion of the marking out stage onwards – of a four light traditional casement frame.

(b) List all the ironmongery, including hinges, that would be required for the window in *(a)* given that all the sashes are to open.

(c) Sketch the treatment given to the horns on the head of a casement frame prior to the frame being "built-in" close to the face of the brickwork.

7. Centres and Formwork

After completing this chapter the student should be able to:

1. State the function and requirements of a centre.
2. Sketch and explain the construction of the various types of centre described.
3. Sketch and describe methods of supporting, striking and easing a centre.
4. Set out the outlines of semi-circular, gothic and segmental arch centres.
5. Calculate the radius of a segmental centre.
6. State the main requirements of concrete formwork.
7. Design and draw forms for simple pre-cast concrete units.
8. Sketch details of *in situ* forms for simple slabs, lintels and beams.

CENTRES FOR ARCHES

Centres are structures built by the carpenter and joiner on which the bricklayer or mason builds an arch. They are thus temporary pieces of work which are removed on completion of the arch, and do not therefore require a smooth fine finish as do pieces of joinery. They are generally made up from sawn timber sections which are nailed together with round wire nails, clenched for strength.

Requirements of a centre

Although an arch centre may be termed a "rough" structure, it nevertheless needs to be properly designed and constructed if it is to serve its purpose effectively. The main points for consideration in the design and construction of a centre are as follows.

(a) The upper surface must conform accurately with the shape of the under surface or "soffit" of the arch required.

(b) It must be strong enough to carry the often heavy load of brickwork or masonry resting upon it without movement or distortion, bearing in mind that it may be required for use several times.

(c) It must be so designed that the upper surface or laggings adequately support the type of masonry or bricks (the voussoirs) which are to form the arch.

(d) It must be so supported during use that it can eventually be removed gently, and eased if necessary, without any jarring or vibration.

Types of centre

The design and construction of an arch centre varies with the span involved, the shape of the arch to be constructed, the width of the soffit, the weight to be carried, and the type of voussoir it is to support.

In this chapter we are concerned with relatively small, simple centres for various types of arch up to about one metre span.

Turning piece

This is the simplest of the centres used in building construction, and is the type used for supporting "camber" and low rise segmental arches in half brick walls.

Fig. 7.1. *Turning pieces.*

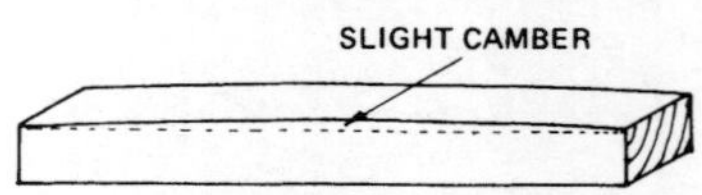

(a) For a camber arch.

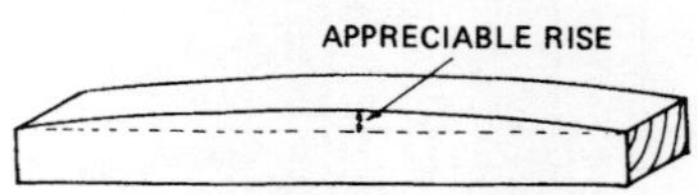

(b) For a segmental arch.

A turning piece for a camber arch is illustrated in Fig. 7.1*(a)* and, as can be seen, is little more than a piece of wood about 75 mm thick and 100 mm deep which has its upper surface very slightly "cambered" (slightly rounded). The amount of camber on such a turning piece is generally about 10 mm rise per metre span. This slight camber gives added strength to the finished arch and allows for a slight settlement during easing and removal of the turning piece.

Another turning piece, this time for a segmental arch, is shown in Fig. 7.1*(b)*. This is similar in construction to the former type but has a more pronounced curvature to the top surface. It has a definite "rise" to produce a true segmental arch.

Rib centre

This type of centre is used for segmental arches of low rise as is the turning piece, but is designed for arches with a wider soffit which would make the simple turning piece impractical or uneconomical.

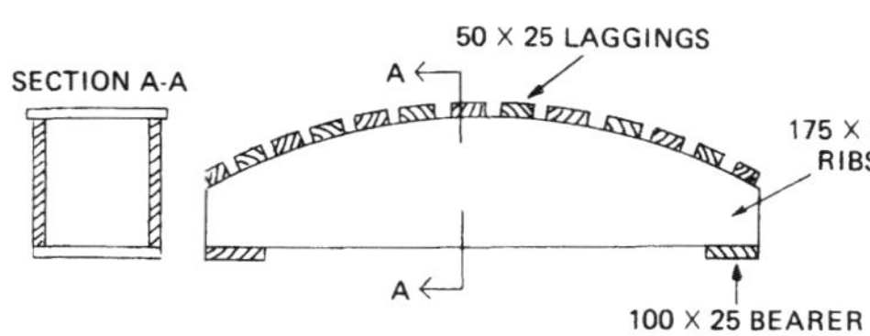

Fig. 7.2. *Rib centre for a segmental arch.*

A typical rib centre is shown in Fig. 7.2 and, as can be seen, is composed of a pair of ribs cut from timber 25 mm to 38 mm thick, connected across the top by laggings and across the bottom by a bearer at either end. The laggings and bearers are nailed to the ribs, their length being about 25 mm less than the width of the arch soffit so they do not interfere with the bricklayer's line and plumb rule during construction of the arch. The sectional size of the laggings is generally about 50 mm × 25 mm, although heavier laggings would be needed where the soffit width is in excess of about 300 mm.

Built-up or laminated centre

Where an arch has a considerable rise, as in the case of semi-circular or Gothic types, it is no longer economical to form the ribs from single pieces of timber, due to the excessive width required. In this case the ribs

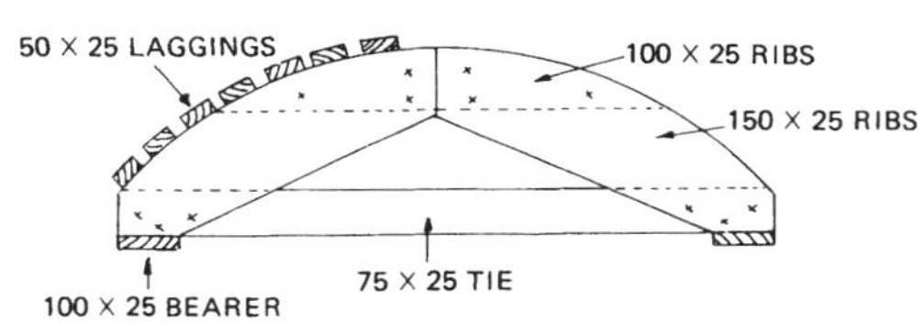

Fig. 7.3. *Built-up centre for a segmental arch.*

are built up from two or more segments, butt jointed together and reinforced with ties, and in larger centres, with a second layer of rib sections nailed to the back of the first with the butt joints staggered. Bearers and laggings are nailed to the built-up sections in the same manner as for the rib centre.

Figure 7.3 shows a built-up segmental centre suitable for an arch of about 750 mm span by 300 mm rise. Figure 7.4 shows a similar type of centre but with a semi-circular outline, whilst Fig. 7.5 shows a larger semi-circular centre of about one metre span. Note that in the latter, the centre is further strengthened by the use of braces. (Braces should always be "normal" to the curvature

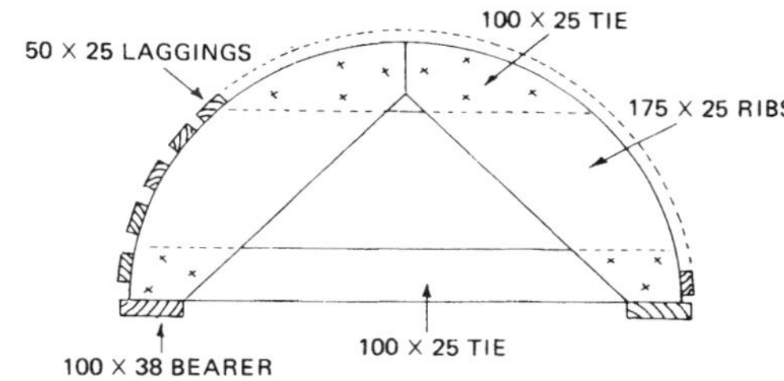

Fig. 7.4. *Centre for a small semi-circular arch.*

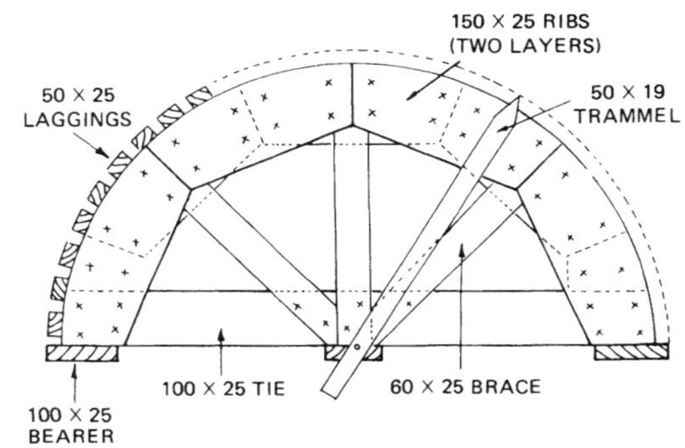

Fig. 7.5. *Semi-circular centre to span 1.000 m.*

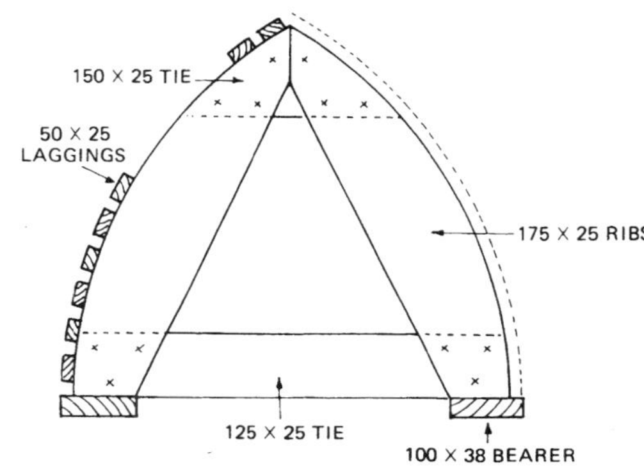

Fig. 7.6. *Centre for an equilateral arch of about 750 mm span.*

Fig. 7.7. *Laggings.*

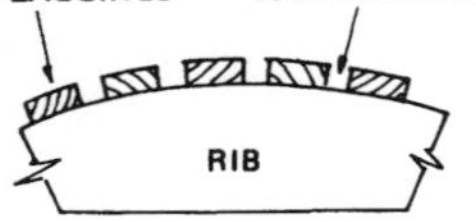

(a) Open lagging for a rough brick arch.

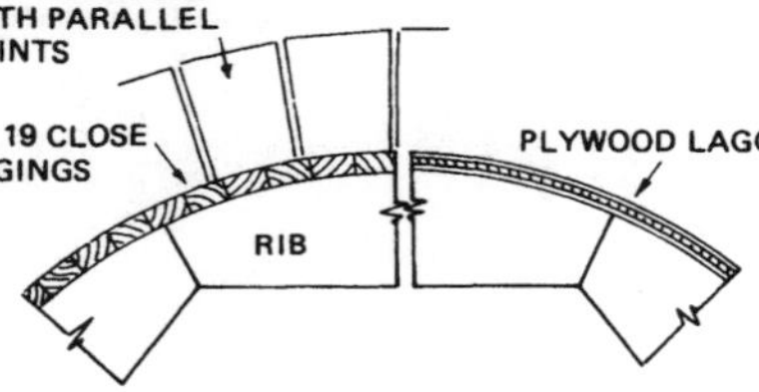

(b) Close lagging for a gauged brick arch.

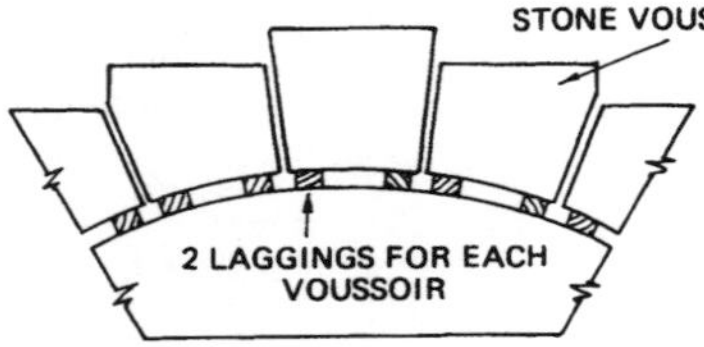

(c) Laggings for a stone arch.

of the arch.) Figure 7.6 shows a centre for an equilateral Gothic arch of about 650 mm span.

Laggings

The type of lagging used in a centre is governed by the type of voussoir used in the building of the arch. Various types are shown in Fig.7.7.

Rough brick arches

Arches built with ordinary bricks thus having tapered mortar joints derive sufficient support from the open laggings illustrated in Fig. 7.7*(a)*.

Gauged brick arches

Also called "axed" arches, these are built with shaped blocks or bricks, and centres for these generally require close lagging to allow for adequate support and fine adjustment of the voussoirs. Two methods of close lagging are shown in Fig. 7.7*(b)*.

Stone arches

Stone arches and arches made from reconstructed stone blocks require two laggings for each voussoir. The laggings must be positioned to support the voussoirs as shown in Fig. 7.7*(c)*.

Trammel rod

The use of a trammel rod is illustrated in Fig. 7.5. This may be required by the bricklayer or mason during the construction of a large or heavy arch and allows him to check for possible distortion of the centre and accurate positioning of the blocks during the building of the arch.

Supporting the centre

Whilst very small arch centres can be adequately supported by a simple prop on either side or even on cleats fixed to the wall, if the arch is of any size or is of considerable weight it should be supported on props *and* folding wedges. This enables the centre to be raised or lowered by fine adjustment to its exact position ready for use. It also facilitates the easy removal of the centre after completion. Figure 7.8*(a)* shows the use of folding wedges in supporting a centre, whilst Fig. 7.8*(b)* shows the use of an adjustable steel prop for the same purpose.

Fig. 7.8. *Supports.*

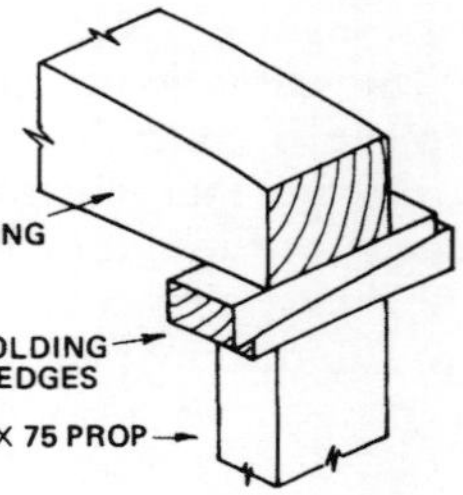

(a) Use of folding wedges to support a turning piece.

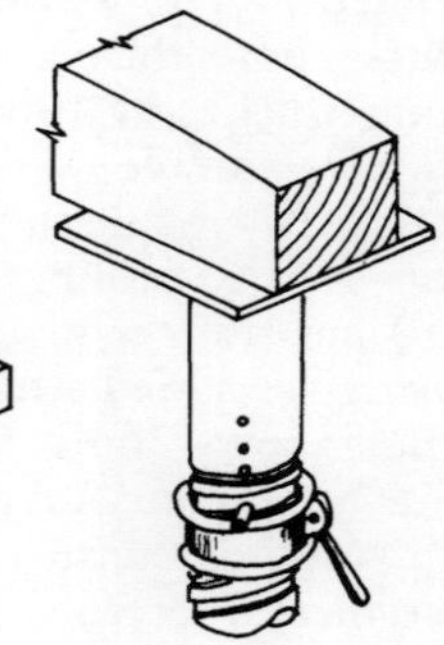

(b) Turning piece supported on a steel prop.

Props, either wooden or steel, should be braced carefully to prevent accidental displacement during the building of the arch.

Striking and easing

The process of removing a centre from under a completed arch is known as "striking", a term which also applies to the removal of formwork for concrete. Striking requires considerable care and must be done without undue knocking or vibration, otherwise a fractured arch may result.

On larger arches the centre is often "eased" before final striking. Easing involves the careful lowering of the centre – perhaps by a millimetre or so – to allow the arch to settle and tighten. Acting upon instructions from the foreman bricklayer, easing may be carried out two or three times (once every 24 hours) until no further settlement takes place, when the centre can be finally removed. Obviously "easement" requires some means of fine adjustment, hence the need for folding wedges or adjustable props.

Setting out arch centres

Semi-circular centres

These require the use of a trammel (*see* Chapter 1), this simply being used in the manner of a large compass to strike the semi-circle. *Do not forget* to allow for the laggings when setting out.

Equilateral-Gothic centres

These are also very simple in outline. The method of setting out an equilateral centre is given in Fig. 7.9*(a)*. Note that the radius for striking the two arcs is equal to the span of the arch.

Segmental arches

These present rather more difficulty than the previous types, but are nevertheless still quite simple. The outline for a camber is obtained simply by bending a timber lath to the outline required and drawing a line along it. This is a quick and effective method where the deviation from a straight line is minimal, as in this case.

For true segmental arches where the rise is significant, a proper geometrical approach is required. If the arch is relatively small, the outline may be drawn with a trammel after first obtaining the radius of the curve as shown in Fig. 7.9*(b)*.

Where larger segmental curves are required it may well be inconvenient to use a trammel, the centre of the circle being inaccessible or the radius too great. In this case the curve may be drawn by using the method shown in Fig. 7.9*(c)*. This method is really a practical use of "loci" – tracing the path of a moving point – and is carried out as follows.

Fig. 7.9. *Setting out arch centres.*

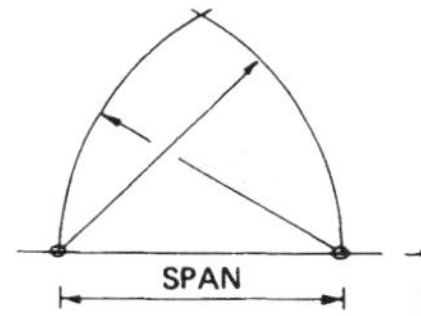

(a) Setting out an equilateral arch.

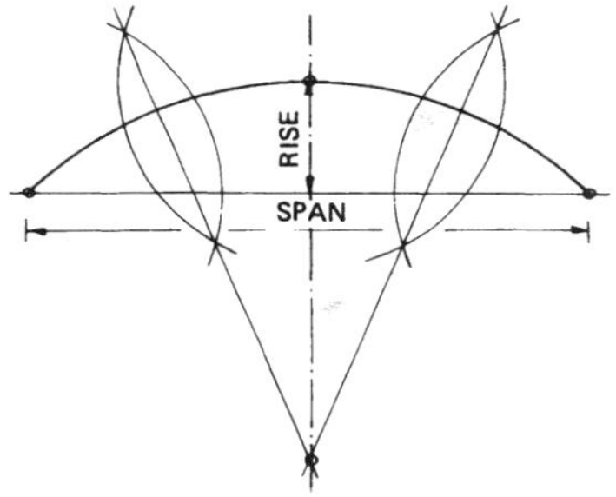

(b) Finding the radius of a segment.

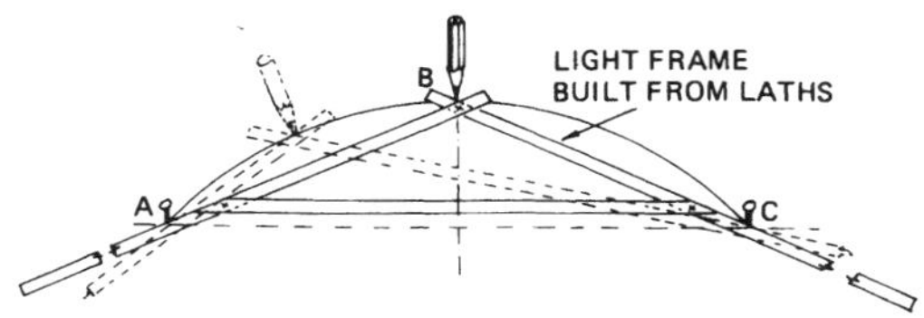

(c) Use of loci in setting out a large segment.

(a) Set out the rise and span of the arch A-B-C.

(b) Drive a small nail into each of the points A-B-C.

(c) Nail three small pieces of lath together to obtain the triangular frame shown.

(d) Remove nail B and substitute a pencil point, sliding the frame along from side to side keeping it in contact with nails A and C to draw the outline.

This is a very quick and efficient way of drawing arcs of large radii.

Calculation of radii

The radius of a segmental curve can be determined by calculation provided the rise and span are known. The calculation involves the use of the "law of intersecting chords", explained in Fig. 7.10 which shows a circle containing two intersecting chords, the four components being identified A, B, C and D. The law of intersecting chords states that A × B = C × D and therefore

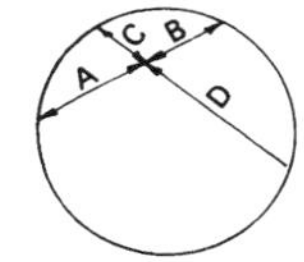

Fig. 7.10. *Intersecting chords A × B = C × D.*

$$\frac{A \times B}{C} = D,$$

so reference to Fig. 7.11 should make it amply clear that where the rise and span of the segment are known, D is the unknown factor which is required, and which, added to C, will give the *diameter* of the circle. The required radius is therefore one half of this.

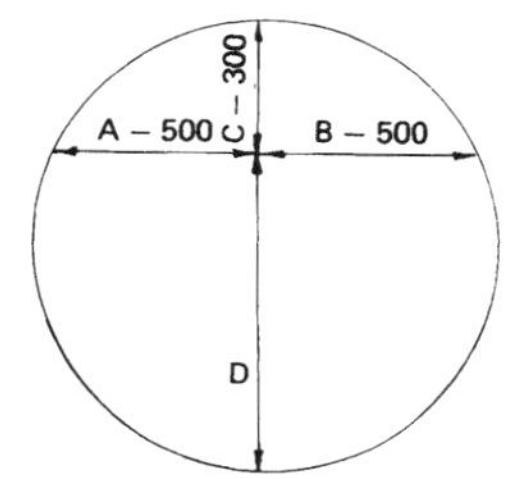

Fig. 7.11. *Diagram showing intersecting chords formed by rise and span of a segment.*

Example

A segmental arch has a span of one metre and a rise of 300 mm. Calculate the radius of the curve. (The problem is shown in diagrammatic form in Fig. 7.11.)

Now:

$$A \times B = C \times D$$

$$\therefore \frac{A \times B}{C} = D$$

Using the dimensions given in Fig. 7.11 the following calculation emerges:

$$D = \frac{500 \times 500}{300} \text{ mm (or } D = \frac{0.5 \times 0.5}{0.3} \text{ m)}$$

$$\therefore D = 833 \text{ mm (or 0.833 m)}$$

The diameter of the circle is therefore:

$$833 + 300 \text{ mm (or } 0.833 + 0.3 \text{ m)}$$
$$= 1133 \text{ mm (or 1.133 m)}$$

The radius of the arch is therefore:

$$\frac{1133}{2} = \underline{566.5 \text{ mm (or 0.5665 m)}}$$

FORMWORK TO CONCRETE

"Formwork" is the term applied to the moulds of all shapes and sizes which are constructed to hold wet concrete in the particular shape required until it has set. It is thus a very important aspect of modern building construction and has become in the last decade or so almost a specialist trade in itself.

Most carpenters and joiners, however, are expected to have a broad working knowledge of the principles involved in formwork (or shuttering as it is sometimes known) in order to carry out work of this nature in instances where the employment of the specialist would be neither economical nor necessary. Like centring, formwork can be regarded as temporary work, not present in a completed building and therefore of no real value in itself – merely a means of enabling concrete work to be carried out. This does not, of course, mean that formwork is inexpensive, of no value or that it need not be looked after and made to last – most often the reverse is the case – simply that it is the *concrete* which the customer is interested in and pays for.

Design and construction of formwork

The following points should be borne in mind when designing and constructing formwork.

(*a*) It should be strong enough to withstand the weight and pressure of the wet concrete, which may be considerable where there is any "depth" and where it is likely to be "tamped", "rammed" or vibrated for consolidation.

(*b*) Joints should be tight enough to prevent loss of fine aggregate.

(*c*) It should be treated with some form of "release agent" (shuttering oil) to ensure that it comes away cleanly without sticking.

(*d*) It should be so designed that it can be "struck" (removed) easily, and in the correct sequence.

(*e*) The face of the formwork must be in keeping with the required finish of the concrete.

Pre-cast concrete

Quite a few of the concrete components in a typical building are *pre-cast*, which means that they are cast in the workshop or factory (and sometimes on site) and then transported or moved into position as they are required. Such components include paving slabs, lintels, fence posts, copings, etc., these being items which can obviously be produced more economically by pre-casting.

Simple slab

Figure 7.12 shows the formwork required for casting a simple concrete slab of about 50 mm thickness. In this case two or more slabs could be cast during the working day

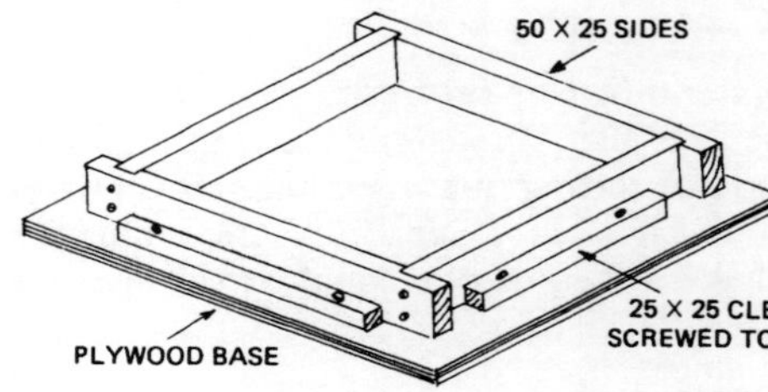

Fig. 7.12. *Form for pre-casting a small concrete slab.*

using the edgings in conjunction with spare base boards. Alternatively, the edge frame can be used without a base board if a good flat concrete floor is available, using polythene sheet to prevent adhesion to the floor.

Lintel

Figure 7.13(*a*) shows a simple box structure for pre-casting a lintel. Note that where several lintels of varying length are required it is only necessary to make the box long enough to cater for the longest lintel, the stop being moved along progressively as the shorter ones are required. Figure 7.13(*b*) shows a casting box which allows for multiple casting of lintels.

Copings

These occur in various shapes and sizes but generally they feature a sloping top surface

Fig. 7.13. *Pre-casting a concrete lintel.*

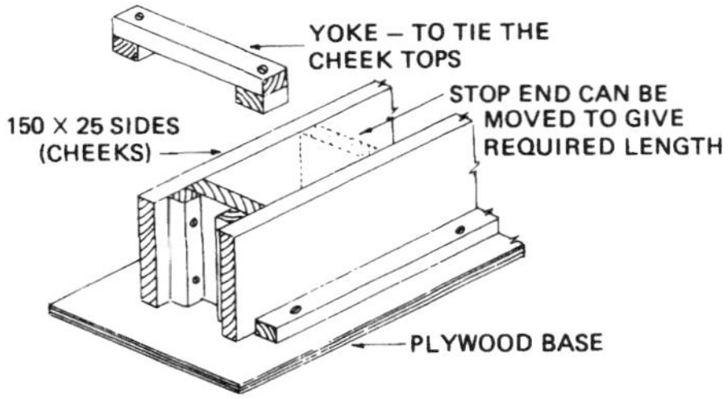

(a) Simple box.

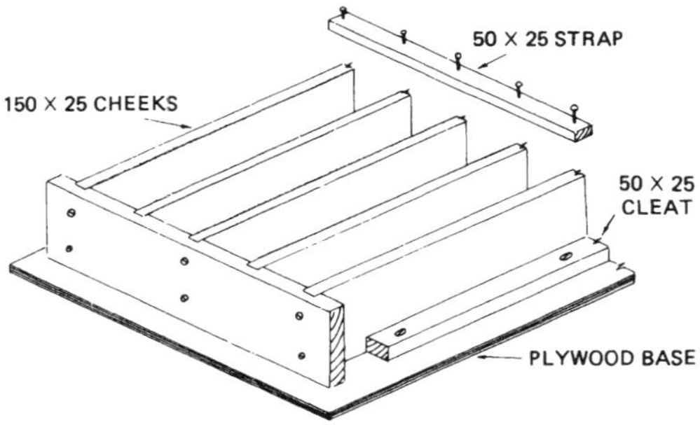

(b) Box for multiple pre-casting of lintels.

– a weathering – and throatings or drips under the front and rear edges.

Figure 7.14 shows details of casting boxes for two such copings. Note that it is often more convenient to cast the concrete upside down as shown. This not only enables the throating to be formed simply by using mild steel rods, but also ensures that the three most important faces of the coping are accurately formed and finished.

In situ **casting**

This term refers to the method of casting concrete in which the required shape is

Fig. 7.14. *Boxes for pre-cast copings.*

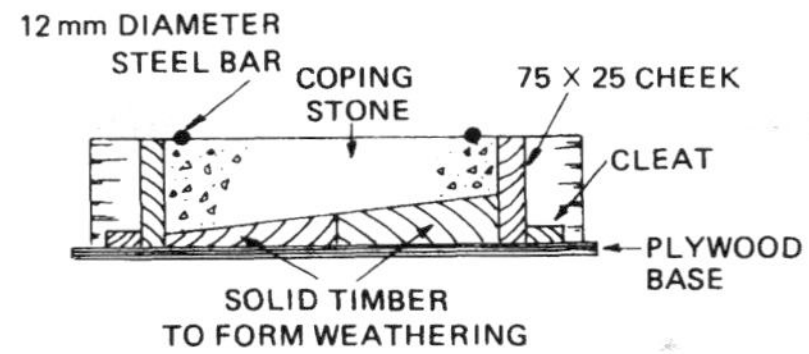

(a) Section through casting box.

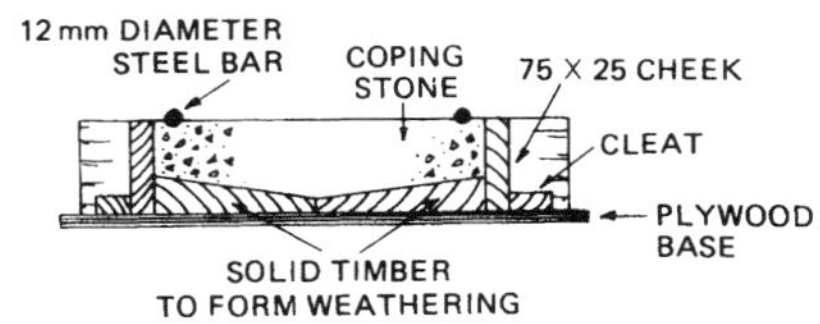

(b) Section through casting box.

formed in its final location, as is usually the case where floors, beams, walls, columns, etc., are concerned. *In situ* concrete, which is invariably reinforced with steel mesh and bars, facilitates the casting of large monolithic structures with a resultant increase in structural strength.

It is in the construction of *in situ* concrete formwork that the greatest demands may be made upon the skill and expertise of the carpenter, such concrete work often forming the main bulk of a building. This volume is concerned only with the simpler types of *in situ* formwork.

Concrete base

Figure 7.15 shows the formwork required for the base of a timber shed or similar structure. It should be realised that even simple jobs like this require the formwork to be accurately squared and levelled.

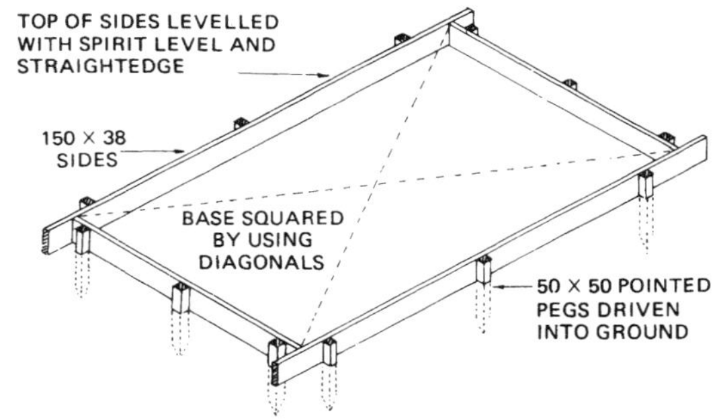

Fig. 7.15. *In situ formwork for a concrete base.*

Beams and lintels

Figure 7.16 shows the formwork required for a plain lintel in a single brick wall. Note that the formwork is supported on props and folding wedges or on adjustable steel

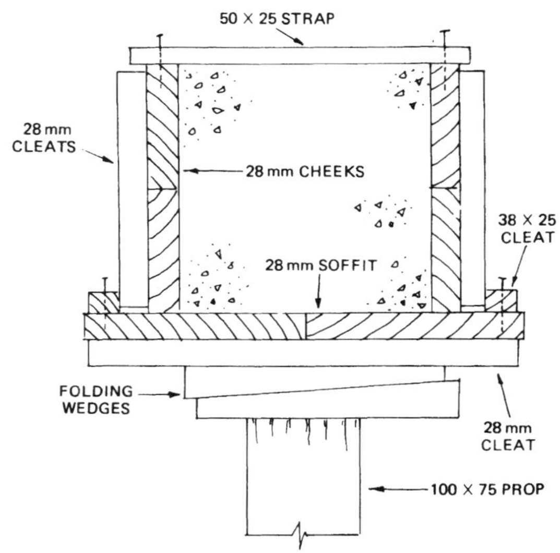

Fig. 7.16. *Formwork for a plain lintel cast in situ.*

props to allow for easing and striking. Where solid boarding is used for the cheeks and soffit, these should be planed to a smooth finish and an even thickness. This latter is an important feature on formwork employing "cleats" to tie the boarding into panels, as any variation in thickness is bound to show on the face of the finished concrete.

Figure 7.17 shows the formwork to a lintel which is to be cast over a doorway where the frame is already positioned. In this case

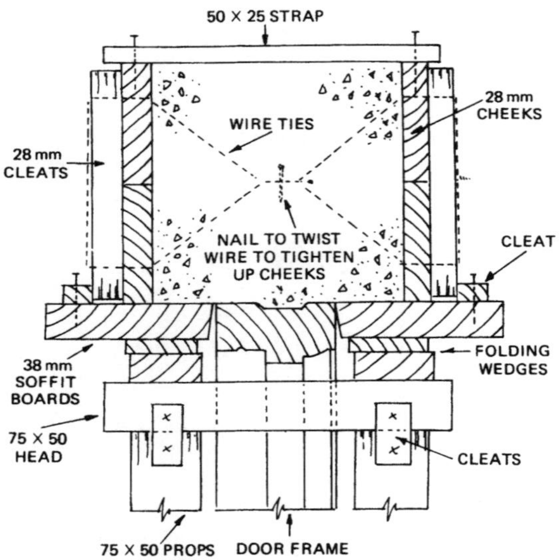

Fig. 7.17. *Formwork for a lintel cast in situ, over a door frame.*

the soffit must be in two separate parts, and therefore the cheeks need to be securely tied to prevent the soffit spreading when the concrete is poured.

Figure 7.18 shows the formwork to a concrete beam of fairly substantial proportions.

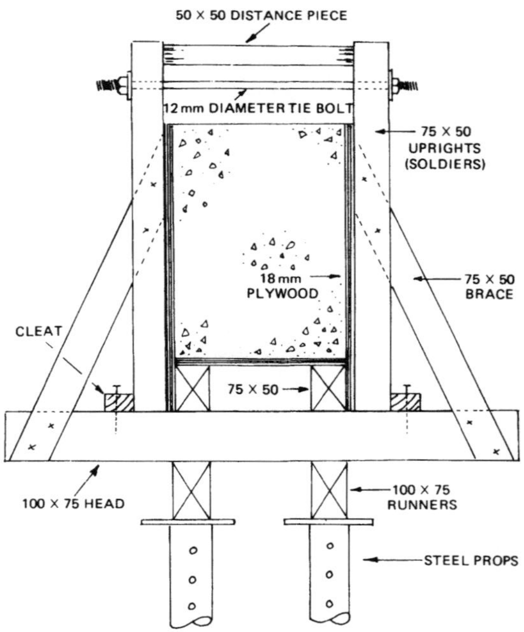

Fig. 7.18. *Formwork to large beam cast in situ.*

In this instance use has been made of plywood sheeting for the beam cheeks and soffit, and therefore cleats are not required. It is, however, necessary to "stiffen" the plywood to prevent bulging between supports. The diagram shows the use of steel tie bolts and distance pieces to form the tie across the top of the beam. Steel bolts should always be well greased when used for this purpose, especially if, as sometimes happens, they pass through the concrete and must therefore be removed during striking.

Striking beam formwork

It is considered good practice to remove the formwork to the cheeks of the beam within a day or so of pouring in order to allow air to come into contact with the concrete and thus hasten its curing. The soffit of the beam, however, must be left in position for some time, to support the concrete while its strength is developing. This may take from seven to twenty-eight days depending upon the loading to which the beam is subjected.

Care of formwork

The materials used in the formwork are, like most other building materials used today, quite expensive. However, the initial cost of the material can often be offset by the number of times it can be used. It becomes important therefore to handle, store and use the materials and "made up" shutters carefully. Forms should not be deliberately ill treated, stacked in a way which will allow them to distort, or used for other purposes. The best time to clean formwork is immediately after striking, when any "green" concrete can be easily removed. After cleaning, re-oil and stack carefully in "sets" ready for further use.

FURTHER READING

Snow, Sir Frederick. *Formwork for Modern Structures.* (Chapman & Hall Ltd.)

SELF-TESTING QUESTIONS

All the information required to answer the following questions is contained within this chapter. Attempt each section *as fully or as briefly* as the question demands, and then check your answers against the information given in the chapter.

1. State: *(a)* the purpose of a centre; *(b)* the main requirements of a centre.

2. *(a)* Illustrate with detailed sketches: *(i)* a turning piece; *(ii)* a rib centre.

(b) Describe the type of arch for which *(a)(i)* and *(ii)* would normally be required.

3. Draw, to a scale of 1:5, centres suitable for the following arches:

(a) a Gothic (equilateral) arch of 650 mm span and a soffit one brick wide;

(b) a semi-circular arch of 1.000 m span and a soffit one brick wide.

4. Sketch details to show the type of lagging required for:

(a) a rough brick arch;

(b) a gauged brick arch;

(c) a stone arch with shaped voussoirs.

5. *(a)* Describe with sketches two ways in which a centre may be supported during the construction of an arch.

(b) Explain what is meant by "easing a centre".

(c) Calculate the radius of a segmental arch which has a span of 800 mm and a rise of 200 mm.

6. *(a)* State the main requirements of formwork for concrete.

(b) Name the two methods which are used to cast concrete components, giving a typical example of the components formed by each.

7. Sketch dimensioned details of the formwork required to cast a "batch" of concrete lintels 1.100 m long by 150 mm deep by 100 mm wide.

8. Sketch dimensioned details of the formwork required to cast ***in situ:***

(a) a lintel over a doorway in a wall one brick in thickness;

(b) a large beam 500 mm deep by 300 mm wide.

9. Show by means of a sketch how the formwork to a rectangular concrete ground slab would be: *(a)* supported, *(b)* squared, and *(c)* levelled.

8. Timber Ground Floors

After completing this chapter the student should be able to:

1. State the main functions and requirements of a timber ground floor.
2. Draw a cross section through a timber ground floor, naming the various members.
3. State the main requirements of the Building Regulations with reference to the construction of timber ground floors.
4. Explain the importance of dpc's, under floor ventilation and honeycombed sleeper walls.
5. Name and define the features of four types of "flooring" material.
6. Outline the procedure for laying and levelling ground floor joists.
7. Outline the procedure for laying t & g and secret nailed floorboards.

FUNCTIONAL REQUIREMENTS OF A FLOOR

The primary function of a floor is obviously the provision of a firm level surface, capable of adequately and comfortably supporting the occupants, furniture and contents of a room within a building.

Secondary but nevertheless important aspects include the following.

(a) A reasonable degree of thermal insulation. This function is performed satisfactorily by a timber ground floor.

(b) Prevention of rising damp. This particular requirement in the case of timber ground floors is a most important aspect of the floor substructure.

(c) Strength and stability. Suspended timber ground floors, properly constructed, perform extremely well in this respect since it is relatively easy to provide sufficient support to the floor structure to make a firm, stiff and lightweight floor of economical proportions.

Resistance to fire and sound insulation are also functions of a floor, but these apply more logically to upper than to ground floors.

SUSPENDED TIMBER GROUND FLOORS

The general construction of a suspended timber ground floor is shown in section in Fig. 8.1. It will be seen that the structure consists of flooring, either floorboards, plywood or chip-board sheets, nailed to light weight beams known as joists. The joists, for a ground floor are spaced at centres of 400 mm to 600 mm, depending upon the thickness/strength of the flooring being supported and the probable loading on the floor itself. Ground floor joists are relatively light in section, 100 X 50 being commonly used for this purpose, sufficient strength and stiffness being obtained by intermediate supports in the form of "sleeper walls" which are built for this purpose.

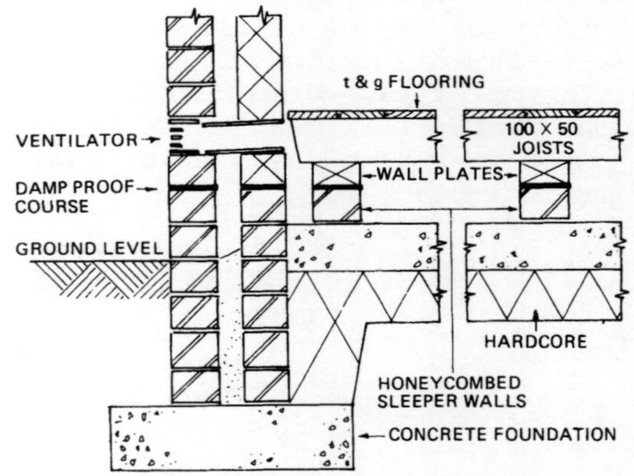

Fig. 8.1. *Part section through a timber ground floor.*

Building Regulations

The Building Regulations lay down certain requirements with respect to the construction of ground floors as follows.

" 1. Such part of a building (other than an excepted building) as is next to the ground, shall have a floor which is so constructed as to prevent the passage of moisture from the ground to the upper surface of the floor.

2. Any floor which is next to the ground shall be so constructed as to prevent any part of the floor being adversely affected by moisture or water vapour from the ground.

3. No hardcore laid under such floor shall contain water-soluble or other deleterious matter in such quantities as to be liable to cause damage to any part of the floor.

"These requirements shall be deemed to be satisfied if:

(*a*) the ground surface is covered with a layer of concrete not less than 100 mm thick composed of cement and fine and coarse aggregate conforming to BS 882 Part 2, in the proportions of 50 kg of cement to not more than 0.1 m^3 of fine aggregate and 0.2 m^3 of coarse aggregate, properly laid on a bed of hardcore consisting of clean clinker, broken brick or similar inert material, free from water soluble sulphates or other deleterious matter in such quantities as to be liable to cause damage to the concrete;

(*b*) the concrete is finished with a trowel or spade finish and so laid that its top surface is not below the highest level of the surface of the ground or paving adjoining any external wall of the building;

(*c*) there is a space above the upper surface of the concrete of not less than 75 mm to the underside of any wall plate, and of not less than 125 mm to the underside of the suspended timbers, and such space is clear of debris and has adequate through ventilation;

(*d*) there are "damp proof courses" in such positions as to ensure that moisture from the ground cannot reach any timber or other material which would be adversely affected by it."

Points to note

Damp proof courses (dpc's)

These are impervious materials used at the base of walls to prevent dampness and moisture rising up above this point into the walls and floors. A dpc should be at least 150 mm above ground level and is a very necessary component in any wall or floor in direct contact with the ground or indirectly in contact via an adjoining permeable structure (*see* Fig. 8.2).

Materials used for dpc's include bitumen, pitch polymer, lead, copper, mastic asphalt, slate and engineering bricks.

Wall plates

These are normally lengths of 100 × 50 or 100 × 75 softwood laid on top of the sleeper walls. Their purpose is twofold:

(*a*) to provide a firm seating for the ground floor joists;

(*b*) to distribute the loads imposed by the individual joists along the length of the sleeper walls.

Lengthening joints in a wall plate are made by means of a half-lap as shown in Fig. 8.3.

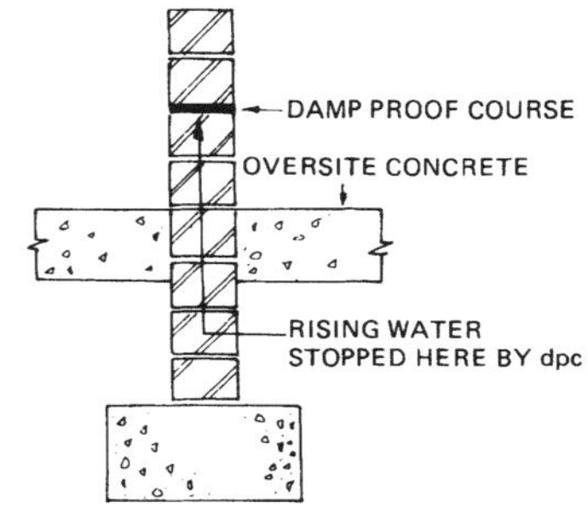

Fig. 8.2. *Function of damp proof course.*

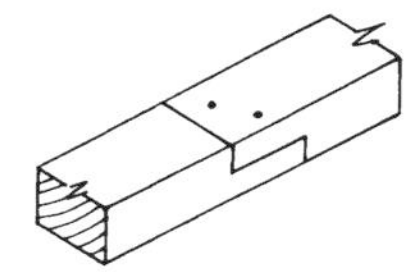

Fig. 8.3. *Half lap joint in a wall plate.*

Sleeper walls

Since sleeper walls are in direct contact with the oversite concrete, and thus indirectly in contact with the ground, a dpc is always required below the wall plate as shown in Fig. 8.4. Sleeper walls are utilised to reduce

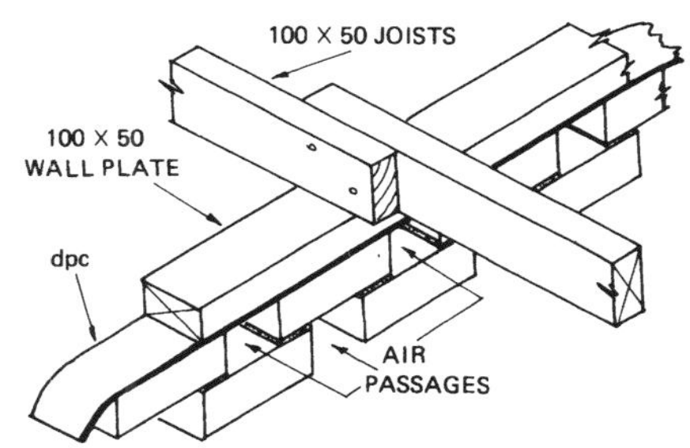

Fig. 8.4. *Sketch showing joists supported by a honeycombed sleeper wall.*

the span of the ground floor joists and thus keep the depth of these to within reasonable limits. Normal spacing for sleeper walls is between 1.300 m to 2.000 m centres.

To allow free passage of air below the floor, sleeper walls are of honeycombed construction as shown in Fig. 8.4.

Joists

For ground floors these are usually of 100 × 50 or 125 × 50 softwood, spaced at 400 mm to 600 mm centres depending upon the type of flooring to be used. They should be laid cambered edge upwards so that any tendency to sag results in a straight joist, and should be held in position by skew-nailing to the wall plates.

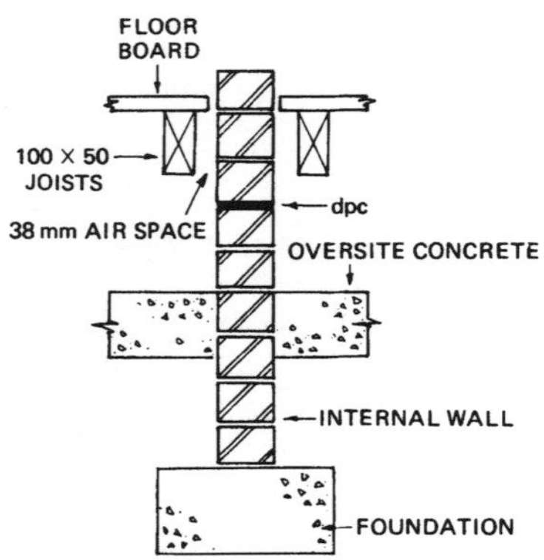

Fig. 8.5. *Sketch showing joists alongside a wall.*

It is important to ensure that the *tops* of the joists are all carefully levelled by inserting packings where necessary between the joist and the wall plate. Failure to do this will result in an uneven floor surface and a springy floor.

Lengthening joints for ground floor joists are made simply by "passing the ends" and spiking them together over a sleeper wall as shown in Fig. 8.4.

It is good practice to allow a flow of air around the ends of joists by leaving a space between wall and joist end as shown in Fig. 8.1. Similarly, joists which lay alongside a wall should be kept 38 mm or so away from the wall for the same purpose, as shown in Fig. 8.5.

Underfloor ventilation

This is essential to prevent damp, humid conditions arising below the floor which might ultimately lead to fungal decay of the

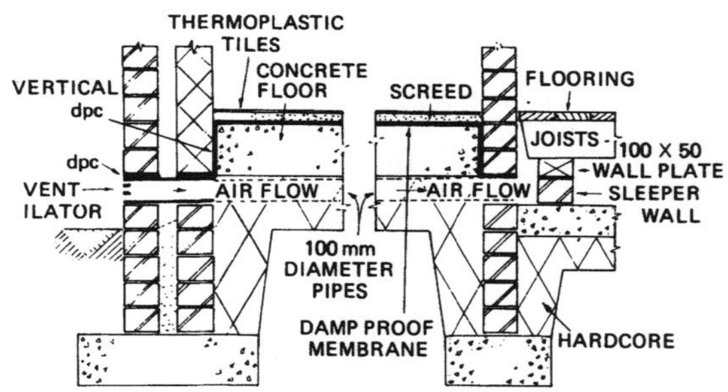

Fig. 8.6. *Section showing ventilation below a solid ground floor.*

timber. This is accomplished by means of air bricks or ventilators built into the outside walls as shown in Fig. 8.1. Air is thus able to flow freely through the under floor space, passing through the spaces in the honeycombed sleeper walls to keep the timber well aired.

Where a suspended timber floor adjoins a solid floor, the necessary air flow can be obtained by means of air ducts as shown in Fig. 8.6.

Hearths

Where a hearth occurs in a timber ground floor, it is essential for the timbers to be so positioned that they do not present a fire hazard.

Certain requirements are therefore laid down in the Building Regulations with respect to the construction of such a hearth and these must be strictly observed. The relevant regulations are as follows.

Constructional hearths for Class 1 appliances

NOTE: The Class 1 appliance referred to in this section means a solid fuel or oil burning appliance with an output not exceeding 45 kW.

"1. Any constructional hearth serving a Class 1 appliance shall –

(*a*) be not less than 125 mm thick;

(*b*) if it adjoins a floor constructed wholly or partly of combustible material, or if combustible material is laid on the hearth as a continuation of the finish of the adjoining floor, in accordance with the provisions of paragraph 2, be so constructed that any part of the exposed surface of the hearth, which is not more than 150 mm, measured horizontally, from the said floor or combustible material is not lower than the surface of the floor, and not lower than the remainder of the exposed surface of the hearth; and either

(*c*) (if it is constructed in conjunction with a fireplace recess) –

(*i*) extend within the recess to the back and jambs of the recess;

Fig. 8.7. *Joist trimming around a hearth.*

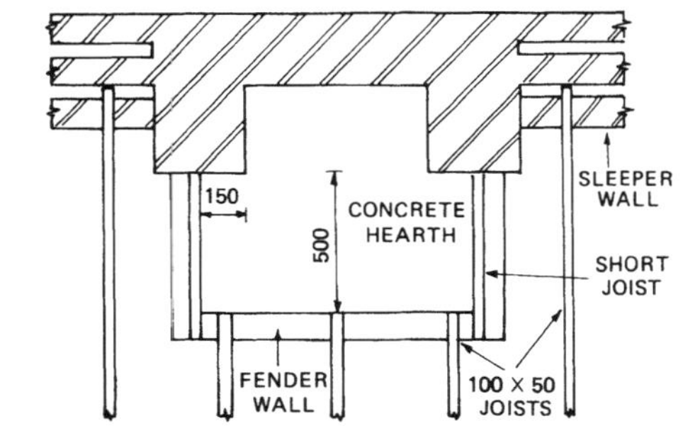

(a) In conformity with the Building Regulations.

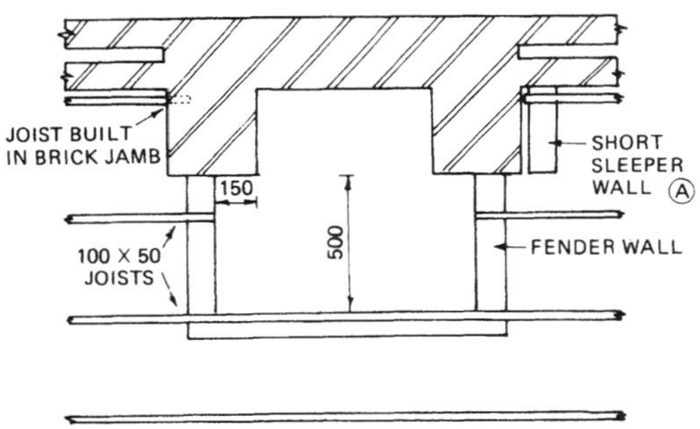

(b) In conformity with the Building Regulations with floor joists parallel to the hearth.

(ii) project not less than 500 mm in front of the jambs; and

(iii) extend outside the recess to a distance of not less than 150 mm beyond each side of the opening between the jambs; or

(d) (if it is constructed otherwise than in conjunction with a fireplace recess) be of such dimensions as to contain a square having sides measuring not less than 840 mm."

The essence of the foregoing regulations, as far as the timberwork in a ground floor is concerned, is explained in Fig. 8.7*(a)*. It will be seen that the concrete hearth extends, in conformity with the regulations, 500 mm beyond the front of the brick jambs and 150 mm on either side. No part of the floor structure must encroach on this area and, therefore, the ground floor joists are cut short (trimmed) to fit up to the hearth, but not into it. The ends of the floor joists so trimmed rest on a "fender wall", built to support and contain the hearth as shown. The fender wall is built up to the same level as the sleeper walls and supports a wall plate, directly below which is the necessary dpc. Note the short length of joist on either side of the hearth also resting on the fender wall which serves to provide a fixing for the ends of the floor boards adjoining the hearth.

Figure 8.7*(b)* shows a similar ground floor hearth, but with the floor joists running parallel to the hearth. Note that the critical hearth dimensions remain the same as in the previous instance, the ends of the trimmed joists again resting on the fender wall. The two trimmed joists alongside the external wall may well be built into the brick jambs with a 100 mm bearing as shown, but are often supported on short sleeper walls as shown at A, this method being equally satisfactory and more straightforward in practice.

Floorings

This term refers to the various types of material which may be fixed to the top of the joists to provide a smooth, level floor surface.

Tongued and grooved (t & g)

Such floorboards are one of the most commonly used flooring materials, varying in thickness from 16 mm to 22 mm (depending upon the spacing of the joists) and from 100 mm to 150 mm in width. Generally, the wider floorboards are quicker to lay since they cover a greater area per board, but the narrower boards tend to be more satisfactory in minimising the effects of possible shrinkage, and are therefore to be preferred in most instances.

Floorboards are fixed to the joists by nailing, two nails being driven through the edges of the boards into each joist as shown in Fig. 8.8. "Heading joints" – end joints in

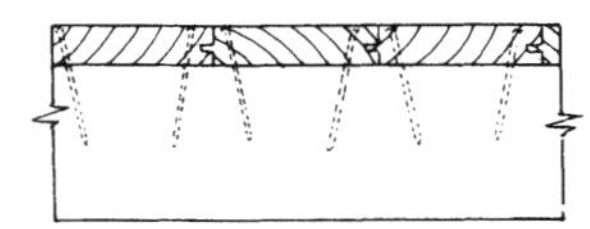

Fig. 8.8. *Nailing t & g flooring.*

floorboards – are either plain butt joints, slightly undercut to ensure a tight surface fit, or splayed as shown in Fig. 8.9*(a)* and

Fig. 8.9. *Nailing heading joints.*

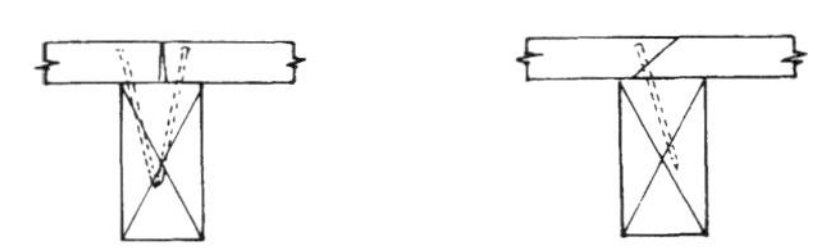

(a) Butt joint. *(b) Splayed joint.*

(b) respectively. Heading joints of either type should be made on a joist and arranged randomly over the floor area, care being taken to ensure that no two heading joints occur consecutively on any joist.

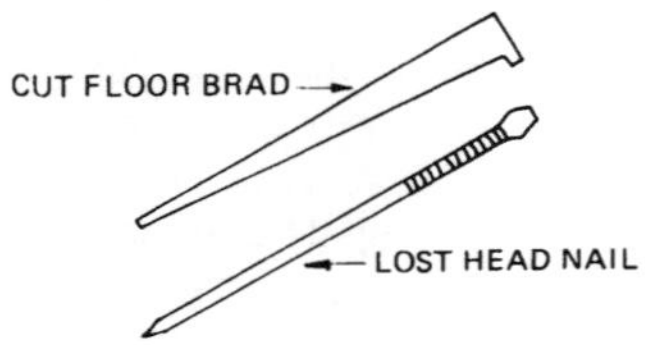

Fig. 8.10. *Flooring nails.*

Tongued and grooved floorboards are fixed with either cut floor brads or lost head nails as illustrated in Fig. 8.10.

Secret nailed floorboards
This type of floorboard is usually quite narrow, never more than 100 mm wide, and has a nominal thickness of 25 mm (20-22 mm finish). The board is designed in such a way that the heads of the fixing nails are concealed, as can be seen in Fig. 8.11, making this type of board ideal for hardwood strip floors or high quality softwood floors. Secret nailed boards must be laid and fixed one at a time, since each board has to be nailed through the tongue as shown. Fine, round wire nails about 56 mm long are suitable for this purpose. Heading joints in secret nailed floorboards usually have machined ends of similar section to that of the edges, thus avoiding the necessity for nailing. These joints need not be made on a joist as in t & g boarding but should nevertheless be staggered over the floor area.

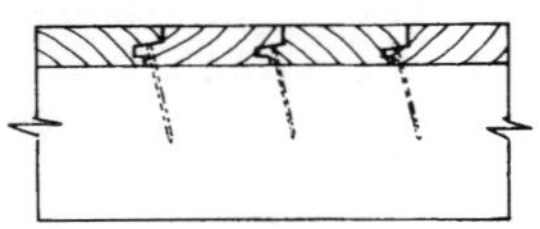

Fig. 8.11. *Secret nailed flooring.*

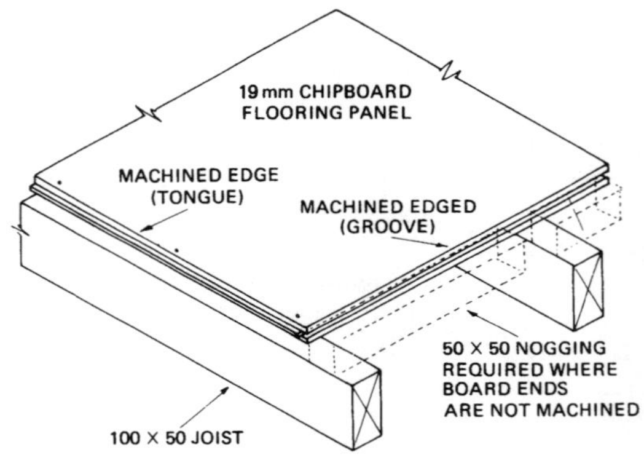

Fig. 8.12. *Chipboard flooring panels.*

Plywood and chipboard flooring
Both these materials are available for flooring purposes made in a special quality known as "flooring grade". These sheets are 19 mm thick and in various widths and lengths to suit the spacing of the joists. Boards specially made for this purpose have tongued and grooved ends and sides as shown in Fig. 8.12 and should be laid with their long edges in the centre of a joist. Heading joints at right angles to the joists do not require "noggings" or bearers unless plain edged boarding is used, in which case noggings should be fixed as indicated in Fig. 8.12 to give support to the sheet ends.

NOTE: Tongued and grooved boarding in either strip or sheet form is much preferred to square edged since it not only prevents the passage of dust and draughts, but also enables each board to derive a certain amount of support from the ones on either side, i.e. the tongues and grooves help to spread and distribute point loadings on the floor.

Plywood and chipboard flooring sheets are fixed with lost head nails about 56 mm long (2½ times the board thickness).

LAYING AND LEVELLING GROUND FLOOR JOISTS

This job is undertaken by the carpenter either when the walls of the building have been built to wall plate level (if the ends of the joists are to be built into the brickwork) or at a much later stage if the joists are supported entirely by sleeper walls, the latter method generally being much preferred since the work can then be done after completion of the roof and therefore under cover. Also, since the joists are laid just prior to boarding, there is less risk of a joist becoming displaced by other workmen.

Laying the joists
Figure 8.13 shows the plan of part of the ground floor of a dwelling house, the position of the sleeper walls and joists being clearly indicated. Taking this particular floor plan as a typical example of such a ground floor, the procedure for laying the joists would be as follows.

(a) Cut and fit the wall plates to the top of all the sleeper walls.

(b) Assist the bricklayer to bed the wall plates on the dpc, keeping a constant check on the straightness and level of the top surfaces. This should be tested with a spirit level and long straightedge, high spots being brought into line with a few taps with a heavy hammer.

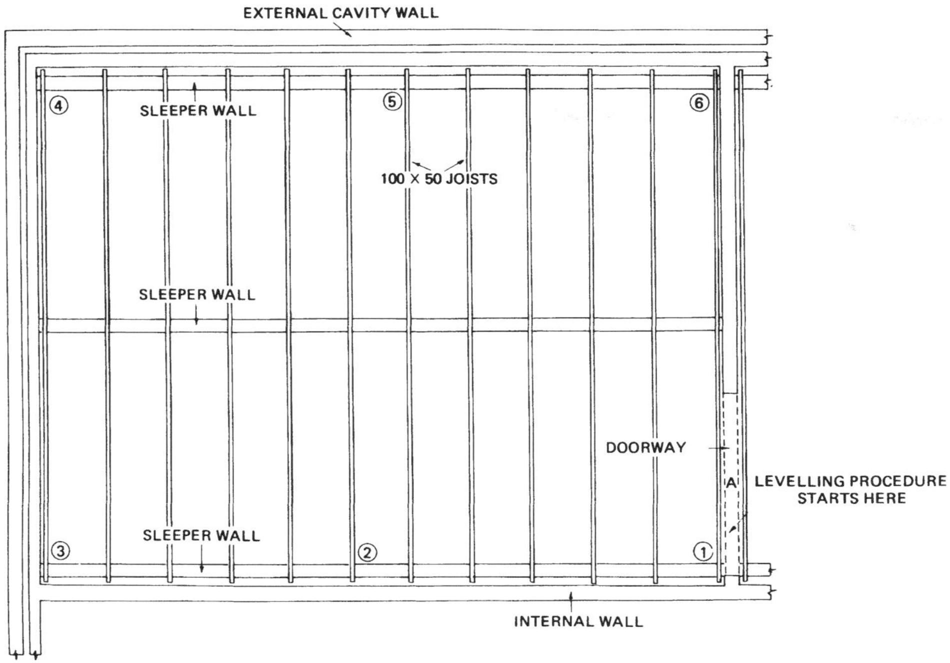

Fig. 8.13. *Plan of a part of a ground floor showing procedure for levelling joists.*

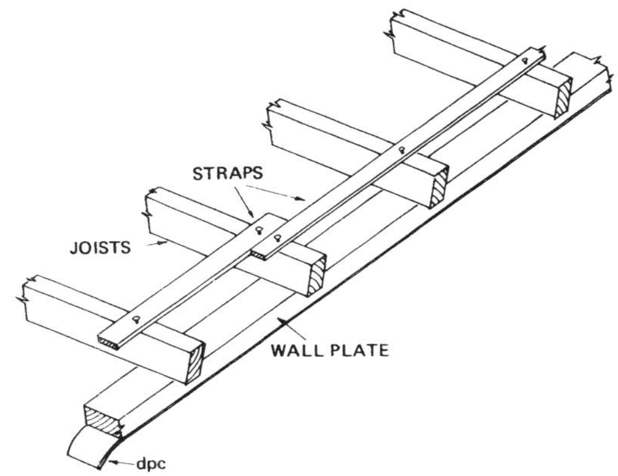

Fig. 8.14. *Sketch showing joists held in position by means of temporary straps.*

(c) Mark the positions of the joists on the two outside wall plates, cut joists to length and place in their respective positions, cambered edge upwards.

(d) Temporarily fix the joists into their final position by means of scrap timber laths (known as joist straps) nailed to the top edges as shown in Fig. 8.14. If the joists are long or badly "bowed", as is often the case, an intermediate strap across the centre of the joists is worth while to straighten them out.

NOTE: Care should be taken to ensure the straps are not in the way of any walls which are to be carried up. If the ends of the joists are to be "built in" they should receive two good flowing coats of creosote or other suitable preservative.

Levelling the joists

This is a simple but important operation, and one which may well have a profound effect upon the stability and levelness of the finished floor. It involves bringing the top surface of all the joists into a single, accurate, horizontal plane, and is necessary because of the inevitable variation in the width of sawn timbers — in this case, the joists.

Referring again to Fig. 8.13, the first decision to be made is the point from which the levelling is to commence — in other words, the establishing of a "datum" from which the level will be taken. Almost invariably, this datum is determined by the position of the bottom of the door jambs (or hardwood threshold) on one of the external door frames. Do not forget to allow for the thickness of the flooring when establishing this!

In this particular instance, we will assume the joists have already been levelled up to point A and the starting point is therefore in the doorway. The levelling procedure will then be as follows.

(a) Place the spirit level in the doorway to ensure that the end of the joist at Point 1 is

Fig. 8.15. *Levelling the joists.*

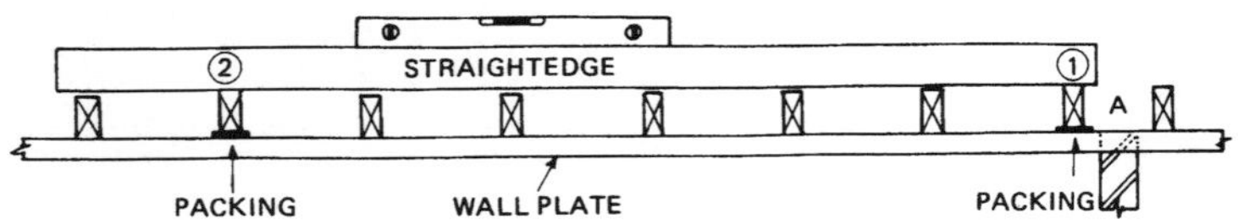

(a) Across the joists.

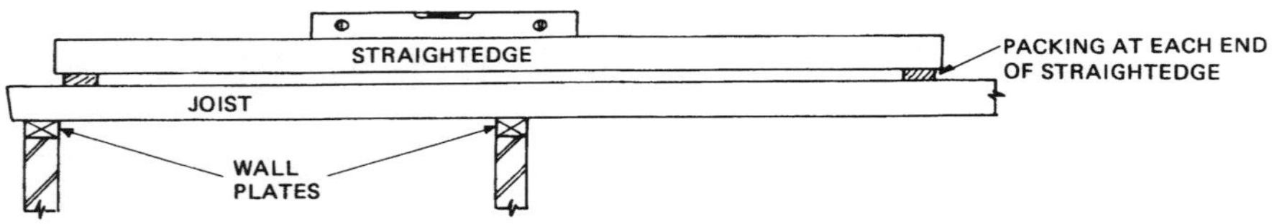

(b) Along a joist.

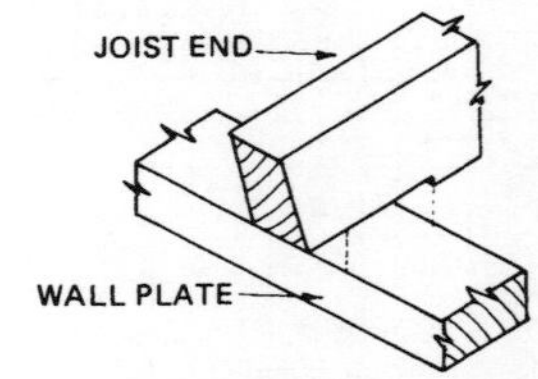

Fig. 8.16. *Treatment for a joist which is too high.*

level with the joists in the adjoining room. If the joist here is low, as it may well be, this should be packed up by inserting one or two small pieces of dpc under the joist until correct. Tap the joist end firmly with a hammer to ensure a firm seating and then place the straightedge alongside the wall, spirit level on top, and level from Point 1 to Point 2 as shown in Fig. 8.15*(a)*, again inserting packing under the joist if necessary. Make a pencil mark on the top of the joist at Point 2 for future reference. Now reverse the straightedge and spirit level and repeat the procedure from Point 2 to Point 3.

(b) Reverse the straightedge and spirit level once again and level from Point 3 to Point 4. When levelling *along* a joist it is necessary to place two packings of equal thickness under the straightedge, one at either end as shown in Fig. 8.15*(b)*, in order to raise the straightedge above the joist straps and to nullify the effect of any camber on the joist itself.

(c) Reverse the straightedge and spirit level once again and level along the far wall to Point 5, and then again from Point 5 to Point 6.

(d) Finally repeat Stage *(b)* from Point 6 to Point 1 – the starting point. If the levelling procedure has been correctly carried out, Points 1 and 6 should be level, any variation indicating an error in the procedure or in the spirit level itself, and requiring a second run through.

NOTE: It is essential to reverse the spirit level and straightedge on each consecutive run in order to cancel out any slight error in the spirit level or straightedge which would otherwise be cumulative.

(e) The spirit level may now be dispensed with, all that remains to be done being the truing up of the intermediate joists by placing the straightedge across the previously levelled (and marked) joist ends and inserting packings as required.

Should one or two of the joists be "high" when levelling through, the only solution is to reduce the depth of the offending joists by cutting the underside away as shown in Fig. 8.16 or, alternatively, by cutting a shallow housing into the top of the wall plate. There is no necessity at this stage to level or pack joists where they lie across intermediate sleeper walls – this is best done during the boarding of the floor as each sleeper wall is reached.

Laying the floorboards

For this, the final part of the floor laying process, the procedure followed depends largely on the type of flooring to be used. If normal t & g flooring is to be laid, then two or three floor cramps, illustrated in Fig. 8.17, will be required, as will also a good straight length of 100 × 50 for use as a cramping piece.

The procedure for laying the boards will then be as follows, illustrated in Fig. 8.18.

(a) Select a long, straight floor board, lay it tongued edge to the wall and nail it securely into place, leaving a 10 mm space between board and wall as shown in Fig. 8.18*(a)*. It is advisable to nail the board only on the edge nearest the wall, as otherwise there may be difficulty in inserting the tongue on the next board. If the room is long, requiring a heading

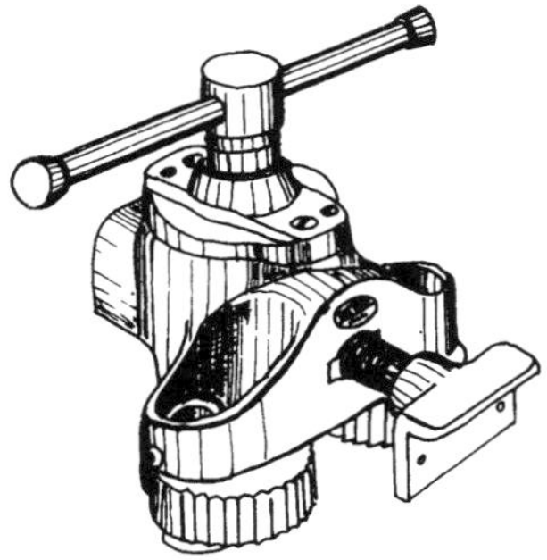

Fig. 8.17. *Flooring cramp.*

Fig. 8.18. *Laying the floorboards.*

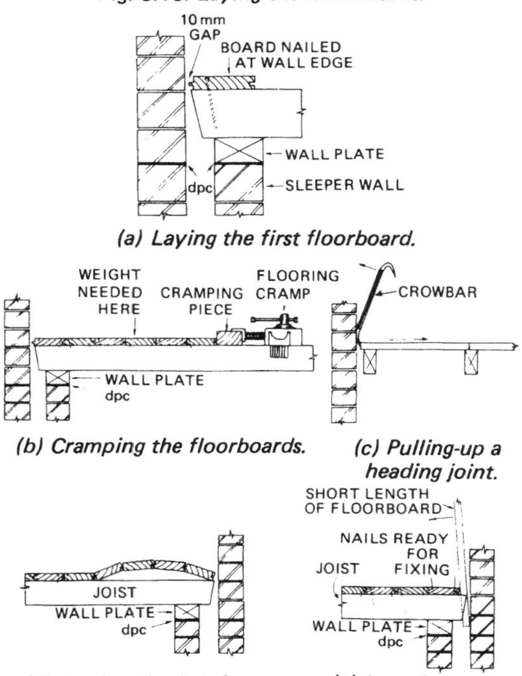

(a) Laying the first floorboard.

(b) Cramping the floorboards.

(c) Pulling-up a heading joint.

(d) Laying the last few floorboards by "folding-in".

(e) Levering up the last board.

joint to be made, or if the board is not quite straight, use a builder's line along the edge as a guide when fixing.

(b) Cut and fit four or five floor boards tapping them gently home with the side of a claw hammer, carefully staggering any heading joints. Next, place the cramping piece against the grooved edge of the last board and apply pressure from the floor cramps as shown in Fig. 8.18*(b)*. Do not apply too much pressure with the cramps on this first laying or the joists may be displaced.

(c) Ensure that all the tongued and grooved joints are tightly fitted with no gaps in the shoulders, and then force all the heading joints up tight by levering gently against the wall with a small crowbar as shown in Fig. 8.18*(c)*.

(d) Nail the boards down, using two nails for each board on every joist, leaving the grooved edge of the last board free for easy insertion of the tongue on the next board. Keep the nails neatly in line in the centre of the joists, dovetailing them slightly for better holding power, and avoiding hammer marks which are the hallmark of an amateur. Finally, punch the nail heads below the surface.

(e) Repeat the procedure, laying up to six boards at a time, until a point is reached close to the wall opposite when there is no longer room for the floor cramps to be used.

(f) The last two or three boards can now be fitted and nailed, cramping the joints up tightly either by "folding" the boards into position as shown in Fig. 8.18*(d)*, or by using a piece of floorboard as a lever against the wall as shown in Fig. 8.18*(e)*.

Points to note

(a) The tongues and grooves on a floorboard are not central but are disposed towards the bottom of the board. Take care not to get a board upside down.

(b) When cramping up the floorboards, it is advisable to have some weight – a fellow workman – on the boards to prevent them springing upwards. Take care not to trap the knees between two boards when kneeling on them – open joints between boards under pressure often snap together unexpectedly.

(c) Keep a constant check on the straightness of the boards during cramping. It is all too easy to depart from a straight line along the edge of the boards, especially if a fellow worker tends to cramp his section more tightly – or less so – than oneself.

(d) Take careful note of the position of any wires or pipes which may have been chased into the top edge of the joists, so as to avoid accidental damage when nailing the boards.

(e) Make sure that tools, debris, offcuts of floorboard, etc., are not left under the floor during the laying of the flooring.

Traps

These are floorboards, or sections of floorboards, which require to be left loose so that they may be removed at a later stage to

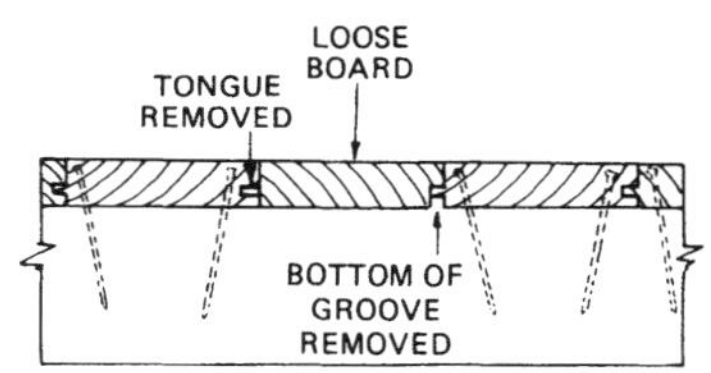

Fig. 8.19. *Forming a "trap".*

enable the plumber or electrician to complete his work below the floor. They are prepared by removing the tongue and lower part of the groove from the edges of a board as shown in Fig. 8.19. The ends of the trap should bear half way on a joist, the board or boards constituting the trap being left unfixed (although they are occasionally secured by screws) and marked for easy identification.

Laying secret nailed floorboards
These boards, because of their sectional shape, must be laid one at a time tongue outwards. They are fixed to the joists by nailing through the splayed tongue as shown in Fig. 8.11. Secret nailed floorboards, or "splay

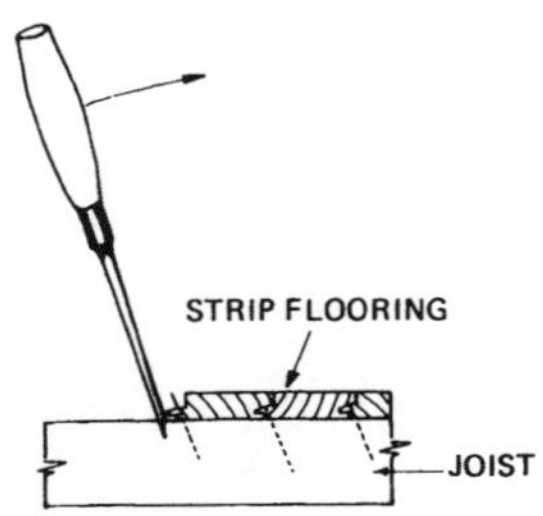

Fig. 8.20. *"Cramping" secret nailed strip flooring using a chisel as a lever.*

rebated, tongued and grooved" as they are sometimes known, are pulled up tightly during laying by levering with a strong chisel as shown in Fig. 8.20, taking care not to damage the tongue in the process.

Care of floors
Newly laid floors, especially hardwood strip floors, should be protected from damage during further building activities by covering the surface with polythene sheet or a layer of clean softwood sawdust or chippings.

Nailing machines
Figure 8.21 shows a manually operated nailing machine designed for the fast and efficient fixing of t & g or secret nailed floor boards. This nailing machine can be fitted with standard (square) or angled base plates to suit whichever type of board is being laid,

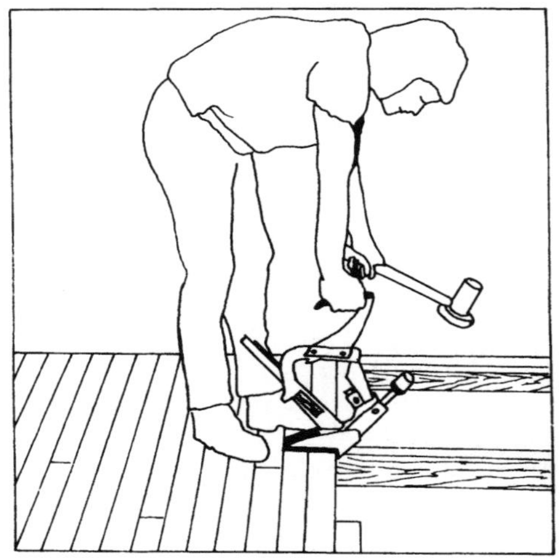

Fig. 8.21. *Using a nailing machine on a hardwood, secret nailed strip floor.*

and can be loaded with either serrated edge or resin coated flooring nails as shown in Fig. 8.22*(a)* and *(b)* respectively, the nails being cohered into strips of up to 100 for fast and easy loading. The nails are driven into the floor by single or multiple blows with the special hammer.

Fig. 8.22. *Flooring nails.*

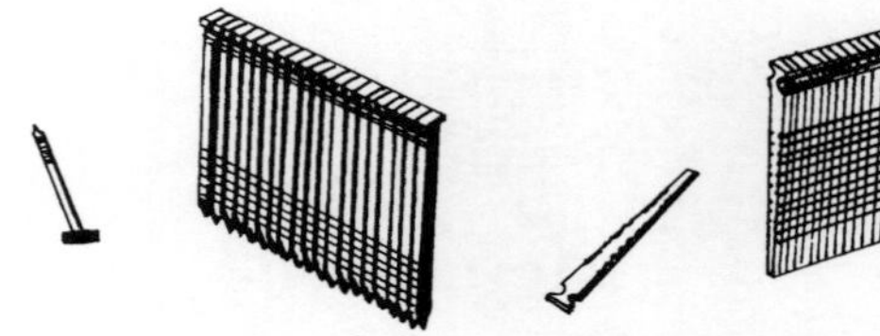

(a) Serrated edge flooring nails. *(b) Resin coated flooring nails.*

A similar type of machine which operates by compressed air and is thus extremely rapid in action is shown in Fig. 8.23.

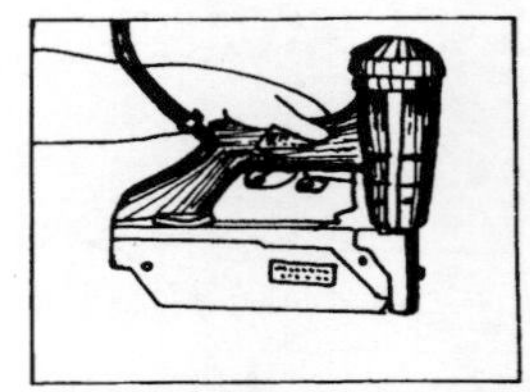

Fig. 8.23. *Pneumatic machine.*

FURTHER READING

British Standards and Codes of Practice

CP 201	Part 2:1972 Wood flooring (board, strip. block and mosaic)
BS 565	Glossary of terms relating to timber and woodwork
BS 1297:1970	Specification for grading and sizing of softwood flooring

Building Research Establishment Digests

No. 18	Design of timber floors to prevent decay

No. 27 Rising damp in walls
No. 77 Damp proof courses
No. 145 Heat losses through ground floors
No. 201 Wood preservation – application methods

Other literature

The Building Regulations 1985 (HMSO)

SELF-TESTING QUESTIONS

All the information required to answer the following questions is contained within this chapter. Attempt each section *as fully or as briefly* as the question demands, and then check your answers against the information given in the chapter.

1. State the main functional requirements of a timber ground floor.

2. Draw to a suitable scale a cross section through a timber ground floor to show: *(a)* the floor boards; *(b)* the joists; *(c)* the sleeper walls; *(d)* the dpc's; *(e)* a means of underfloor ventilation.

3. Explain with dimensioned sketches the critical measurements to which a hearth in a timber ground floor must conform in order to satisfy the Building Regulations.

4. *(a)* Sketch approximately full size sections through:
(i) a t & g floorboard;
(ii) a secret nailed floorboard.
(b) Name two materials (other than *(i)* and *(ii)*) which are used as floorings to floor joists.

5. *(a)* Describe the procedure by which the tops of ground floor joists are levelled.
(b) State the reason for reversing a straightedge and spirit level for each consecutive "run" when levelling floor joists.

6. *(a)* List five important points which need consideration when laying t & g flooring.
(b) Explain the function and sketch the construction of a "trap" in a boarded floor.

9. Construction of Single Roofs

After completing this chapter the student should be able to:

1. State the main functions and requirements of a roof.
2. State the factors which govern the pitch of a roof.
3. Explain the importance of triangulation in a roof structure.
4. Sketch sections and details to show the construction of lean-to, couple, collar-tie and couple close roofs.
5. Sketch details to show the construction of flush, open, closed and sprocketed eaves.
6. Determine by scale drawing, the length, plumb cut and seat cut for a common rafter.
7. Calculate the length of a rafter given the rise and span of the roof.
8. Set out a pattern rafter for a gable roof of specified dimensions.
9. Outline the sequence of operations in the erection of a gable roof.

FUNCTIONS OF A ROOF

The main and most obvious function of a roof is to provide a weatherproof covering to a building, but there are other functions which a roof must serve, not perhaps so fundamental as the former but nonetheless important. These include strength and durability of the structure, prevention of heat loss from the building, resistance to the spread of fire, sound insulation, and lastly the enhancement of the building's appearance. This may well seem a formidable list of requirements, but in fact a well designed and constructed roof fulfils all of these functions quite satisfactorily and yet still remains a relatively simple structure.

DESIGN, PITCH AND COVERINGS

A pitched roof is one which has a slope of 10° or more, anything less than this being described as a flat roof. The pitch of the roof is, to a great degree, dependent upon the type of roof covering and vice versa, the lower the pitch, the longer the roof covering units, and the more closely fitting they must be in order to prevent the penetration of rainwater and snow. Table III gives typical minimum slopes for a variety of roof covering materials.

TABLE III. ROOF PITCH

Material	*Minimum pitch*
Plain tiles	40°
Pantiles	35°
Interlocking tiles	30°
Slates (medium size)	35°
Asbestos slates	35°
Thatch	45°
Shingles	30°
Corrugated asbestos	12°

Since these are minimum pitches it would be a good policy to increase them for extra protection, especially where the roof is in a very exposed position or in a high rainfall area.

The design of a single, pitched roof is relatively simple, the roof comprising a sloping rafter or spar supported at the top by a ridge, or in the case of a lean-to roof, a wall piece, and at the bottom by a wall plate. There are no intermediate supports to the rafter in a *single* roof. The lower end of the roof usually overhangs the wall to a greater or lesser degree and this feature is referred to as the eaves.

The roof covering, the tiles, slates, etc. are laid on battens which are fixed along the roof at right angles to the rafters. A layer of sarking felt is provided under the battens to prevent penetration of any rainwater or snow which may be driven under the roof covering by the wind.

Triangulation of roof members

A roof structure is subjected to various loadings due to the weight of the roof covering itself and to the weight of snow and the effects of wind, the latter two being variable and sometimes considerable.

It should be understood that the increased wind speed and turbulence deriving from a flow of air across a pitched roof can lead to a negative pressure – suction – on various parts of the roof, especially at the eaves.

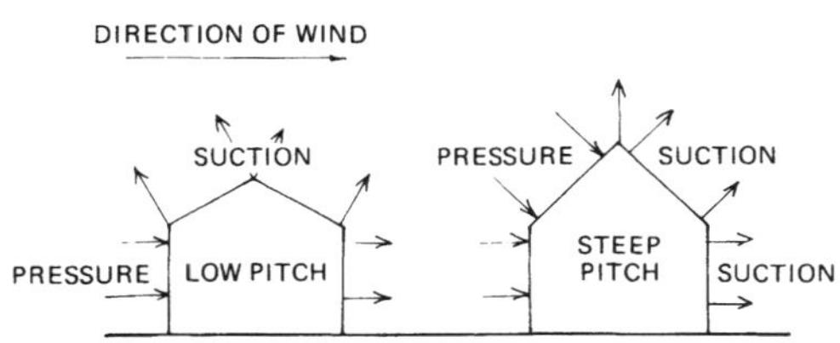

Fig. 9.1. *Effect of wind on roofs.*

This necessitates secure fixing of the roof structure to the walls, and of the roof covering to the structure itself to avoid possible lifting or stripping in high winds. The effect of wind on roofs of varying pitches is shown in Fig. 9.1.

As is well known, a well joined triangular structure is both strong and rigid and resists

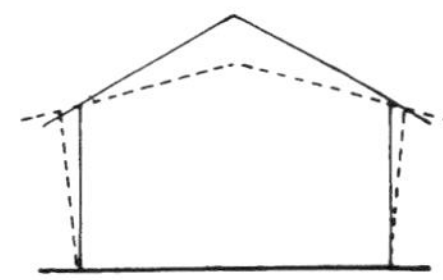

Fig. 9.2. *Untriangulated roof showing tendency of rafters to push walls outwards.*

any tendency to distort, a fact which can be put to good use in the design of a pitched roof. Figure 9.2 shows in diagrammatic form a simple couple roof and, as can be seen, a roof of this type, lacking as it does a horizontal tie to complete the triangle, suffers the inherent weakness associated with such a structure – a tendency to push the top of the walls outwards. This tendency is resisted only by the the walls themselves, which must therefore be very strong and need to increase in thickness in proportion to the span, factors which limit the span to a maximum of about 3 metres.

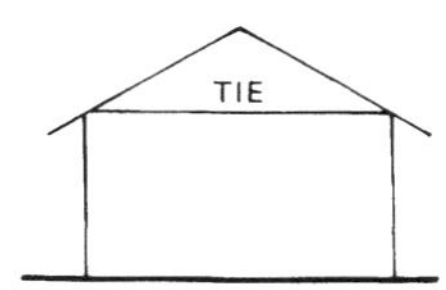

Fig. 9.3. *Triangulated roof giving stability.*

Figure 9.3 shows a much more stable structure incorporating a tie, which effectively resists any tendency on the part of the rafters to spread or slide. Where single roofs are concerned, the only real limitation to the span of the roof is the ability of the rafters to resist a tendency to sag. This can be achieved by increasing the depth of the rafters. This obviously imposes economic limitations and therefore such roofs rarely exceed about 4.500 metres to 5.000 metres span. Beyond this point a double roof would be required, having an intermediate support to the rafters in the form of a longitudinal beam known as a purlin. The construction of double roofs is, however, beyond the scope of this volume.

TYPES OF SINGLE ROOF

The following descriptions and drawings show typical details of the construction of the four most common types of single roof.

Lean-to roof

This type of roof is commonly used on domestic outbuildings and garages for which purpose it is suitable for spans up to 2.500 metres. The construction details of a lean-to roof are shown in Fig. 9.4.

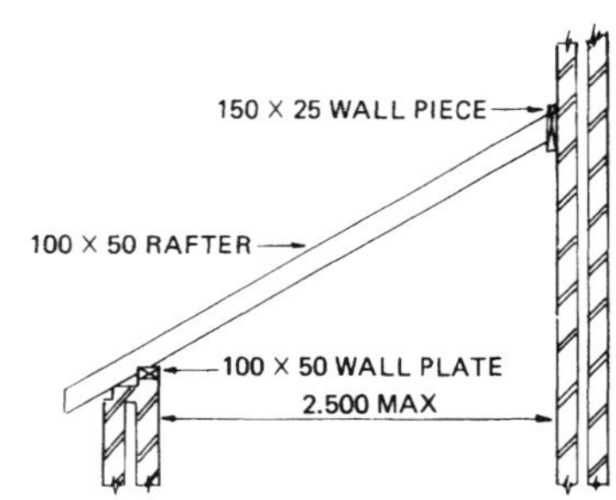

Fig. 9.4. *Lean-to roof.*

Joint at head of rafter

The head of the rafter may be secured in one of several ways, two of which are shown in Figs. 9.5(*a*) and 9.5(*b*).

Fig. 9.5. *Securing the head of rafters.*

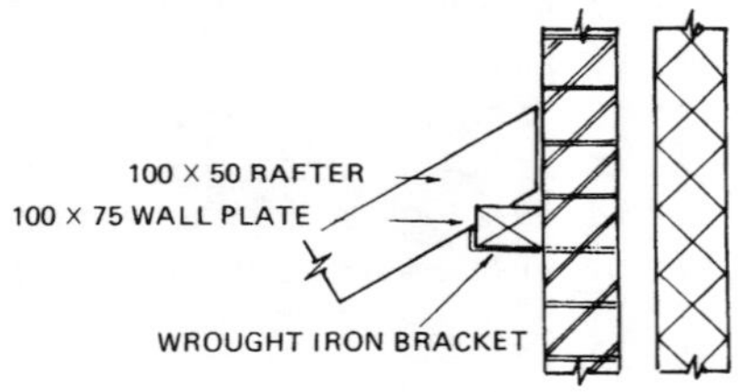

(a) Detail at head of a lean-to roof.

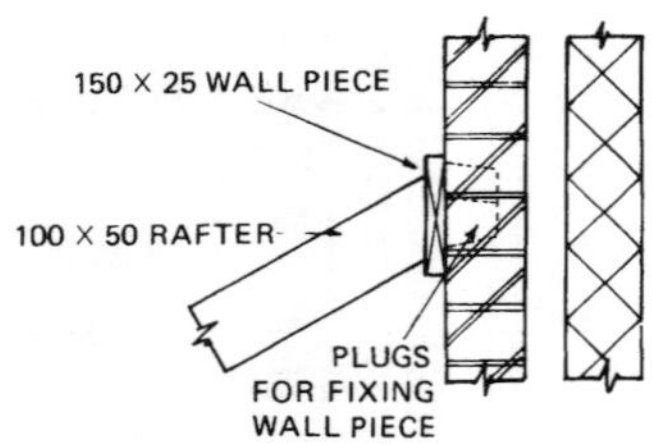

(b) Alternative detail at head of lean-to roof.

Joint at foot of rafter

The foot of the rafter is nailed to a wall plate, a good bearing being obtained by the use of a birdsmouth joint as shown in Fig. 9.6*(a)*. Note that the depth of the birdsmouth should not exceed one third the depth of the rafter. Figure 9.6*(b)* shows a variation of the birdsmouth joint which is useful on half brick walls with flush eaves.

Wall plate

This is usually a length of 100 X 50 or 100 X 75 sawn softwood which is levelled and bedded down on to the wall with mortar. Lengthening joints in a wall plate are plain halvings (*see* Chapter 3) fastened with a pair of nails. The wall plate should be secured to the head of the wall by means of mild steel

Fig. 9.6. *Birdsmouth joints.*

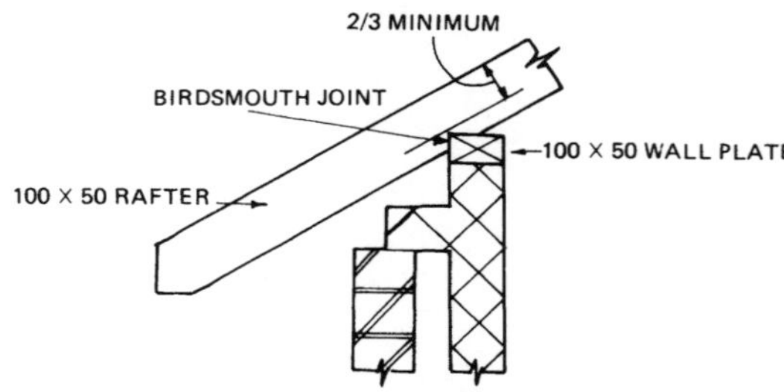

(a) Sketch showing proportion of birdsmouth joint.

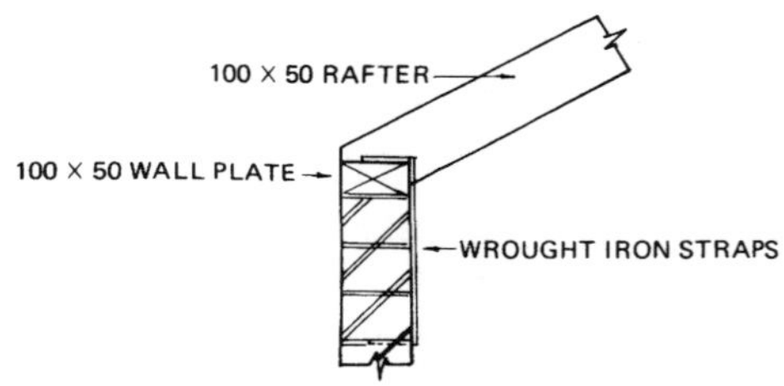

(b) Variation of the birdsmouth joint.

anchor straps built into the wall as shown in Fig. 9.6*(b)*. Note that, as shown in Fig. 9.6*(a)*, the wall plate is usually flush with the inside face of the inside skin of masonry, although in practice this often leads to a crack developing in the plaster at this point due to movement in the timber. The risk of this defect occurring can be offset by the use of a narrow strip of expanded metal fixed to the edge of the wall plate before plastering.

Couple roof

This roof design shown in Fig. 9.7 is comparatively weak because of the lack of a tie and should therefore be restricted to spans of up to about 3.000 metres for the reason pre-

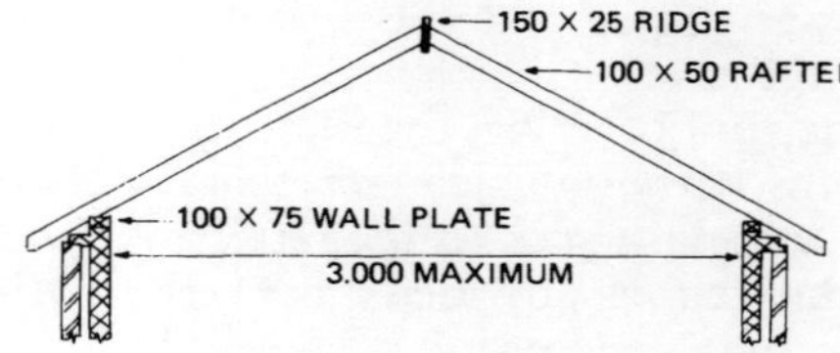

Fig. 9.7. *Couple roof.*

viously explained. It will be appreciated that the outward thrust on the walls is greater with a roof of low pitch than where the rafters are more steeply inclined.

The rafters in this type of roof are erected in pairs or couples, and are held at the top with a ridge board to which they are nailed and by a wall plate at the foot, the joint here being identical with that used on the lean-to roof.

Couple close roof

This roof shown in Fig. 9.8 is a much stronger design than the couple roof owing to the fact that it is completely triangulated by means of a tie or ceiling joist fixed to the

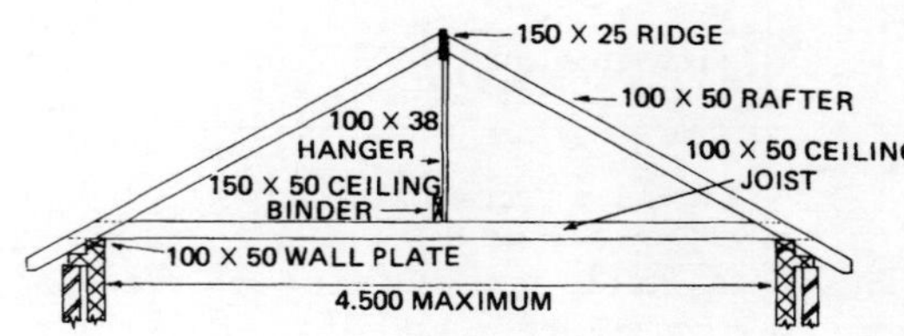

Fig. 9.8. *Couple close roof.*

rafters and wall plate. It is suitable for small domestic purposes up to spans of about 4.500 metres. The joints at the head and foot of the rafters are as previously described.

Hangers and ceiling beams

Whilst the two opposing rafters can be regarded as being separate members, each covering one half of the span, the ceiling joists in the couple close roof will have a tendency to sag unless they receive some intermediate support. Since an internal load bearing wall is unlikely in a building of such small span, this support is generally provided by means of a ceiling beam or binder suspended from the ridge by hangers as shown in Fig. 9.8. The hanger is nailed to the sides of the ridge and ceiling beam. In addition, the ends of the ceiling beam can be built into the gable walls at either end of the building, and possibly supported at mid-point by a cross wall.

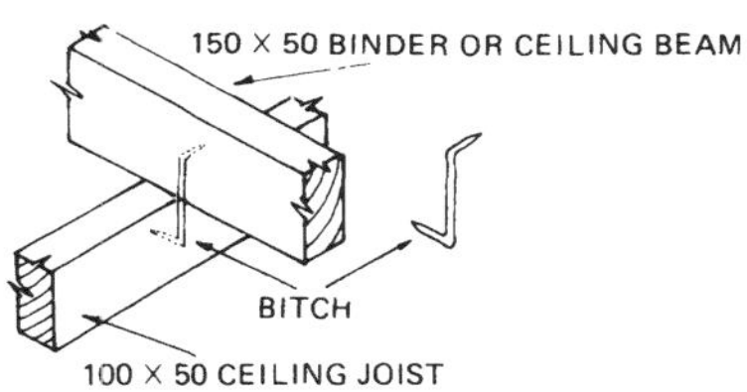

Fig. 9.9. *Method of fastening ceiling joists to binders.*

The ceiling joists are skew nailed to the ceiling beam or, more usually, fixed with "bitches" as shown in Fig. 9.9.

Collar tie roof

These roofs shown in Fig. 9.10 are used where extra headroom is required without the use of additional courses of masonry to raise the wall plates.

The general construction of a collar tie roof is similar to the couple close, the main difference being in the position of the tie

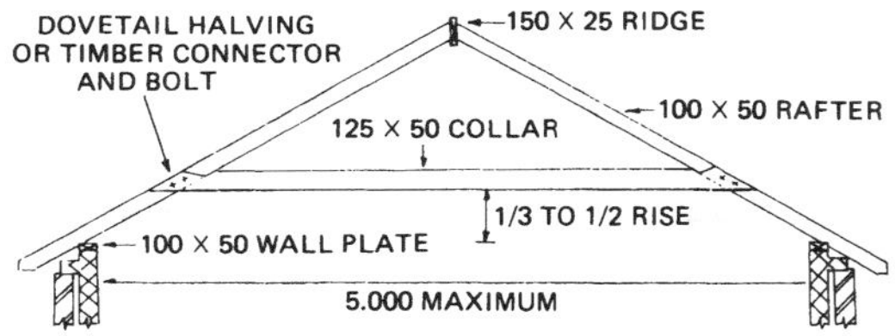

Fig. 9.10. *Collar tie roof.*

(collar) which is raised above the level of the wall plate. The collar is usually placed in the lower third of the roof, but in any case its height should never exceed more than half the rise of the roof.

It should be noted that a collar tie is subjected to greater strain than a normal ceiling joist because of its position in the roof and is generally in considerable tension, although it does, to some extent, give a certain degree

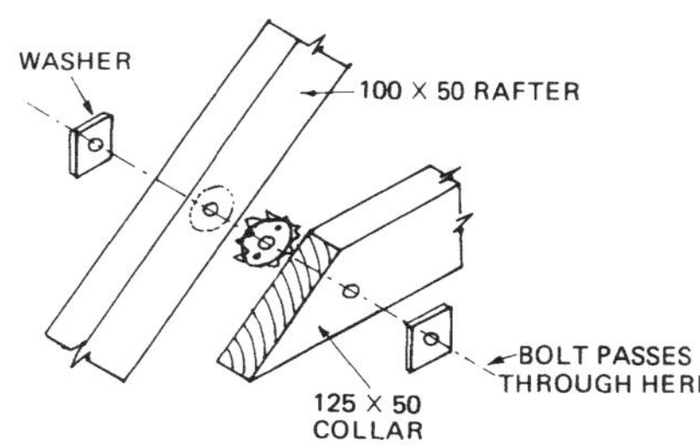

Fig. 9.11. *Use of timber connector.*

of support to the rafters themselves. It is necessary, therefore, to make a strong joint between the collars and the rafters, this being done traditionally with a dovetailed halving joint as shown in Fig. 9.10, but nowadays more often with a bolt and timber connector as shown in Fig. 9.11.

Treatment at eaves

The eaves of a roof – the overhanging portion at the feet of the rafters – can be treated in one of several ways, the choice being mainly a matter of preference in design.

Flush eaves

Shown in Fig. 9.12*(a)* these are used where an overhanging eaves is not desired or perhaps not permitted. Note that the fascia board must be deep enough to cover the ends of the rafter feet. If necessary the foot of the rafter can be cut as shown by the broken line to reduce the width of the fascia board. The function of the fascia board is twofold: *(a)* to provide a neat finish to the edge of the roof, and *(b)* to provide a surface to which the gutter may be fixed.

Tilting fillets are employed, as their name suggests, for tilting the bottom edge of the first courses of slates or tiles to reduce the tendency of water to penetrate between them due to capillary action, but as can be seen from the diagram this purpose is served by the upstand of the fascia board itself. A tilting fillet in this position is therefore used to support the top edge of the fascia board (to which it is nailed) and could, perhaps, be better described as an angle fillet.

Open eaves

Shown in Fig. 9.12*(b)* these project beyond the face of the wall thereby giving it added weather protection. As can be seen from the diagram the rafter feet are exposed to view from below and therefore require to be cleaned up before fixing in preparation for painting. In addition, an eaves lining of hardboard, plywood or asbestos cement sheet is usually fixed to the top surface of the rafters

Fig. 9.12. *Treatment at eaves.*

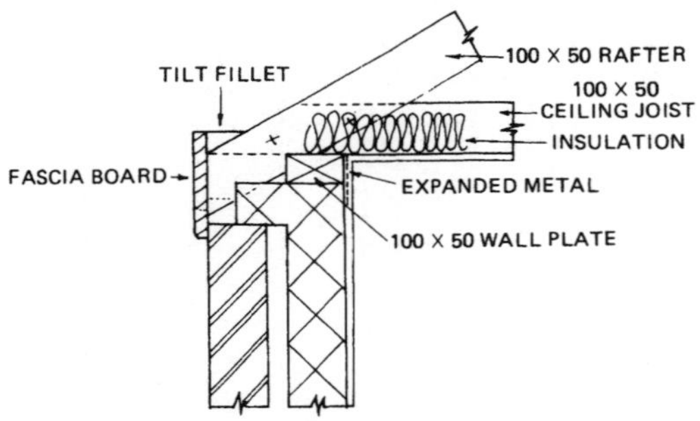

(a) Flush eaves.

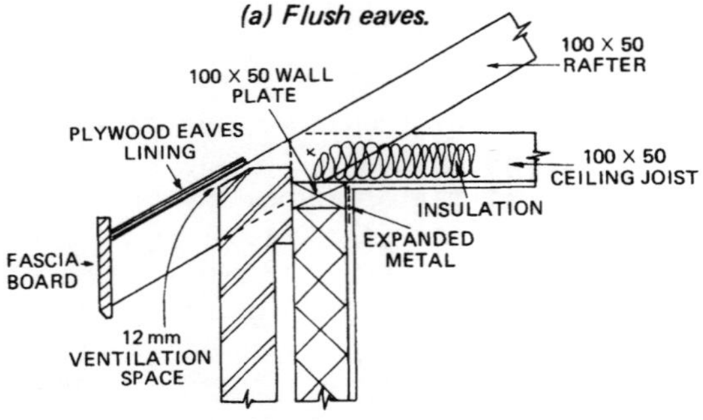

(b) Open eaves.

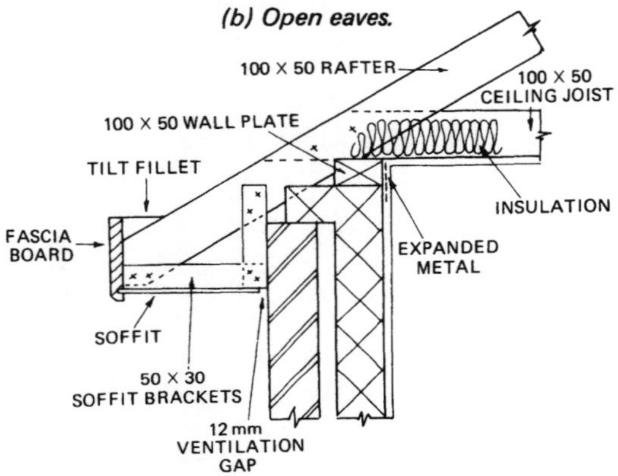

(c) Closed eaves.

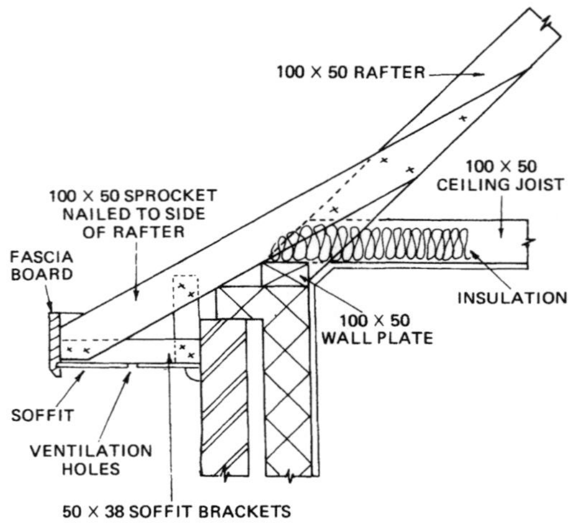

(d) Sprocketed eaves.

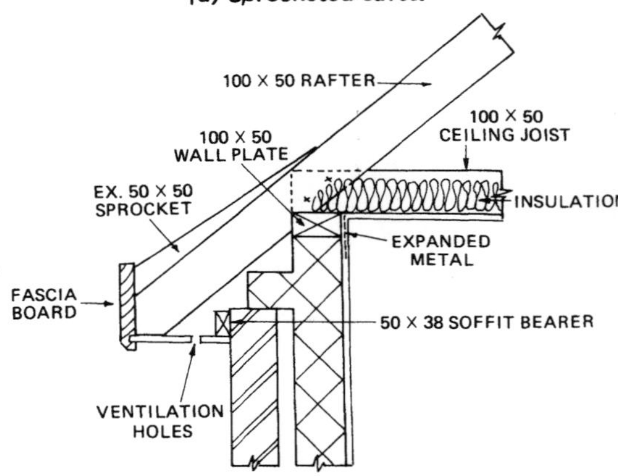

(e) Sprocketed eaves.

as shown to cover the underside of the tiles and sarking felt.

Closed eaves

Eaves of this type incorporating a "soffit" to enclose the underside of the rafter feet are shown in Fig. 9.12*(c)* and are by far the most common form of treatment used on domestic buildings. Treatments for closed eaves vary widely according to the pitch of the roof, the width of the soffit and general appearance required, the details shown being fairly typical.

The soffit, which may be of matchboarding, asbestos cement sheet or plywood, is tongued into a groove ploughed into the back of the fascia board and is further supported by soffit bearers nailed to the side of the rafter feet.

Sprocketed eaves

These are illustrated in Fig. 9.12*(d)* and *(e)* which show methods of fixing the sprockets to the side and top of the rafters respectively, the small tapered sprocket shown in Fig. 9.12 *(e)* being commonly known as a sprocket piece. The use of a sprocket at the eaves gives a pleasing appearance to a steeply pitched roof and tends to slow down the flow of water into the gutter, which on a steeply pitched roof might otherwise in a severe rainstorm be of sufficient velocity as to rush over the edge. A sprocket also acts as a superior form of tilting fillet by tightening the lower edges of the slates or tiles more effectively than a simple tilt fillet.

It will be realised that some form of sprocket can be applied to flush, open or closed eaves, wherever this may be desired.

Treatment at verges

The verge or sloping end of a gable roof requires protecting from the weather no less than the eaves, and where this projects considerably beyond the face of the gable wall,

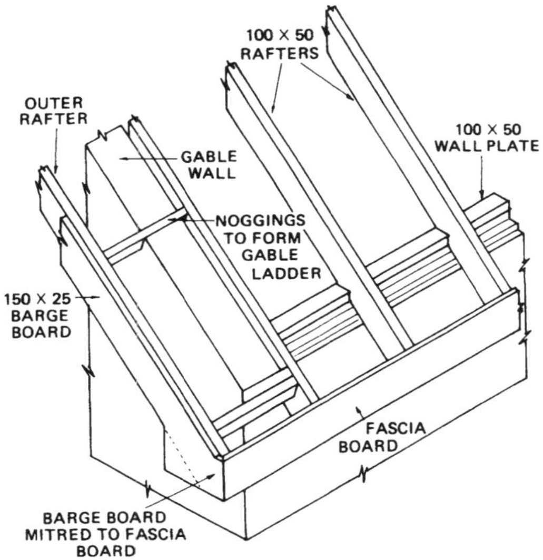

Fig. 9.13. *Sketch showing roof detail at verge.*

the carpenter is usually called upon to fit a "barge board" as shown in Fig. 9.13. The barge board is really a type of sloping fascia as shown in the diagram. Barge boards fit closely to the underside of the overhanging tiles and are mitred together at the ridge and nailed to an outer rafter. This rafter is in turn fixed to short cantilevered nogging pieces, fastened to the last pair of common rafters (the barge couple) and built into the top of the gable wall. The foot of a barge board is often shaped to give a smart appearance and is mitred to fit the end of the horizontal fascia board as shown.

Trimming round chimney stacks

Where a chimney stack penetrates a roof surface, a hole must be formed in the structure

Fig. 9.14. *Trimming around chimneys.*

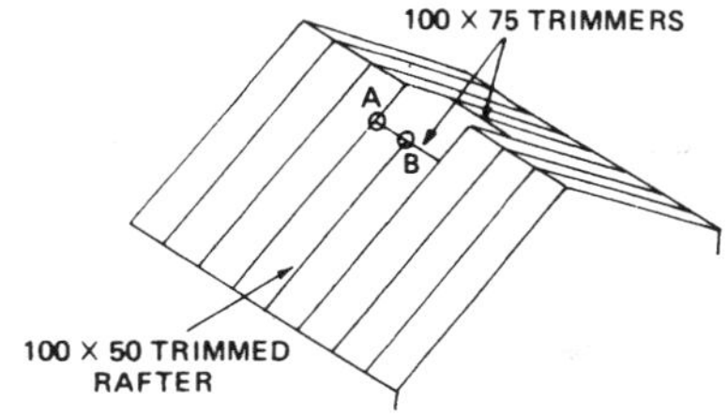

(a) Chimney passing through ridge.

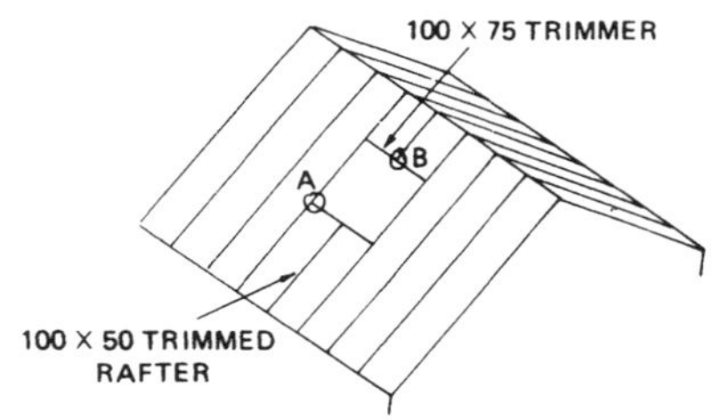

(b) Chimney passing through roof to one side of ridge.

through which it can pass. This procedure is known to the carpenter as "trimming" and must be done in a manner which satisfies the requirements of Part J of the Building Regulations. This part of the Regulations is concerned with the prevention of fire arising as a result of combustible materials (timber) being in close proximity to hot flue gases. The Regulations state that where a flue serves a 4.5 kW appliance, no structural timbers must be placed closer than 40 mm to the outer face of the chimney, unless the flue itself is surrounded by solid non-combustible material 200 mm or more in thickness. The structural timbers referred to here include rafters (and trimming rafters), ridge boards, purlins, wall plates, struts, ties, etc., but not

Fig. 9.15. *Trimming joints.*

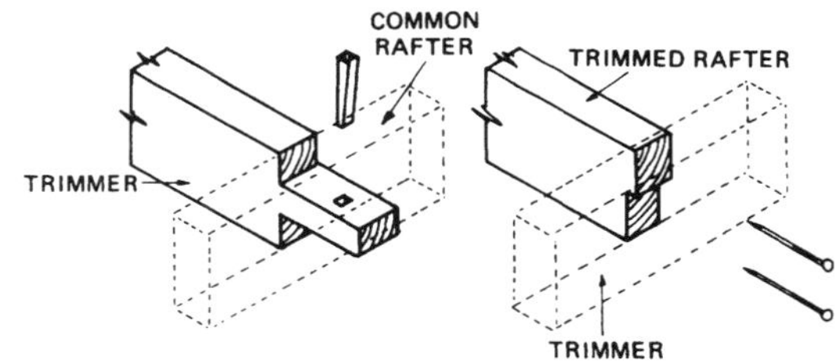

(a) Joint at "A". *(b) Joint at "B".*

the gutter bearers and boards which the plumber requires around the chimney to support his lead flashings.

Figure 9.14 shows the trimming necessary for chimney stacks which pass *(a)* through a pitched roof at the ridge, and *(b)* to one side of the ridge. Trimming joints are shown in Fig. 9.15, although the simple housed joint (Fig. 9.15*(b)*) is nowadays commonly used for both situations. Since the trimming rafters have to carry the trimmed rafters which lie between them, they should be thicker than the common rafters by 12 to 25 mm.

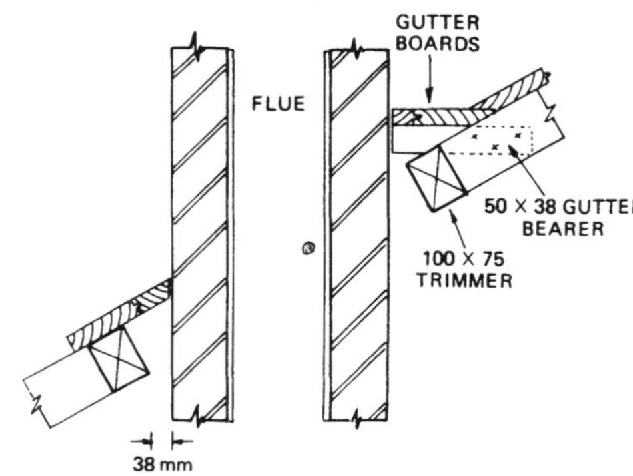

Fig. 9.16. *Vertical section through chimney showing trimming.*

Figure 9.16 is a vertical section through a chimney stack and shows the application of the Building Regulations with reference to the roof trimming and the construction of the gutter behind the chimney.

DETERMINATION OF LENGTHS AND BEVELS FOR RAFTERS

Whilst the cutting and erection of a gable ended roof is a fairly straightforward procedure, the setting out of the pattern rafter (from which all the other common rafters are set out) requires a certain amount of technical knowledge and skill if the cumbersome and time consuming alternative of laying out the members on the floor is rejected. This setting out can be done quickly and efficiently if the terminology and basic principles of roof carpentry are soundly understood.

Roofing terminology

The following terms are illustrated in Fig. 9.17.

Span

This in itself can be a rather confusing term and may be given as:

(a) clear span – the clear distance between the supporting walls;

(b) effective span – the horizontal distance between the centres of the supporting walls.

The span as far as the carpenter setting out the roof is concerned is the horizontal distance between the outside edges of the wall plates, and usually approximates closely but not exactly to the effective span. This dimension should always be taken on site by measuring across the wall plates with a tape rule.

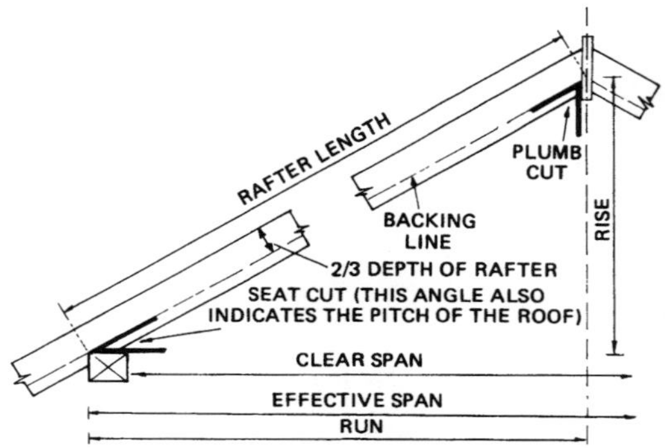

Fig. 9.17. *Roofing terminology.*

The run

The run of a rafter is its length on plan – excluding the eaves. In the case of a simple pitched roof with equally inclined surfaces it can be taken (for a common rafter) as one half of the effective span.

The rise

The rise of a roof is the vertical height to which the rafters rise above the wall plate, measured to the backing line.

The pitch

The pitch of the roof indicates the slope of the rafters and is given in degrees or as a fraction: rise/span. Thus a half pitch roof has a rise of one half of its span, and so on. In the same way we can deduce that a roof having a span of 4.500 m and a pitch of one third has a rise of 1.500 m.

The backing line

This is gauged on the pattern rafter for setting out purposes and intersects the inside corner of the birdsmouth.

Rafter length

This is the measured or calculated length of the rafter discounting the eaves, and is taken along the backing line.

Plumb cut

This is the angle which is marked at the head of a rafter where it abuts the ridge board. Thus a plumb cut is in fact "plumb" (vertical) when the rafter is raised to its correct pitch.

Seat cut

This is the angle between the slope of a rafter and the horizontal top surface of the wall plate. Thus the seat cut is really the "complement" of the plumb cut, i.e. seat cut + plumb cut = 90°.

Rafter lengths and bevels

Having ascertained the pitch of the roof from the drawings or specifications and the span from the building itself, it becomes a

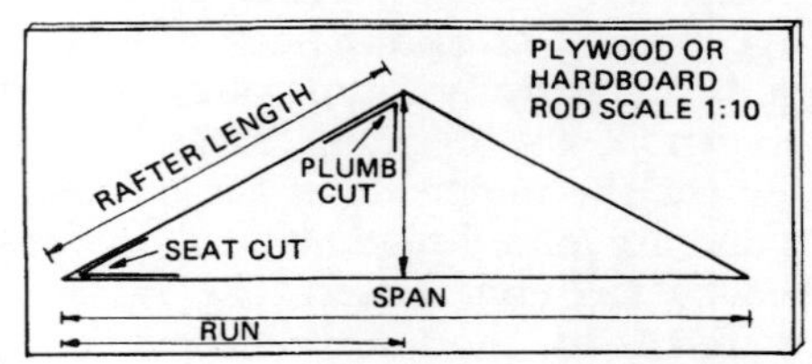

Fig. 9.18. *Setting out to scale to determine rafter lengths and bevels.*

simple matter to set out to a reasonable scale of, say, 1:10 a line diagram of the roof. This is usually drawn on a piece of plywood, hardboard or similar material as shown in Fig. 9.18. The plumb and seat cuts can be marked in and, provided the drawing is accurately done, it is then possible to determine the length of the rafter from the drawing by scale measurement.

Calculation of length

If it is deemed worthwhile to calculate the rafter length, this can be done quite simply by the application of the Pythagorean theorem which states: "The square of the hypotenuse of a right angled triangle is equal to the sum of the squares of the other two sides." The

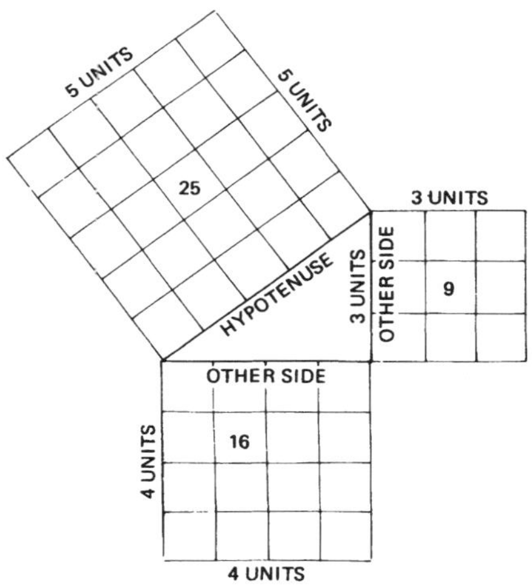

Fig. 9.19. *Graphical explanation of Pythagoras' theorem.*

significance of this theorem to the carpenter is that the "hypotenuse" represents the "rafter length" whilst the "other two sides" are the "run" and the "rise" respectively.

This theorem can be made easier to understand by reference to Fig. 9.19. This shows quite clearly that a right angled triangle having sides of 3.000 m, 4.000 m and 5.000 m has squares on each side as follows:

$$\text{Side A} = 5.000\text{ m} \therefore A^2 = 5 \times 5 = 25\text{ m}^2$$
$$\text{Side B} = 4.000\text{ m} \therefore B^2 = 4 \times 4 = 16\text{ m}^2$$
$$\text{Side C} = 3.000\text{ m} \therefore C^2 = 3 \times 3 = 9\text{ m}^2$$

and therefore:

$$A^2\ (25\text{ m}^2) = B^2\ (16\text{ m}^2) + C^3\ (9\text{ m}^2)$$
$$A^2\ (25\text{ m}^2) = B^2 + C^2\ (25\text{ m}^2)$$

The theroem can be memorised therefore as:

$$A^2 = B^2 + C^2$$

So taking the square root of each side of the equation gives:

$$A = \sqrt{B^2 + C^2}$$

Example 1

A roof has a span of 4.000 m and a rise of 2.000 m. Calculate the length of the rafter. (Note that the *span* is given as 4.000 m; the *run* which we require to solve the problem is therefore one half of this.) The problem can thus be solved as follows:

$$A = \sqrt{B^2 + C^2}$$
$$\therefore \quad A = \sqrt{2^2 + 2^2}$$
$$= \sqrt{4 + 4}$$
$$= \sqrt{8}$$

$\therefore$ Rafter length = 2.828 m

Example 2

A roof has a span of 5.500 m and a pitch of one third. Calculate the rafter length.

$$\text{Rise of roof} = \frac{5.500}{3} = 1.833\text{ m}$$

$$\text{Run of roof} = \frac{5.500}{2} = 2.750\text{ m}$$

$$A = \sqrt{B^2 + C^2}$$
$$\therefore \quad A = \sqrt{(1.833)^2 + (2.750)^2}$$
$$\therefore \quad A = \sqrt{3.360 + 7.562}$$
$$\therefore \quad A = \sqrt{10.922}$$

$\therefore$ Rafter length = 3.304 m

SETTING OUT THE PATTERN RAFTER

Having determined by scale drawing the plumb and seat cuts for the common rafter, and its length by scale drawing and/or calculation, the pattern rafter should be set out (*see* Fig. 9.20) as follows.

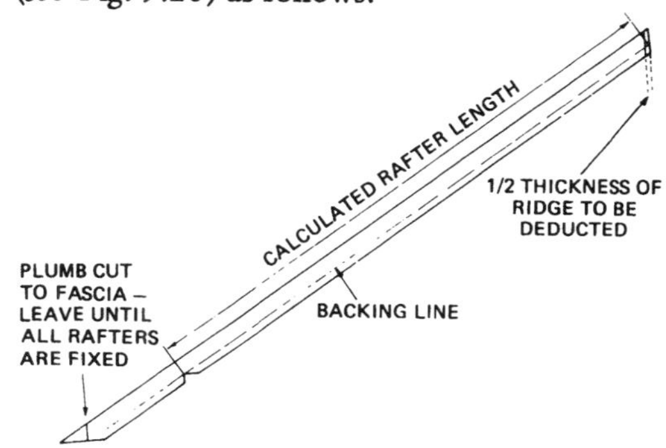

Fig. 9.20. *Setting out the pattern rafter.*

(*a*) Select the top edge of the rafter (the straightest possible length of rafter material should be used for this) cambered edge upwards.

(*b*) Mark the plumb close to the top end using a sliding bevel set accurately to the angle on the drawing.

(*c*) Gauge the backing line along the rafter – two thirds of its depth – from the top edge.

(d) Measure the rafter length along the backing line starting from the intersection of the plumb cut and backing line.

(e) Mark the birdsmouth joint using sliding bevels set to plumb and seat cuts, the internal corner of the birsdmouth being placed at the marked rafter length.

(f) Add to the foot of the rafter whatever dimension is required for the eaves (mark off with the same two bevels).

(g) Deduct half the thickness of the ridge board (measured at right angles to the plumb cut).

The pattern rafter should now be checked over for accuracy and cut to the lines marked (plumb cut, birdsmouth and horizontal eaves cut). A second rafter should now be cut from this pattern and the two rafters can then be tried in position to check that the rafters sit neatly on the wall plate, and butt accurately against the ridge board (hold an offcut of similar thickness to the ridge between the rafters for this purpose). If everything is correct, mark the pattern rafter so that it can be identified for further use.

NOTE: The plumb cut to the back of the fascia board at the bottom of the rafter is best left unsawn until all the rafters are fixed.

ERECTING THE ROOF

Safety

Whilst the erection of a small gable roof is a reasonably simple and straightforward matter, it is as well to bear in mind that it involves working on a scaffold, usually at a considerable height, and requires the use of ladders, trestles and perhaps some form of temporary platform. Statistics prove that a high proportion of accidents occurring in the building trades are due to workmen falling from buildings under construction or to objects falling upon them from above. The job of erecting a roof, therefore, should be undertaken in the full knowledge that accidents can occur unless sensible and reasonable safety precautions are observed (*see* Chapter 12). The competent, intelligent craftsman never takes unnecessary risks in any situation, especially when working at height.

Procedure

The first step is to cut, prepare and bed the wall plates on the top of the wall. Here the carpenter must work alongside the bricklayer, and he should make sure the wall plates are level, straight and parallel.

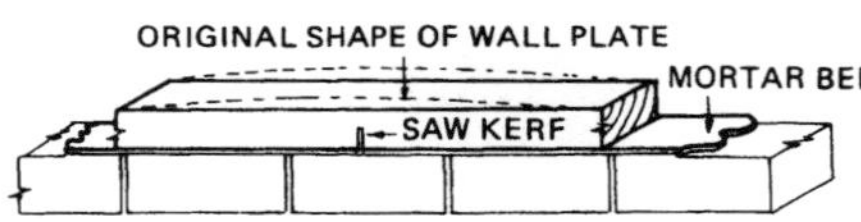

Fig. 9.21. *Method of flattening at excessively cambered wall plate.*

Where a wall plate has an excessive camber, this would be relieved by sawing through part way from the underside as shown in Fig. 9.21. Temporary timber straps fixed from one wall plate to the other can be used where necessary to pull them straight if they show any tendency to bow inwards or outwards. The wall plate should be fastened down with anchor straps as previously mentioned.

The position of all the rafters should now be marked on the top surface of the wall plates, keeping as near as possible to the specified centres and allowing the required distance between rafters and chimney stacks where these occur. The number of common rafters required should be noted.

The next step is to cut all the remaining rafters, using the pattern rafter only to mark them, and making sure that any tendency towards a camber on the rafters is put to the best advantage — camber upwards! Any variation in the width of a rafter should be nullified by working only from the top edges. Having cut the required number of rafters these can be leaned against the scaffold — in a safe manner — plumb cuts upwards, in their approximate position so they may be conveniently reached from the scaffold as required.

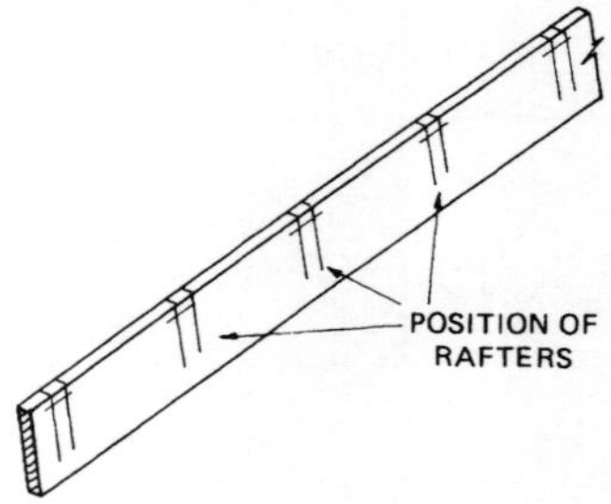

Fig. 9.22. *Ridge board marked out ready for erection.*

The ridge board can now be marked out by placing it on top of one of the wall plates and taking the position of the rafters directly from this. Run a gauge or pencil line along the top of the ridge to mark the position of the top of each rafter (*see* Fig. 9.22) and set on one side ready for use.

Where the roof incorporates ceiling joists

these should be fixed first, nailing them to the wall plates with 100 mm or 125 mm oval wire nails, *and ensuring they are fixed at the side of the rafter positions.* As with rafters, these should also be fixed camber upwards. Where a partition wall occurs, ceiling joists will be required on either side of it so as to give a fixing point for the plaster board ceiling. (The spacing of ceiling joists is more critical than that of the rafters since the plaster boards which will eventually be fixed to them are supplied to suit a particular spacing.)

The ceiling beams or binders can now be positioned and fixed using nails and/or ceiling bitches to fasten them to the ceiling joists and placing packings of waste dpc under the bearings on gable or partition walls as required to obtain a completely flat ceiling (a very slight upward camber is acceptable here, but a sag is not).

It should now be possible to erect a staging of scaffold boards across the ceiling beams so that access can be gained to the ridge position.

If the ceiling shows any tendency to sag under the weight of the workman it should be propped from below. Also – and very important – on a large, high roof the use of proper scaffold should be considered for this purpose. This, however, is not normally warranted on a roof of the scale described here.

It is now possible to fix the rafters into position using 125 mm oval wire nails at the wall plate. A further nail 150 mm long may be driven down through the top of the rafter into the wall plate for extra strength. The top of the rafter is nailed to the ridge board

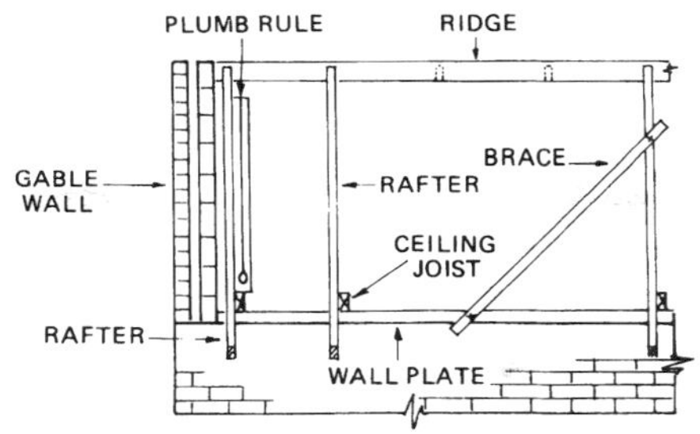

Fig. 9.23. *Plumbing the rafters.*

with 75 mm oval wire nails, a somewhat tricky procedure until a pair of rafters has been fixed at each end of the roof to support the ridge.

It will be appreciated that the initial stages of erection really requires three pairs of hands – one at the foot of each rafter, and one at the ridge. From here onwards, the job can be carried out comfortably by two craftsmen.

The rafters can now be plumbed as shown in Fig. 9.23 and braced back to the wall plate to keep the pairs of rafters stable. The remainder of the rafters are fixed in a similar manner, keeping an eye on the ridge board to make sure it is straight. The rafters should always be fixed in pairs, one opposing the other.

Finally, cut and fit any trimming and trimmed rafters around the chimney stack where applicable.

If barge boards are to be fitted, the outer rafters should be fixed at this stage, cutting and fixing noggings between inner and outer rafters to form gable ladders as shown in Fig. 9.13.

To prepare the rafter feet for the eaves, the exact projection from the face of the wall should be measured at each end of the roof and a line pulled through tightly, ensuring the plumb cut on the feet of all the rafters is exactly in line. A short plumb rule or sliding bevel should be used to mark each plumb cut from the line.

Next fit the soffit bearers, again using the line to ensure a straight soffit (the soffit bearers should be slightly below the level cut on the rafter feet), and fix the soffit boarding, allowing this to project beyond the rafter foot to fit into the groove in the back of the fascia board. Heading joints in the soffit should be made on a soffit bearer.

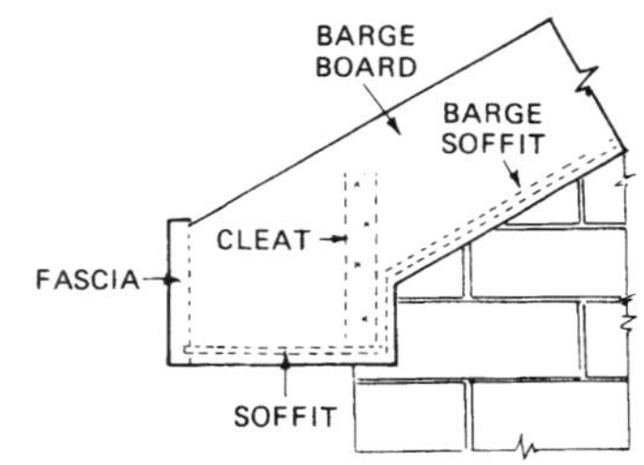

Fig. 9.24. *Boxing-in the foot of the barge boards.*

All that now remains to be done is the cutting and fixing of the barge boards, verge soffit below the barge boards and the fascia board. If the verge soffit is tongued into the back of the barge boards, it is well to cut and fix this first, nailing it to the noggings already described and making a neat joint where it connects with the eaves soffit as shown in Fig. 9.24. The barge boards can now be cut and shaped – mitring the intersection at the top of the gable and mitring and shaping the foot as shown in Fig. 9.13.

Fix the barge boards to the outer rafters with 63 mm oval wire nails. Finally cut and fit the fascia boards, making a neat mitred joint at the intersection with the barge boards, and fix them securely to the rafter feet with 63 mm oval wire nails. Where heading joints occur because of the length of the fascia, these should be splayed, the joint being made on one of the rafter feet.

FURTHER READING

British Standards and Codes of Practice

CP 112	Parts 2 and 3 The structural use of timber
BS 565:1972	Glossary of terms relating to timber and woodwork
BS 5803	Thermal insulation for pitched roof spaces in dwellings Part 1:1979 Specification for man-made mineral fibre thermal insulation mats

Building Research Establishment Digests

No.156	Specifying timber
No.180	Condensation in roofs
No.181	Timber – efficient and economic use
No.201	Wood preservatives – application methods

SELF-TESTING QUESTIONS

All the information required to answer the following questions is contained within this chapter. Attempt each section *as fully or as briefly* as the question demands, and then check your answers against the information given in the chapter.

1.*(a)* State the main functions of a roof.

(b) List six different roof covering materials and state the recommended pitch for each.

(c) Describe the forces which may act upon a roof structure and which have to be taken into account at the design stage.

(d) Explain what is meant by "triangulation" in roof design. State the importance of triangulation and give an example of its application.

2. *(a)* Make rule-assisted sketches to illustrate the general construction of the following types of roof: *(i)* lean-to; *(ii)* couple; *(iii)* collar-tie; *(iv)* couple close.

(b) Define clearly what is meant by the term "single roof".

3. Sketch details to show the following:

(a) the fixing of a rafter to the ridge;

(b) the proportion of a birdsmouth joint;

(c) the joining of a wall plate in its length;

(d) the function of "hangers" and ceiling beams or "binders";

(e) two methods of joining a collar to the rafters.

4. Make rule-assisted sketches to show the general construction of: *(a)* open eaves; *(b)* closed eaves; *(c)* flush eaves; *(d)* sprocketed eaves.

5. *(a)* Sketch details to show the "trimming" required where a chimney stack passes through a timber roof structure.

(b) Explain briefly the Building Regulations which govern the roof trimming around a chimney serving a Class 1 appliance.

6. Explain by means of diagrams the following terms applied to roofs: *(a)* effective span; *(b)* run; *(c)* rise; *(d)* pitch; *(e)* backing line.

7. *(a)* Determine either by calculation or scale drawing the rafter length of a gable roof which has a span of 5.500 m and a pitch of one third.

(b) Set out to a scale of 1:5 a pattern rafter for the roof in 7*(a)*, taking into account the following information:

(i) Rafters – 100 × 50 mm;

(ii) Ridge – 150 × 25 mm;

(iii) Width of eaves – 500 mm, measured horizontally from the outer face of the wall plate.

8. List the sequence of operations in the following stages of constructing a couple close roof of about 3.500 m span (assume there is no trimming or barge boards required);

(a) determination of lengths and bevels;

(b) setting out and cutting the rafters;

(c) erection of the rafters;

(d) fixing a fascia board for flush eaves.

10. First and Second Fixing

After completing this chapter the student should be able to:

1. List under separate headings the activities generally referred to as first and second fixings.
2. List the tools commonly employed for first and second fixing.
3. Sketch, name and describe the use of the various types of patent fixing devices available.
4. List the sequence of operations involved in the fixing of door linings, skirting boards, grounds and window boards.
5. List the sequence of operations involved in fixing door stops, architraves, skirting boards (second fixing) and shelves.
6. List the tools required and the sequence of operations involved in fixing latches, locks and postal plates.
7. Describe the function and fitting of a threshold to an internal door.

Among the many aspects of work with which the site carpenter is concerned, one of the most important is the fixing inside a building of the various pieces of joinery and finishings such as doors, door linings, door furniture, skirting boards, architraves and shelving. This work is often referred to as the first and second fixing – first fixing embracing work carried out *before* the building has been plastered, and second fixing that which is carried out *afterwards*.

In both first and second fixing, the main points for consideration by the craftsman are as follows.

(a) fixings, whether by means of nails, screws or patent fixing devices, should be firm and secure.

(b) Methods of fixing should be neat and unobtrusive so as not to spoil or detract from the appearance of the finished work.

(c) Considerable care must be taken to ensure that the members are fixed plumb, level, square and straight as the case may be.

(d) When fixing members to unplastered walls, allowance must usually be made for the thickness of the plaster which is to be eventually applied. This is normally in the region of 12-15 mm. Since the plasterer will almost invariably use the carpenter's first fixing as a guide to his plasterwork, it follows that any inaccuracy here is likely to lead to similar errors in the plastering.

FIXING TOOLS AND DEVICES

Plugging chisel

This tool, shown in Fig. 10.1, is used to chop out mortar joints in the brickwork

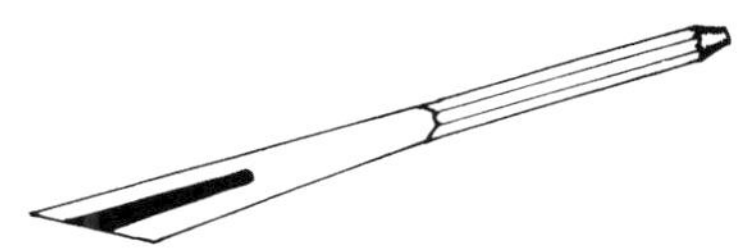

Fig. 10.1. *Plugging chisel.*

ready to receive a wooden plug. The plugs are shaped as shown in Fig. 10.2 so that they

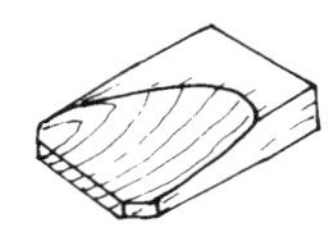

Fig. 10.2. *"Chopped plug".*

wedge very tightly when driven into the chopped out mortar joints. They are then sawn off to a predetermined length, often level with the finished plaster line as shown

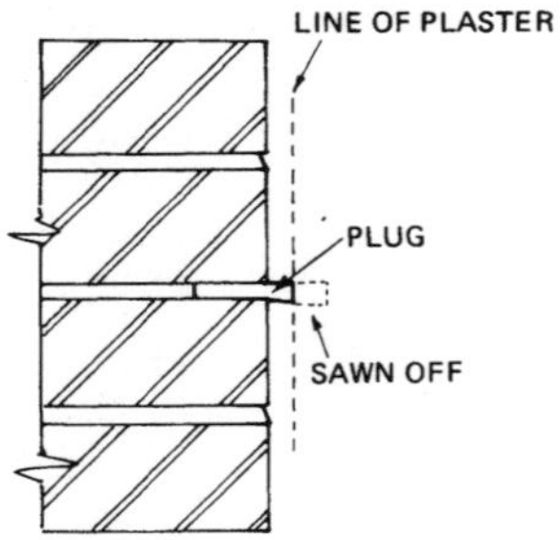

Fig. 10.3. *Trimming off a wooden plug.*

in Fig. 10.3. Fixings into this type of plug are usually made with nails which should be dovetailed slightly, as shown in Fig. 10.11, to give increased holding power.

Carpenters axe

This tool, shown in Fig. 10.4, is used mainly

Fig. 10.4. *Carpenters axe.*

to chop the wooden plugs shown in Fig. 10.2 but also has many other uses. A sharp axe in skilled hands is a valuable tool for the rapid removal of waste material and many other forms of rough trimming.

Star drills

Shown in Fig. 10.5, these are percussion tools used to cut round holes in brickwork and masonry to receive either round wooden plugs or one of the patent fixing devices shown in Fig. 10.9. Star drills are available

Fig. 10.5. *Star-drill.*

in a range of sizes from 9 mm to 25 mm in diameter.

A small but very useful version of the star drill is shown in Fig. 10.6, this tool being generally known as a jumper. Jumpers comprise of a handle into which any one of the various size drill bits supplied can be fitted. In use the jumper is driven into the brick,

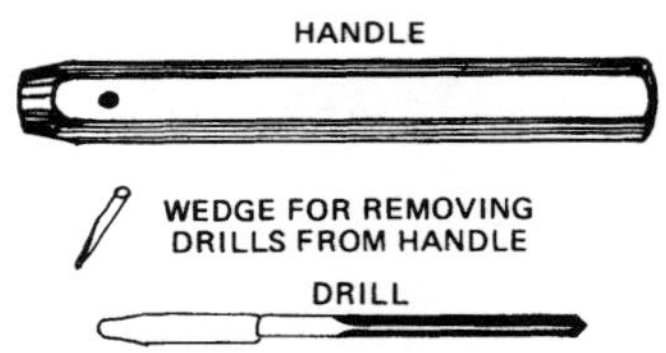

Fig. 10.6. *Rawlplug "jumper".*

stone or concrete with rapid hammer blows while it is turned continuously to cut a hole into which a wood, fibre, metal or plastic plug can be inserted. The small wedge shown in the figure is used for removing the drill bits from the handle.

Masonry drills

These extremely useful items of equipment shown in Fig. 10.7 are available in a wide range of sizes (diameter and length) and have tungsten carbide inserts in the cutting edges, enabling them to be used to drill holes in brick, stone and concrete. Normally used in a slow speed electric or pneumatic drill they facilitate the cutting of holes in the fabric of a building with an absolute minimum of vibration and damage, and are thus the

Fig. 10.7. *Masonry drill.*

obvious choice for obtaining a fixing in a plastered, decorated wall. Certain types of masonry drill are especially robust for use in rotary/percussion drills, which are a decided advantage when drilling into very tough materials.

Masonry nails

These are specially toughened and tempered nails which can be driven directly into relatively soft brick or stone with a hammer, thus dispensing with the need for drilling and plugging. Masonry nails of this type are useful where no great stress is likely to be applied to the fixture, but need to be used with some caution where loadings are considerable. They are less effective on very hard, tough backgrounds.

Ballistic fixing tools

Tools of this type employ explosive cartridges to drive specially toughened nails or bolts into brick, stone, concrete and even steel backgrounds, and are thus extremely powerful and useful tools. Figure 10.8 shows

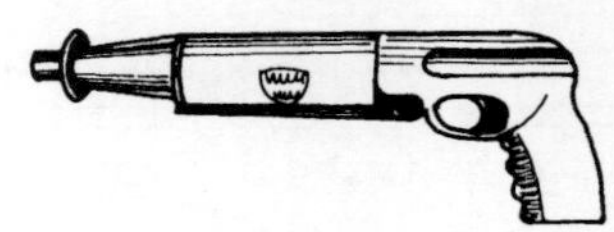

Fig. 10.8. *Ballistic fixing tool.*

a typical type of ballistic tool. In use, the penetration of the nail into the background can be regulated by the strength of the cartridge used, this being denoted by its colour. Table IV shows the colour coding generally recommended for various types of background. To some extent the choice of cartridge is a matter of trial and error, often being determined by a few practice shots.

TABLE IV. COLOUR CODING OF CARTRIDGES

Colour of cartridges	*Strength*
Brown	Extra low
Green	Low
Yellow	Low/medium
Blue	Medium
Red	Medium/high
White	High
Black	Extra high

The modern types of ballistic fixing tool employ an adjustable gas escape in the breech mechanism, enabling a single cartridge type to be used over a whole range of possible situations. Ballistic fixing tools provide a quick, easy and secure means of fixing into tough backgrounds but are costly in use and are therefore best suited where more conventional types of fixing would prove difficult or laborious. Some care is needed to avoid undue damage to the timbers being fixed: this can be reduced by interposing a scrap piece of thin plywood or hardboard between the barrel and the work to absorb powder burns.

IMPORTANT NOTE: It should be constantly borne in mind that ballistic tools can be extremely dangerous if used improperly or without due care – not only to the operator but to any other persons in the vicinity. They must only be used therefore by either a competent person or under his supervision. The competent person should ascertain that neither he nor other operators suffer from colour blindness, as this could adversely affect the choice of cartridge used. In the interests of safety, persons using ballistic fixing tools are advised to wear a safety helmet, eye shields and ear muffs, and to ensure that cartridges are stored in a cool, dry place – under lock and key.

Patent fixing devices

Figure 10.9 shows a selection of some of the various fixing devices obtainable. Their uses are described below.

Fig. 10.9. *Patent fixing devices.*

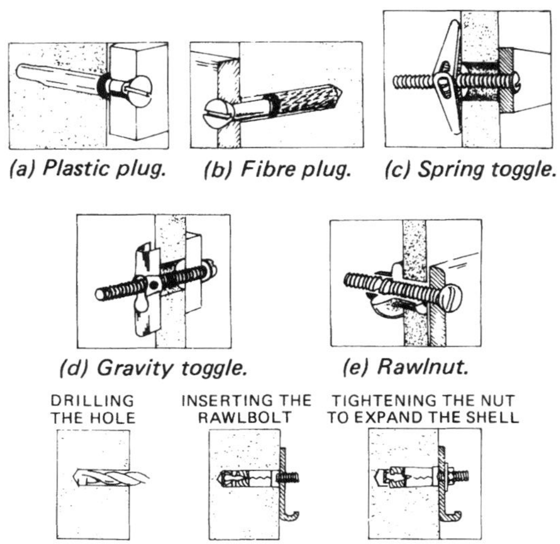

(a) Plastic plug. *(b) Fibre plug.* *(c) Spring toggle.*

(d) Gravity toggle. *(e) Rawlnut.*

(f) Rawlbolt – showing stages in use.

Plastic and fibre plugs
(*See* Fig. 10.9*(a)* and *(b)*.) These are inserted into holes drilled into the background and are intended for use with screws. As the screw is driven into the plug, this is expanded within its housing, gripping tightly and providing a good secure fixing for light to medium weight use. Available to suit screws from gauge 6 to 16.

Spring and gravity toggles
(*See* Fig. 10.9*(c)* and *(d)*.) These are very useful for obtaining fixings in pot bricks, hollow walls, plaster boards, etc. The device is inserted through a hole drilled in the skin. It then opens up, either by means of a spring or by gravity, to give a very secure fixing for light to medium weight purposes.

Rawlnuts
(*See* Fig. 10.9*(e)*.) These are used for fixing to relatively thin membranes such as asbestos cement or metal sheet. Inserted through a hole of suitable size, the tightening of the screw expands the rubber grommet causing it to grip tightly within its housing. Rawlnuts are especially useful where a watertight fit is required.

Rawlbolts
(*See* Fig. 10.9*(f)*.) These are available in small to quite large diameters and are used for heavy duty fixings into brick, stone, concrete, etc. Inserted into a pre-drilled hole of correct diameter, the device is spread by the tightening action of the nut (or bolt head) and gives an immediate and extremely powerful grip.

FIRST FIXING

As previously explained, this term refers to the fixing of frames, linings, grounds, etc., prior to the internal plastering.

Door linings or casings

These lightweight internal frames are fixed into position by nailing through the jambs into pads, plugs or fixing blocks in the walls.

Fixing to blocks and pads

Where fixing blocks or pads are utilised, these will have been built into the door opening

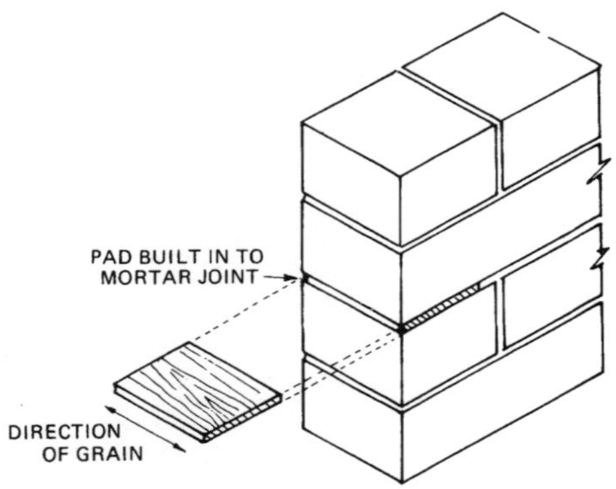

Fig. 10.10. *Pad for fixing of door and window frames.*

(by the bricklayer) as shown in Fig. 10.10 – three or four being required for each jamb.

Invariably the opening will have been built with 20 mm or so clearance so that the lining fits loosely into the opening with room for adjustment when plumbing and aligning with the walls. Since the opening is oversize, it is necessary to insert packings between the jambs and the brickwork as shown in Fig. 10.11. This packing must be

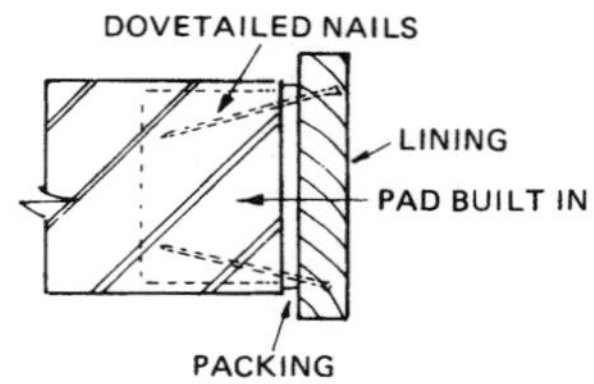

Fig. 10.11. *Fixing into a pad.*

carefully done to ensure the lining jambs are both plumb and straight. Normal practice is to fix one jamb first, making sure it is plumb, straight and correctly positioned with respect to the line of the eventual plaster, and then fix the second jamb, sighting through the frame to ensure that it is exactly parallel to the first. Linings of this type are normally fixed with 100 mm oval wire or cut clasp nails, although screws may be used occasionally.

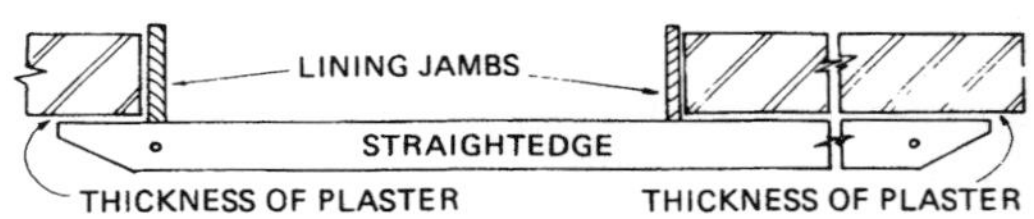

Fig. 10.12. *Lining-up a door lining.*

When fixing a door lining to an unplastered wall, care must be taken to ensure that the jambs are accurately lined up with the wall as shown in Fig. 10.12. This should be tested with a straightedge during the fixing as illustrated.

Fixing to plugs

Where there are no fixing blocks or pads available, the opening will have to be "plugged", using the plugging chisel to chase out four mortar joints on either side. Suitable plugs are chopped from scrap timber and

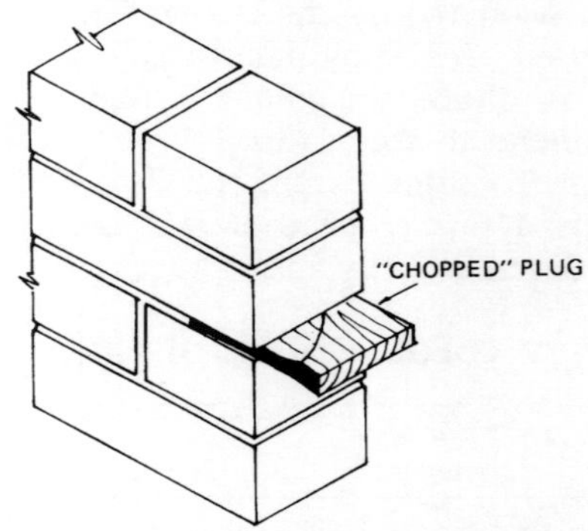

Fig. 10.13. *Sketch showing plug driven into mortar joints.*

driven into the slots as shown in Fig. 10.13. The ends of the plugs can then be sawn off to a plumb line, taking care to keep the sawn ends exactly in line and square to the opening as shown in Fig. 10.14. The lining can now be pushed into the prepared opening, tested for accuracy and fixed with 100 mm oval wire nails. When fixing the lining it is good practice first to make a temporary fixing with a single nail at the top and bottom of each jamb, leaving the nail heads protruding. Then test the frame for accuracy and correct position, and, if all is well, drive in the remaining nails. This method of fixing allows for adjustment to be made easily before the fixing has progressed too far. The carpenter should realise that unless the jambs are exactly plumb and straight, there will be difficulty when hanging the door; a little care at this stage may save considerable time later on.

NOTE: Braces and stretchers should not be removed from the lining until the fixing is complete.

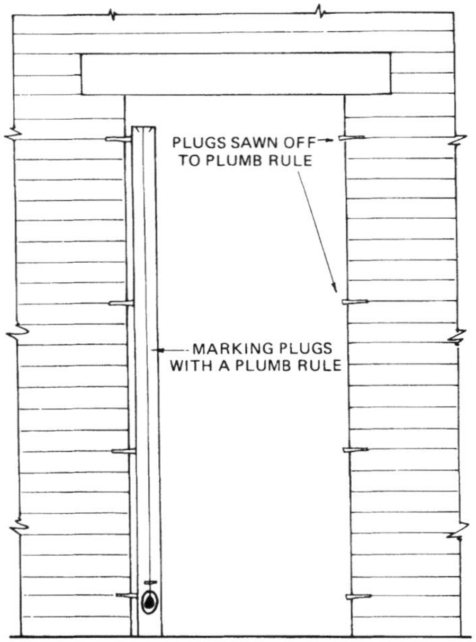

Fig. 10.14. *Plugging the joints for fixing a door lining.*

Fig. 10.15. *Fixing skirting boards.*

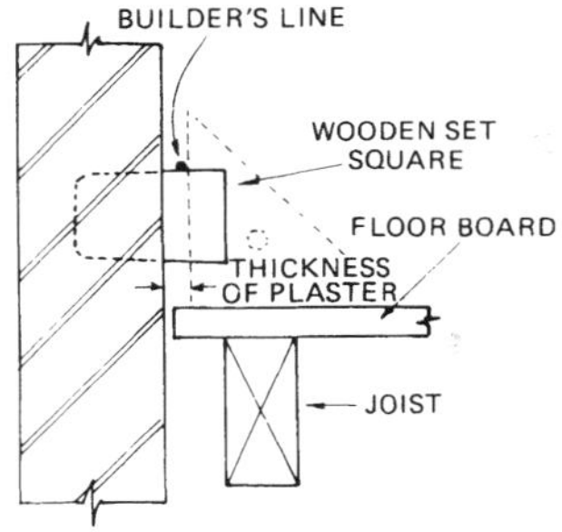

(a) Trimming plugs.

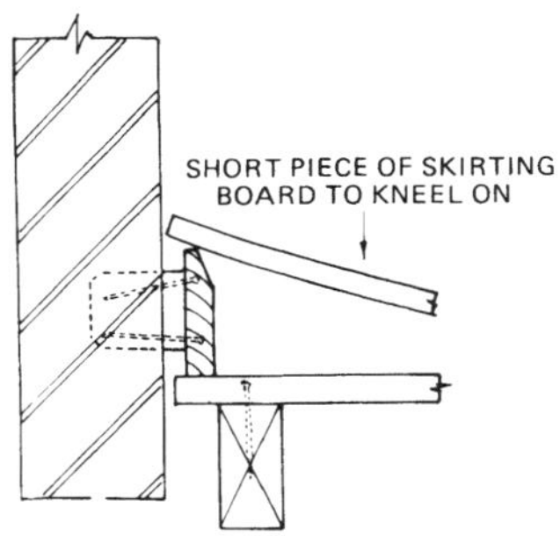

(b) Fixing skirting boards.

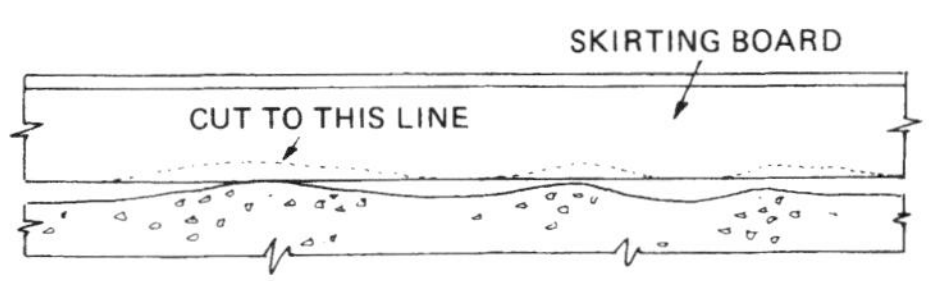

Fig. 10.16. *Scribing skirting board to an uneven floor.*

Fig. 10.17. *Corner joints on skirtings.*

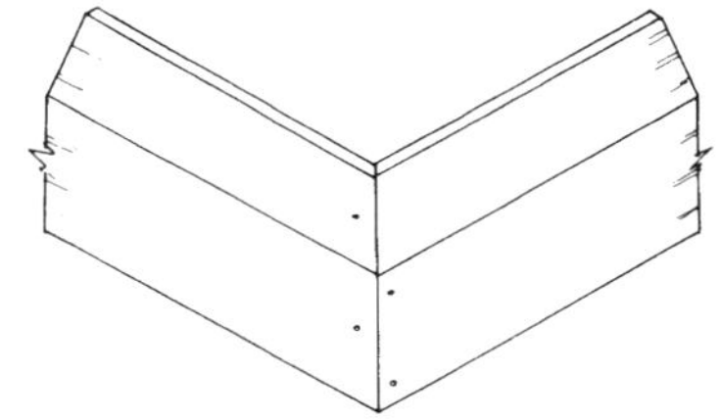

(a) External corner – mitred and pinned.

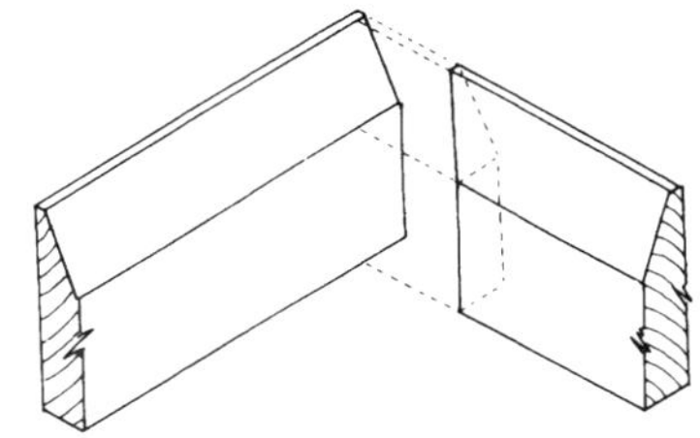

(b) Internal corner – scribed.

Skirting boards (first fixing)

When the fitting of skirting boards is done as part of the first fixing, the walls must first be plugged all round at intervals of 2-3 bricks.

The plugs, driven in tightly, are marked off to exact length using a line and try square to ensure the cut ends are plumb and in line as shown in Fig. 10.15*(a)*. The plugs must be sawn off to leave a projection equal to the thickness of the plaster to be applied.

Skirting boards are fixed by driving two 60-75 mm oval wire nails into each plug as shown in Fig. 10.15*(b)*. Note the use of the short piece of floorboard or similar material which the carpenter kneels on during the fixing. This ensures the skirting board fits tightly to the floor. If the floor is very uneven, as may happen on occasions, the bottom edge of the skirting board must be scribed to fit, using a rip saw, plane or axe to cut away the waste as shown in Fig. 10.16.

Corner joints

External corner joints on a skirting board are plain mitres, cut with the panel saw using a mitre box to obtain the correct angle. The mitre is nailed as shown in Fig. 10.17*(a)*.

Internal joints should be formed by scribing (whenever possible) as shown in Fig. 10.17*(b)*, a scribed joint being easy to cut and satisfactory in service since it is less likely to open up due to shrinkage than a mitred joint.

Provided that the skirting boards are nicely plumb on the face – as they should be if the plugs have been accurately sawn off – scribing can be done by simply cutting a mitre across the moulded top edge of the skirting and squaring this down to the floor line as

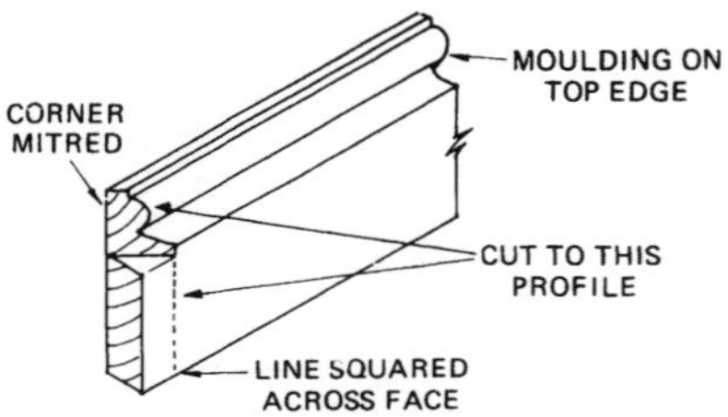

Fig. 10.18. *Scribing a skirting board.*

shown in Fig. 10.18. It is then a simple matter to cut to the profile so obtained using a coping saw for the moulded portion and a panel saw for the straight.

When fixing skirting boards in a normal room, it is usual to start near a doorway and work round the room and thus back to the

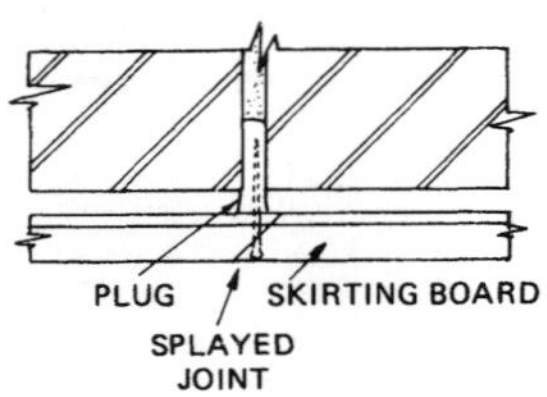

Fig. 10.19. *Splayed heading joint for skirting board (plan).*

doorway. In this way there will normally be only one joint to cut on each piece of skirting board. On long walls it will usually be necessary to joint the skirting boards in their length, this being done by the use of a splayed joint – over a plug – as shown in Fig. 10.19. At doorways the ends of the skirtings should be trimmed off square, allowance being made for the width of the architrave and its margin as shown in Fig. 10.20.

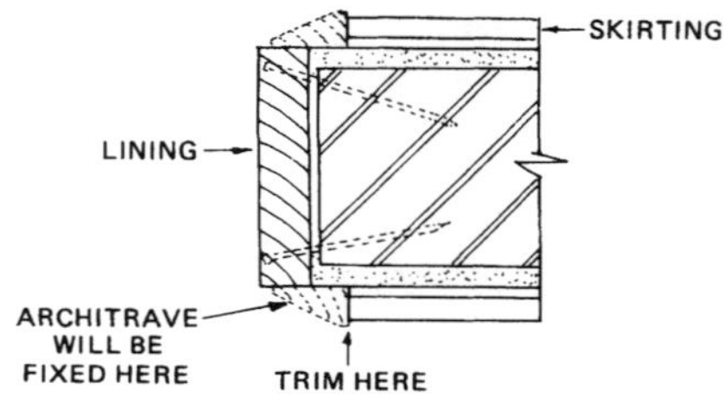

Fig. 10.20. *Horizontal section through door lining showing finish of skirtings.*

Use of grounds

In good class work, and especially where hardwood skirting boards are concerned, grounds may be fixed to the wall as shown in Fig. 10.21. Grounds must be fixed carefully, keeping them straight and plumb with their face flush with the plaster line. Grounds are fixed by nailing into wooden plugs in a similar manner to fixing skirting boards. (The plugs must be cut off flush – or nearly so – with the face of the wall.) The purpose of the ground is threefold:

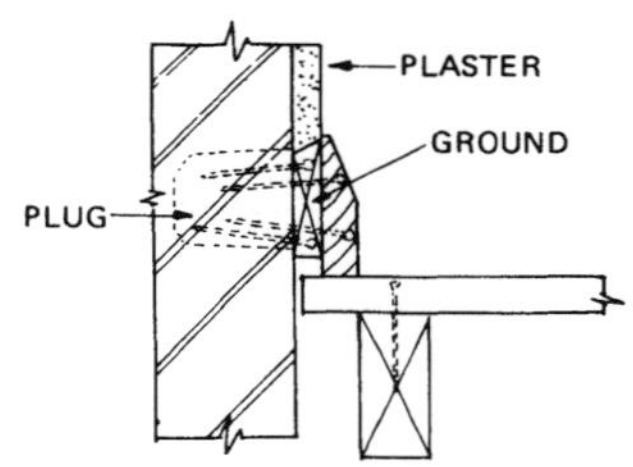

Fig. 10.21. *Use of grounds for skirting boards.*

(a) to provide a line for the plasterer to work to;

(b) to provide a continuous fixing – where necessary – for the skirting board;

(c) to enable the skirting board to be fixed as a second fixing, thus avoiding any possible damage during plastering.

The top edge of the ground is normally bevelled as shown in Fig. 10.21. This gives a better joint between the ground and the plaster. The ground should be so positioned that its top edge is cloaked by the skirting board when this is eventually fixed.

Window boards

Window boards are used to provide a neat finish to the inner skin of brick or blockwork on the inside of the window cill. Since they are usually fixed before the plastering is carried out, they also form part of the first fixing.

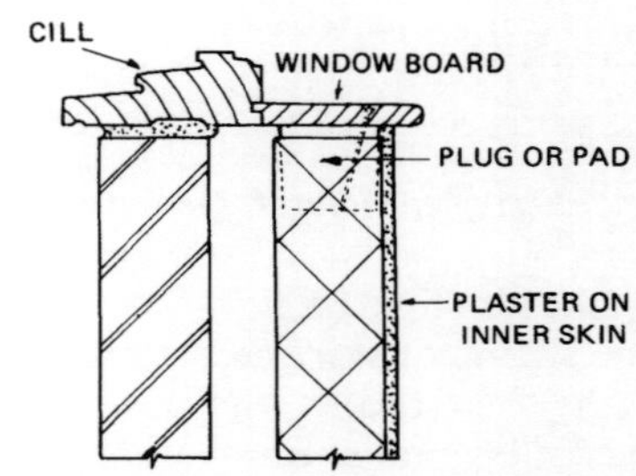

Fig. 10.22. *Section through cill and window board.*

A typical window board is shown in section in Fig. 10.22, and as can be seen it is nailed to plugs driven into the inner skin of masonry. The back edge of the window board is tongued into the cill, allowing a certain amount of movement in the timber without a gap occurring. This is an important feature of window board design since many modern buildings equipped with central heating have radiators fixed below the windows. The window board is therefore subjected to a constant flow of hot dry air which dries the moisture from the window board, tending to cause shrinkage.

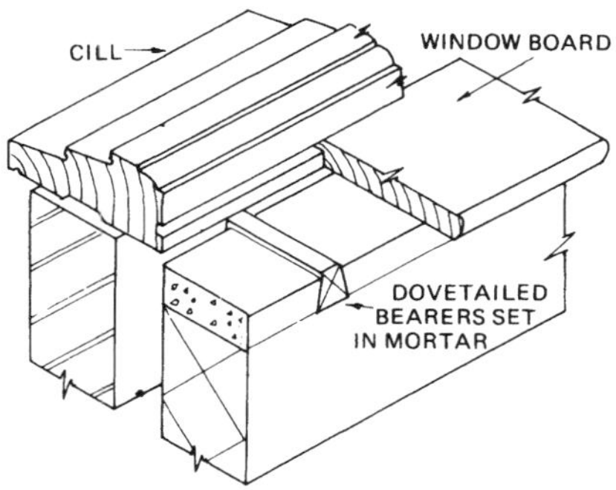

Fig. 10.23. *Method of fixing window board.*

Window boards may also be fixed to dovetailed blocks or bearers bedded on to the inner skin as shown in Fig. 10.23. This method is very suitable where wide or hardwood window boards are to be fixed by slot screwing (*see* Chapter 3) to allow the board to move marginally. The front edge of the

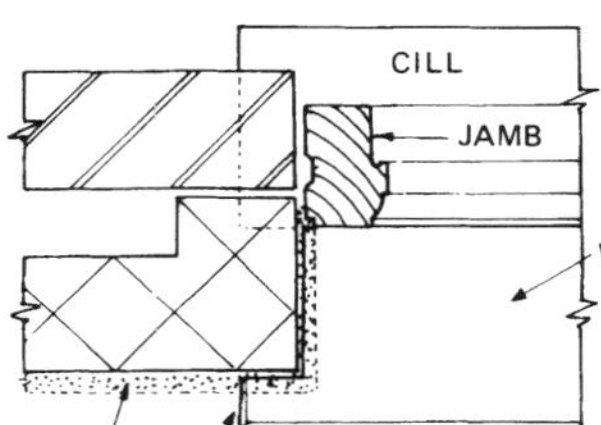

Fig. 10.24. *Horizontal section through jamb, showing returned end of window board.*

window board is "nosed", as are the returned ends which should be cut to run past the brick jambs as shown in Fig. 10.24.

SECOND FIXING

As stated previously, this term refers to the fixings and finishings carried out after plastering is completed.

Architraves and door stops

Architraves are used to cover the joint between the plaster and the frame on internal door linings as shown in Fig. 10.25, and are

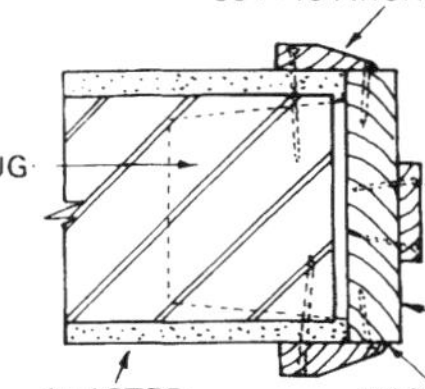

Fig. 10.25. *Horizontal section through jamb of door lining showing finishings.*

best fixed after the door has been hung (*see* Chapter 5) as otherwise they tend to get in the way during this operation. Architraves are nailed to the edges of the lining and to the ends of the skirting boards. The top corners should be neatly mitred and pinned as shown in Fig. 10.26. On occasions, the mitres are cut in the workshop by machine, and the architraves delivered to the site in sets ready for fixing. When fixing architraves, care should be taken to keep the margins parallel – an error of one or two millimetres here being very obvious.

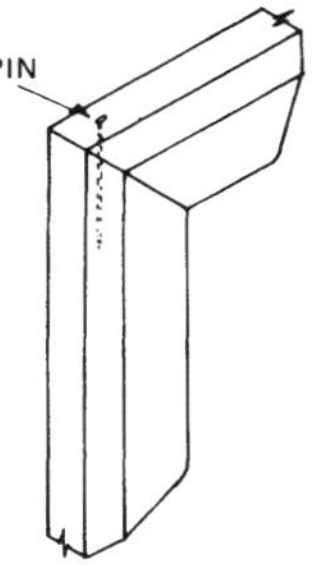

Fig. 10.26. *Mitred corner of architrave.*

Planted door stops are nailed to the inside face of the door lining after the door has been hung, that on the head being cut and fixed first and those on the jambs last. The nails – usually 38 mm oval wire – should be staggered and slightly dovetailed. It will be found most convenient to fit the door stops after having fitted the lock or latch as this holds the door closed whilst they are positioned and fixed. Door stops should not be fitted too tightly against the face of the door – allowance must be made here for the thickness of several coats of paint which will eventually be applied, otherwise the door may become "hinge-bound".

Skirting boards (second fixing)

Where the fixing of skirting boards is carried out as part of the second fixing, there are several possible variations in procedure, the method adopted depending on the quality of the work and also upon the type of background involved.

Fixing to brick or block

In this case, the wall will have already been plugged during first fixing, the ends of the plugs having been sawn off to the line of the finished plaster (or possibly just below). In any case, the position of the plugs should be marked on the floor so they may be easily found when nailing back the skirting boards to the plastered wall. The procedure for cutting, fitting and fixing the skirting is exactly as previously described under first fixing, with perhaps one slight difference — the skirting boards in this instance must be cut to dead length between plaster angles; they cannot run right up to the brick or block wall as they do in the former instance.

Whilst this method of fixing is quite common and reasonably satisfactory, the finished skirting boards are rarely as plumb, straight, and accurately fitted to the wall as when grounds are used.

Fixing to grounds

The provision of grounds makes for an extremely effective and superior way of fixing the skirting boards. It is also, of course, considerably more costly due to the extra work and materials involved and can probably only be justified on the best quality work or with a hardwood skirting. The fixing and purpose of the grounds has already been described.

When fixing skirting boards to grounds, the nails should be driven through the skirting and ground and into the plugs. Short pieces of skirting and returned ends may be fixed directly to the grounds. It will be obvious to the reader that the use of grounds renders the fitting and fixing of skirting boards a relatively simple and straightforward operation.

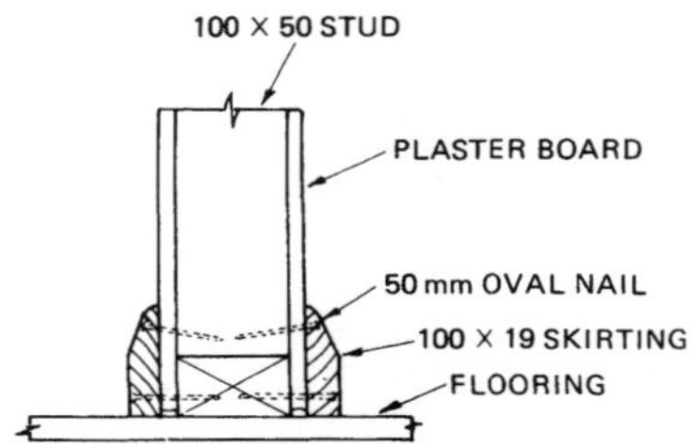

Fig. 10.27. *Fixing skirting to timber studding.*

Fixing to timber studding

Internal, non load-bearing walls are often built in the form of stud partitions, covered on either side with plaster board sheets. The plaster boards are skimmed with plaster about 2 mm thick or left unplastered and finished with a dry lining technique. Whichever is the case, the skirtings are fitted after the plaster boarding has been fixed and finished.

Fixing to timber studding is a relatively simple operation since no plugging is necessary, the skirting being nailed into the cill or vertical studs as shown in Fig. 10.27. Internal and external corners are formed in exactly the same way as in the previous case.

Shelves

Shelves are a simple but important aspect of the second fixer's work and may be either solid or slatted as shown in Fig. 10.28. Solid shelving up to about 225 mm wide by 25 mm thick is the more common type for general storage purposes. Slatted shelves are used mainly in airing cupboards or in commercial store-rooms and warehouses, either to encourage air circulation or simply for reasons of economy.

Fig. 10.28. *Types of shelving.*

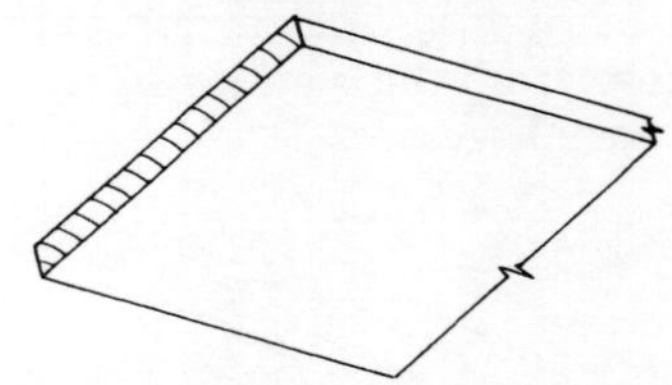

(a) Solid.

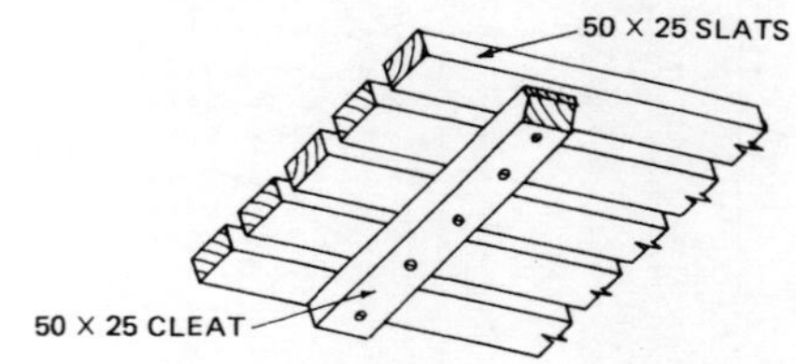

(b) Slatted.

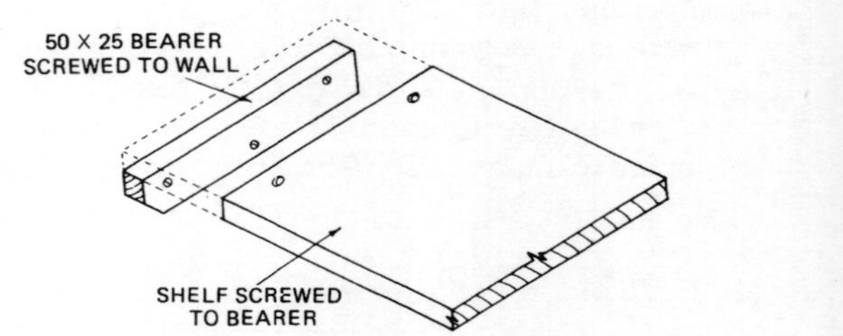

Fig. 10.29. *Supporting the end of a shelf.*

Shelves are generally supported by means of shelf brackets, but where the end of a shelf abuts a wall, a cleat as shown in Fig. 10.29 can be used. Long shelves obviously

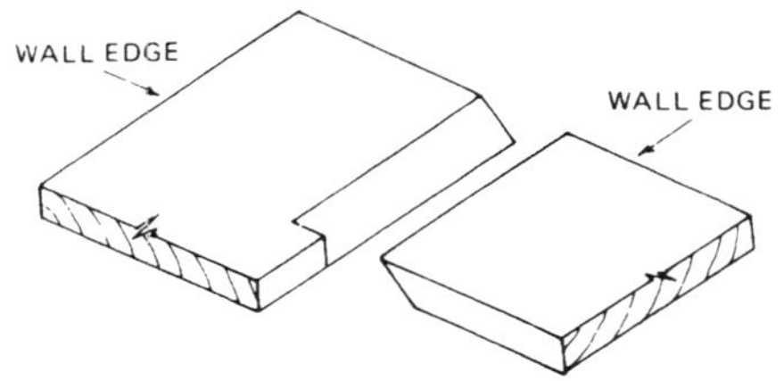

Fig. 10.30. *Ship-lap joint for internal junction between shelves.*

require more shelf brackets than do shorter ones, the brackets being spaced at about one metre centres, although this can vary depending upon the load the shelf has to carry.

Figure 10.30 shows a form of ship-lap joint which is used to join shelves at an internal angle.

Fixing the shelf brackets

Where single shelves are concerned, it will generally be found best to screw the shelf brackets directly to the wall using a suitable sized masonry drill.

Where multiple shelves are required, however, it will be found more convenient to fix vertical battens to the wall and screw the shelf brackets to these as shown in Fig. 10.31. This considerably reduces the time spent drilling and plugging the wall for individual brackets. Note that where this method is adopted it is necessary to cut a recess in the back of each shelf to fit round the batten (this is not required on the top shelf which normally extends over the top of the batten). The use of wall battens may be essential where the shelves are to be fixed to a stud partition as it is unlikely that the position of the studs will coincide with the ideal position for the brackets.

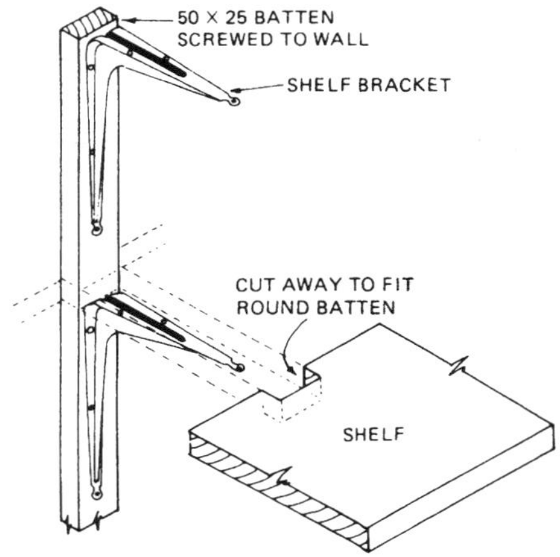

Fig. 10.31. *Use of shelf brackets.*

Points to note

Whilst the fixing of shelving is a relatively simple undertaking, the following points are worthy of note.

(*a*) Careful setting out is required! A level line should be marked around the walls indicating the top surface of each shelf.

(*b*) Brackets/wall battens should be carefully plumbed and spaced.

(*c*) The bearers and battens should be cut, drilled, countersunk and fixed in sequence so as to avoid undue waste of time through repetition.

(*d*) Shelf brackets should be fixed to the underside of the shelves before screwing to the wall – it is difficult and often dangerous to attempt to fasten shelves to brackets from below.

(*e*) A long bladed screwdriver is a decided advantage when fixing wide shelves as this will give a firmer grip in the screw slot and will also avoid wear and tear on the knuckles.

Fitting door furniture

This aspect of second fixing always requires a certain amount of care and precision to ensure a neat appearance and smooth, correct functioning of the ironmongery involved.

Mortice locks and latches

The first step is to open the door to an angle which gives good access to the edge to be worked on, and hold it in this position with a small wedge driven under the bottom edge. The mortice for the lock or latch can now be marked out on the edge of the door, and cut out using a brace and bit to remove as much of the waste as possible as shown in Fig. 10.32(*a*). The mortice is then cleaned out with a mallet and chisel until the lock or latch can be inserted easily, right up to the face plate. A fine line is then scribed around the edge of the face plate, and the recess carefully cut out using a mallet and sharp chisel as shown in Fig. 10.32(*b*). (It will be found helpful to mark the vertical lines for the face plate with a cutting gauge.) The position of the spindle and key hole should now be marked carefully on both faces of the door, allowance being made for any "lead" planed on the door edge (*see* Fig. 5.26). These can be cut out using a brace and bit, chisel and key hole saw and working from both sides of the door. It should now be possible to insert the lock/latch and examine the spindle/key holes for correct alignment. If all is well, the face plate can be secured with the two screws supplied.

Fig. 10.32. *Fitting mortice locks and latches.*

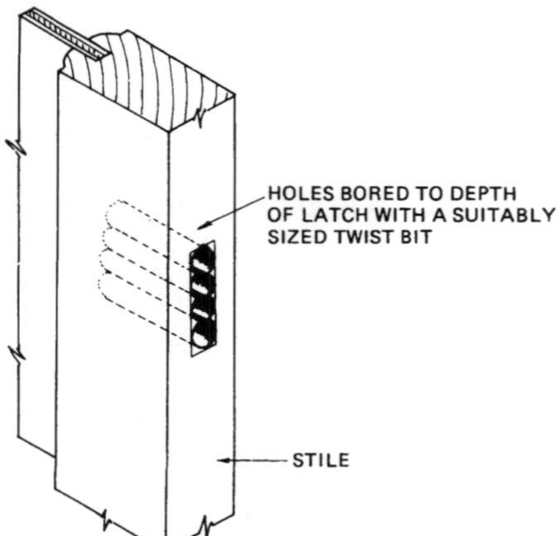

(a) Sketch showing quick removal of waste when fitting a mortice lock/latch.

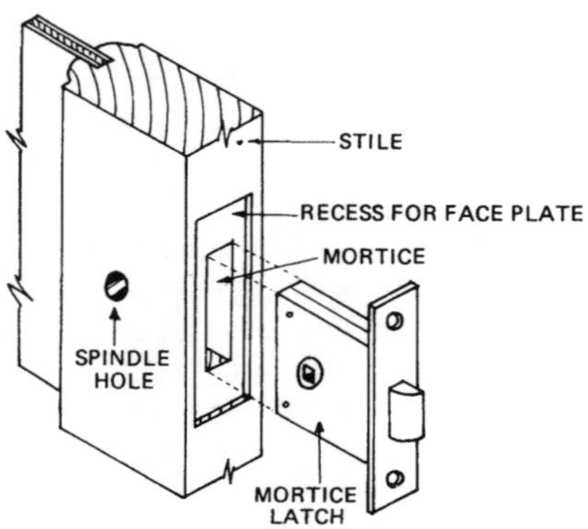

(b) Mortice and recess completed ready for fixing latch.

NOTE: It may be necessary to reverse the latch to suit the "hand" of the door before the lock/latch is fixed permanently.

To fit the handles, a spindle should be inserted through the door and the handles pushed on to this when the screw holes can be started with a bradawl. When doing this, make sure that the handle is accurately centred, testing for free play from side to side and up and down, and setting the handle to a mid-position. The handle faceplates can then be screwed to the door using the screws provided, taking care not to damage the slots in the process. Screw slots should be left vertical for the sake of neatness.

To fit the keeper, rub the tip of the latch with a soft pencil and close the door a time or two, to mark its position on the jamb. Hold the keeper in position, scribe round it with a sharp pencil and cut out the recess to the required depth. Replace the keeper and mark the position of the mortices for the snib and bolt. Chop these out with a mallet and chisel and then screw the keeper into position. Finally, test the door for correct and easy self-closing action of the lock/latch, making sure the handles and key work easily and smoothly from both sides when the door is closed.

Points to note

(a) Whilst the lock/latch should hold the door firmly closed without any obvious sloppiness, it should not hold the door too tightly against the stop. Should the latter be the case, the keeper must be removed and eased on the inside of the slot with a fine file until the closing action is correct. Failure to do this will make the handles and key stiff to turn – a situation which will be aggravated when the door and frame are painted.

(b) Horizontal mortice locks (*see* Chapter 5) usually penetrate some distance into the end grain of the lock rail. A mortice lock chisel will be required to cut the mortice in such cases. Also the key hole for this type of lock will require a separate escutcheon as shown in Fig. 10.33.

Fig. 10.33. *Keyhole escutcheon.*

(c) Whilst a spindle hole can reasonably be cut oversize to give plenty of clearance, key holes should not be too sloppy, otherwise insertion of the key into the lock will be an annoying matter of hit or miss.

(d) For the sake of neatness and convenience in use, take care to ensure that all handles are placed at the same hand height, as far as is possible, throughout the building. (Position of locks, postal plates, hinges, etc., are given in BS 459.)

Postal plate

The fitting of a postal plate requires the cutting of a fair sized hole through the rail of the door. This should be marked out carefully, allowing a few millimetres clearance all round the flap, and cut out using a brace and bit and a pad saw (*see* Fig. 10.34). The rounded corners of the hole will need to be cleaned out with a sharp chisel, and the completed hole should be cleaned up with a piece of glasspaper wrapped around a small flat stick. Mark the holes for the fixing screws and bore these out with a twist bit of appropriate size. Having fixed the postal plate to the door, it is advisable to stand back a metre or two and "eye it in". An error of a millimetre or so here is very obvious.

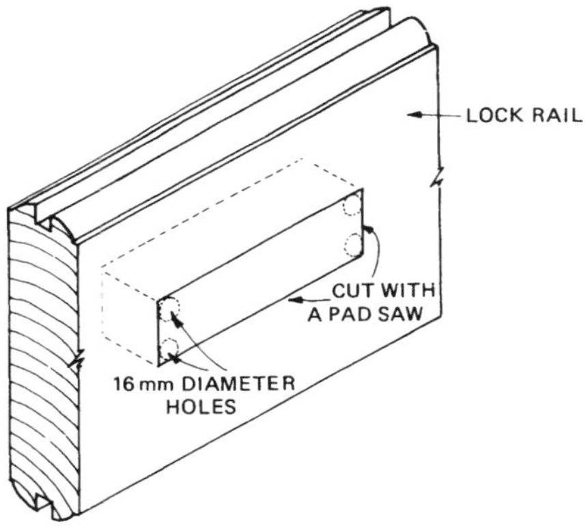

Fig. 10.34. *Cutting a lock rail to receive a postal plate.*

Thresholds

A hardwood threshold of similar section to that shown in Fig. 10.35*(a)* will often be required where an internal door has to open over a fitted carpet. To fit such a threshold it is necessary to mark carefully the bottom of the door *on both sides* to indicate the amount to be removed. (The door should be fitted to within 2-3 mm of the threshold.) The waste portion must be cut off the door in a manner which will not damage the bottom edge, flush doors, especially veneered types, being particularly prone to damage of this sort. To avoid spoiling the door in this way, it is advisable either to cut from both sides with a tenon saw or to first cut the lines deeply with a marking knife or chisel before using a panel saw at a relatively low angle.

The threshold must be cut into the door opening to fit neatly around the door stop as shown in Fig. 10.35*(b)*, and screwed down with brass countersunk screws.

Fig. 10.35. *Fitting a threshold.*

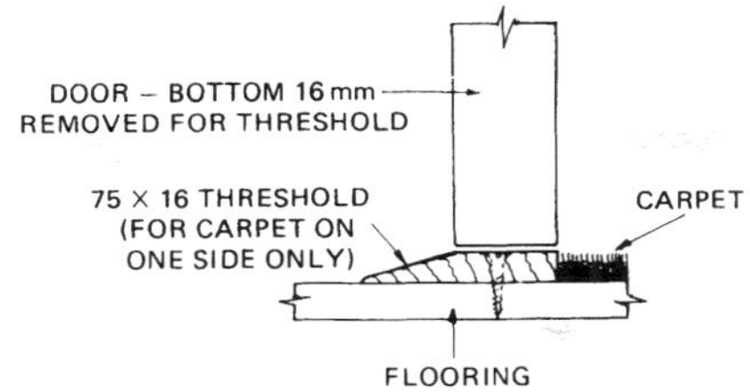

(a) Hardwood threshold fitted under an interior door.

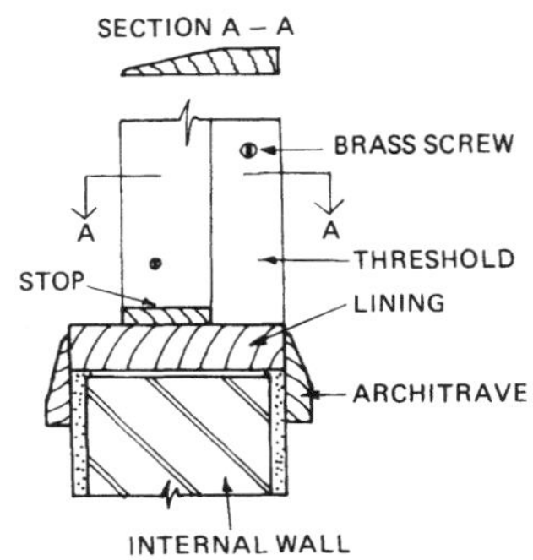

(b) Section through jamb of lining showing fitting of threshold.

FURTHER READING

British Standards and Codes of Practice

CP 151	Doors and windows including frames and linings Part 1:1957 Wooden doors
BS 565:1972	Glossary of terms relating to timber and woodwork
BS 584:1967	Specification for wood trim (softwood)
BS 1186	Specification for quality of timber and workmanship in joinery Part 1:1971 Quality of timber Part 2:1971 Quality of workmanship
BS 1567:1963	Wood door frames and linings.
BS 4787	Internal and external wood doorsets, doors and casement frames Part 1:1980 Specification for dimensional requirements

Building Research Establishment Digests

No. 27	Rising damp in walls
No. 163	Drying out buildings
No. 211 No. 212	Site use of adhesives

SELF-TESTING QUESTIONS

All the information required to answer the following questions is contained within this chapter. Attempt each section *as fully or as briefly* as the question demands, and then check your answers against the information given in the chapter.

1. *(a)* State what is broadly meant by the terms "first" and "second" fixing.

(b) List three activities carried out by a carpenter under the headings: *(i)* first fixing; *(ii)* second fixing.

(c) List the tools which are used by the site carpenter for first and second fixing operations, giving an example for the *use* of each.

2. With reference to ballistic fixing tools state:

(a) the purpose of colour coding of cartridges;

(b) how the cartridges should be stored;

(c) the protective equipment that should be worn by the user of a ballistic fixing tool;

(d) who is allowed to use a ballistic fixing tool;

(e) the advantage of using a ballistic fixing tool that has an adjustable gas escape.

3. List four patent fixing devices, giving an example of where each might be used to advantage.

4. Briefly describe the procedure for fixing an internal door lining making use of: *(a)* fixing blocks or pads; *(b)* chopped wooden plugs.

5. By means of sketches, show the joints used to join skirting boards at: *(a)* an internal corner; *(b)* an external corner; *(c)* mid point of a flank wall.

6. *(a)* Explain the reasons for using "grounds" for fixing purposes.

(b) Show the use of grounds applied to skirting boards.

(c) Show by sketches a method of fixing a window board.

7. *(a)* Show by means of sketches: *(i)* a slatted shelf; *(ii)* a solid shelf. Give an example of where each type would be most suitable.

(b) Sketch the joint used to join solid shelves at an internal corner.

(c) Show by means of sketches a means of supporting a shelf: *(i)* at mid-point along a wall; *(ii)* where it abuts a wall.

8. *(a)* Define with sketches: *(i)* a mortice latch; *(ii)* a keyhole escutcheon.

(b) List the tools required to fit a mortice latch into a door stile.

9. *(a)* State the function of a hardwood threshold fitted under the bottom edge of an internal door.

(b) Sketch a section (approximately full size) through a threshold to a doorway which has a thick carpet fitted closely up to one side.

11. Machine Woodworking

After studying this chapter the student should be able to:

1. List the functions of a woodworking machine.
2. List the "common sense" behavioural requirements of persons using woodworking machines.
3. Explain the "general" requirements (all machines) which are mandatory under the Woodworking Machines Regulations 1974.
4. Identify and state the specific functions of the following machines;

(a) hand fed circular saw;
(b) dimension saw;
(c) surface planer;
(d) thicknesser;
(e) chain and/or hollow square chisel morticer.

5. Sketch and name the guards and safety devices associated with the use of the overhand planer, indicating all critical dimensions.
6. Sketch details and describe the function of the various types of circular saw blade and teeth.
7. Explain with sketches the cutting action of the chain and hollow square chisel morticing machines.
8. Calculate from given data:

(a) the pitch of cutter marks on machine planed timber;
(b) the peripheral speed of a circular saw blade;
(c) the pitch of circular saw teeth.

FUNCTIONS OF A WOODWORKING MACHINE

The natural evolution and development of industry taken along with the ever increasing need for economy of labour has brought about a situation whereby the greater part of all joinery, cabinet making and similar work is nowadays carried out by machine rather than by hand.

Generally speaking, wood machining on this scale is undertaken by the specialist – the wood machinist who is a skilled craftsman in his own right. However, such is the broad nature of the construction industry that greater and greater use is being made of woodworking machinery, both *on site* and in the workshop, and inevitably this leads to the necessity of the hand woodworker to turn towards the machine, either to assist or to actually use it himself. In many of the smaller workshops, the use of elementary machinery is expected as a matter of course, and it is therefore absolutely essential that the woodworker should understand the working of these machines, be fully aware of the dangers which using them introduces, and, most important of all, know how to use the machines in such a way that the risk of injury, either to himself or to others, is reduced to the absolute minimum.

It cannot be too strongly stressed that the proper and correct use of woodworking machinery, far from detracting from a craftsman's skills, actually adds to them. Furthermore, it is true to say that the most efficient and highly skilled operative is the "safe operative" – the worker who takes no risks, understands his job and produces the best work.

The functions of a woodworking machine may be listed as follows:

(a) to increase production;
(b) to decrease costs;
(c) to reduce manual labour;
(d) to make precise and repetitious work easier.

Safety

The following common sense safety precautions should be learned and understood by all who are involved in any way with the use of woodworking machinery.

(a) Guards are required by law on most machines – make sure they are used, fitted securely and positioned correctly.

(b) Keep fingers well away from moving cutters and blades of any description.

(c) Use push sticks, push blocks and jigs to hold short awkward pieces of timber – keep such implements to hand.

(d) Stop the machine before making adjustments; isolate it before carrying out any maintenance or remedial work.

(e) Keep the machine bed or table clear of obstructions such as spanners, spare pieces of timber, etc.

(f) Ensure that all cutters, blades, fences and lock nuts are secure before starting the machine.

(g) Never assume that a machine which has been left for any length of time has not been altered or tampered with. Check before starting.

(h) Avoid wearing loose, flapping clothes – sleeves, neckties, etc. – or confine them in a suitable protective garment. Long hair should be confined under a suitable cap.

(i) Stop the machine when the work is completed. Do not leave a machine running unattended.

(j) Do not abuse a machine with work for which it is not intended or which is beyond its capacity.

(k) Ensure that machinery is well maintained, cutting edges kept sharp and accessories in good condition.

(l) Keep the area around a machine clean and tidy. Littered floors can be very dangerous.

(m) Do not operate any machine without being fully trained and confident in its use, except under expert supervision.

(n) Behave sensibly in a machine workshop – skylarking and practical joking have no place here.

(o) Stop the machine at the first sign of any unusual behaviour and investigate or report the occurrence.

(p) Operatives should be familiar with and *fully understand* the Woodworking Machines Regulations. These must be displayed in the machine shop.

(q) Take no risks or chances with the machinery. Concentrate on the job in hand and do not allow yourself to be distracted – or to distract other operatives.

If the foregoing precautions, many of which are dealt with in greater detail in the following pages, present a formidable and disturbing list of behavioural requirements, take comfort from the fact that these are all "common sense" aspects of the job (most are mandatory), and that many thousands of machine woodworkers carry out their work happily and efficiently with no more risk of personal injury than any other person employed in the construction industry.

WOODWORKING MACHINES REGULATIONS 1974

The following is an abridged and simplified version of the Woodworking Machines Regulations 1974, Part 2, and are general regulations which apply to *all* woodworking machines. Individual regulations applying to specific machines will be outlined and explained as the relevant machine is dealt with.

It should be noted that the Woodworking Machines Regulations 1974 were laid before Parliament on 10th June 1974 and are therefore *legal requirements*. They should be read carefully, and the implications understood fully, by *all* who use woodworking machines.

Guards

The cutters of every woodworking machine shall be enclosed by a guard or guards to the greatest extent that is practicable having regard to the work being done, unless the cutters are in such a position as to be as safe to every person employed as they would be if so enclosed. All guards shall be of substantial construction. No person shall, while the cutters are in motion, make any adjustment to the guard or to any part of the machine unless such an adjustment can be made without danger. At all times when the cutters are in motion, all guards and safety devices shall be kept constantly in position and properly secured and adjusted.

Machine controls

Every woodworking machine shall be provided with an efficient device or devices for starting and stopping the machine and the control of the device or devices shall be in such a position as to be readily operated by the person operating the machine.

Working space

There shall be provided around every woodworking machine sufficient clear and unobtructed space to enable the work being done at the machine to be done without risk of injury to persons employed.

Floors

The floor or surface around every woodworking machine shall be maintained in a good and level condition, free from chips and other loose material, and shall not be allowed to become slippery.

Temperature

In that part of the room or other place (not in the open air) in which a woodworking machine is being used the temperature must not be allowed to fall below 13° Celsius (this applies to the machines described in this chapter). Heating appliances used in the machine shop area must not be such that there is any likelihood of accidental ignition of any material within that area by reason of contact with the heating element or flame. No method of heating shall be employed which results in the escape into the air of any fumes of such a character and to such an extent as to be likely to be injurious or offensive to persons employed in the machine shop area.

Training

No person shall be employed at any kind of work on a woodworking machine unless he has been sufficiently trained at machines of a class to which that machine belongs in the kind of work on which he is employed, except where he works under the adequate supervision of a person who has a thorough knowledge and experience of the working of the machine and of the matters specified in the following paragraphs.

Every person, while being trained to work at a woodworking machine, shall be fully and carefully instructed as to the dangers arising in connection with such machine, the precautions to be observed, the requirements of the regulations which apply and, in the case of a person being trained to operate a woodworking machine, the method of using the guards, devices and appliances required by these regulations.

No person who has not attained the age of 18 years shall operate any circular sawing machine, any sawing machine fitted with a circular blade, and planing machine for surfacing which is not mechanically fed, or any vertical spindle moulding machine, unless he has successfully completed an approved course of training in the operation of such machine, save that where required to do so as part of such a course of training, he may operate such a machine under the adequate supervision of a person who has a thorough knowledge and experience of the working of the machine and of the matters specified in the previous paragraph.

Duties of persons employed

Every person employed shall, while he is operating a woodworking machine, use and keep in proper adjustment the guards and devices provided in accordance with the regulations, and use the spikes, push sticks, push blocks, jigs, holders and back stops provided in accordance with the regulations.

It is the duty of every person employed by the occupier of a factory and trained in accordance with the regulations, who discovers any defect in any woodworking machine, or in any guard, device or appliance provided in accordance with the regulations, or who discovers that the floor or surface of the ground around any woodworking machine is not in good and level condition or is slippery, to report the matter without delay to the occupier, manager or other appropriate person.

Lighting

The lighting whether natural or artificial for every woodworking machine shall be sufficient and suitable for the purpose for which the machine is used, and the means of artificial lighting for every woodworking machine shall be so placed as to prevent glare and so that direct rays of light do not impinge on the eyes of the operator while he is operating such a machine.

Noise

Where any factory or part thereof is mainly used for work carried out on woodworking machines the following provisions shall apply to that factory or part thereof.

Where on any day any person employed is likely to be exposed continuously for 8 hours to a sound level of 90 dB or greater, such measures as are reasonably practicable shall be taken to reduce noise to the greatest extent which is reasonably practicable, and suitable ear protectors shall be provided, made readily available and used by the person for whom they are provided.

SAWING MACHINES

The two sawing machines illustrated and described in this chapter are those which are most likely to be met with in the average joiner's workshop, or which the young craftsman is most likely to use himself or assist in using.

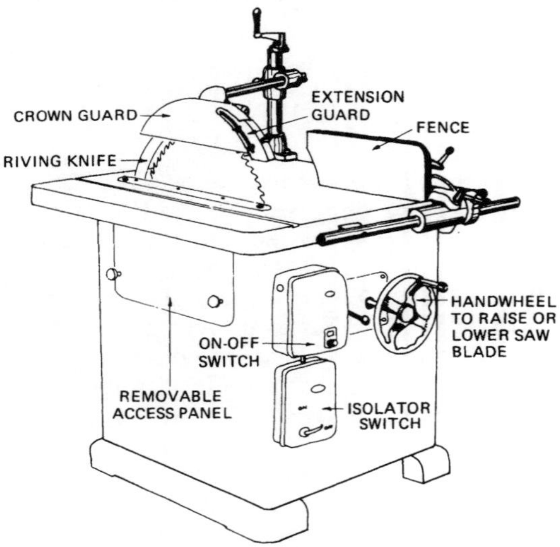

Fig. 11.1. *Circular saw bench (hand fed).*

Circular saw bench (hand fed)

This machine is shown in Fig. 11.1. and comprises a cast iron table mounted on a steel main frame. The main frame houses the electric motor which drives the plate saw via multiple vee belts. The plate saw is secured to a ball bearing spindle by means of a large locking nut which has a left hand thread, this latter feature ensuring the saw does not work loose in use. The following are features worth noting.

(a) The adjustable crown guard. This can be raised and lowered to suit the work being done and the projection of the saw blade above the table. It is fitted with an adjustable front extension piece which should be positioned close to the timber being cut.

(b) The riving knife. This protects the back edge of the saw blade, and since it must be slightly thicker than the plate saw itself it helps to prevent the saw kerf closing and binding on the saw blade.

(c) The fence. This can be set and locked in any position to provide a guide for the timber being fed into the saw.

(d) The handwheel. This is used to raise and lower the saw blade and riving knife to the required projection above the table.

(e) The on/off switches. In the form of push buttons these are conveniently placed for the operator.

Use of the machine

The machine illustrated is a hand fed saw-bench, the timber being fed into the saw by hand, and is a general purpose saw used mainly for ripping and deeping as shown in

Fig 11.2. *Sawing operations.*

(a) Ripping (flatting).

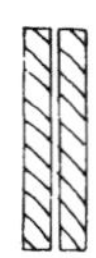

(b) Deeping.

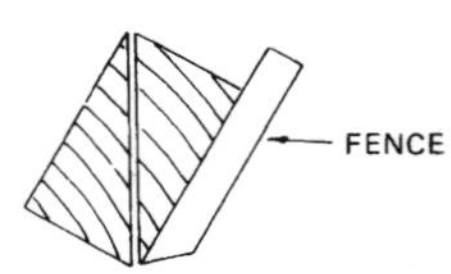

(c) Splay cutting.

Fig. 11.2*(a)* and *(b)*. The machine can also be used for splay and arris cutting up to about 45° by tilting or "canting" the fence as shown in Fig. 11.2*(c)*.

The correct and proper position of the fence is illustrated in Fig. 11.3. Note that the curved edge on the back of the fence should be set just level with the gullets of

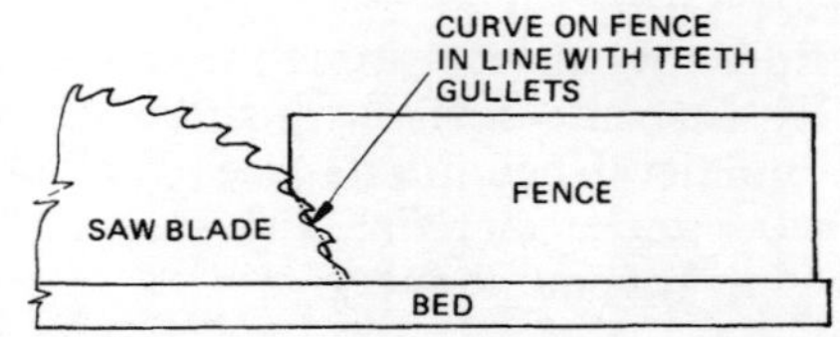

Fig. 11.3. *Correct position of fence.*

the saw teeth to prevent possible binding between saw blade and fence.

Figure 11.4 shows a push stick used to push the end of a piece of timber past the

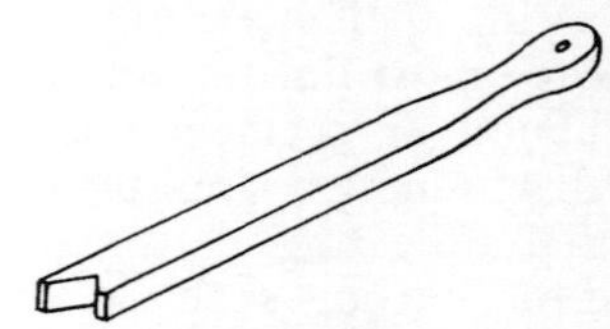

Fig. 11.4. *Push stick (500 mm long).*

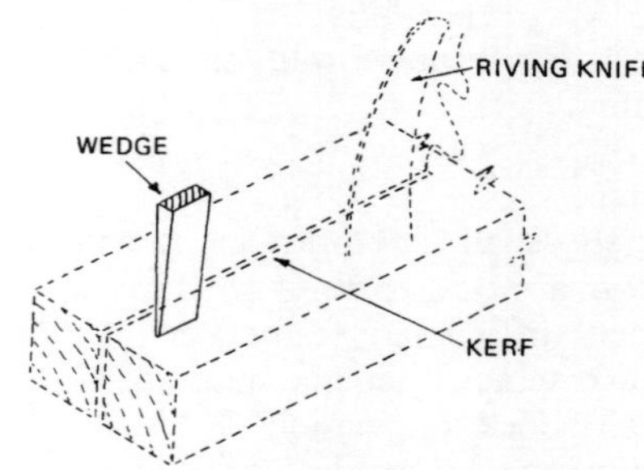

Fig. 11.5. *Wedge used to keep kerf open on case hardened wood.*

saw blade without bringing the fingers into close proximity to the blade. The fingers should never be nearer than 300 mm to the moving saw blade. Push sticks should always be conveniently to hand when using the saw.

Figure 11.5 shows a small wooden wedge which is used by the person pulling off behind the saw, the wedge being driven into the saw kerf to keep it open if the timber shows any tendency to bind (case hardened timber tends to do this – *see* Chapter 2).

Saw blades
These take varying forms depending upon the work being done, the variation being either the shape of the saw teeth or the section of the saw blade.

Fig. 11.6. *Saw teeth.*

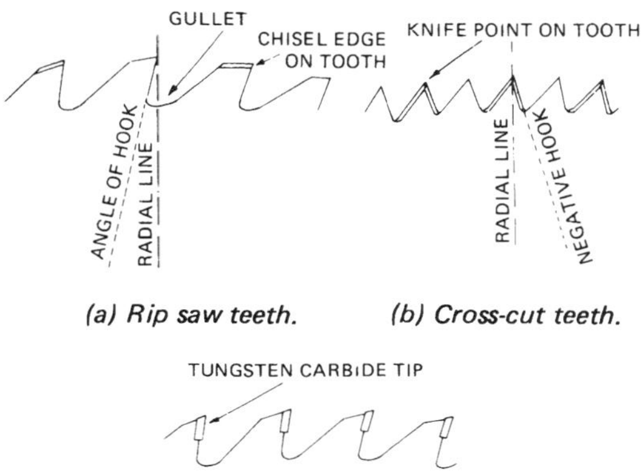

(a) Rip saw teeth. *(b) Cross-cut teeth.*

(c) Tipped teeth.

Figure 11.6*(a)* shows the shape of saw teeth designed for ripping softwood. Note that these teeth are fierce and have a considerable hook (the angle between the front edge of the tooth and a radial line touching the tip of the tooth). Teeth designed for ripping hardwood are generally similar in shape to the former, but have less hook.

Figure 11.6*(b)* shows the shape of teeth designed for cross cutting, these having a "negative hook" and being generally similar in shape to cross cut teeth on a hand saw. Many modern saw blades have tungsten carbide inserts on the tips of the teeth as shown in Fig. 11.6*(c)*. These saw teeth hold a sharp edge for a much longer period of time than do ordinary saw teeth.

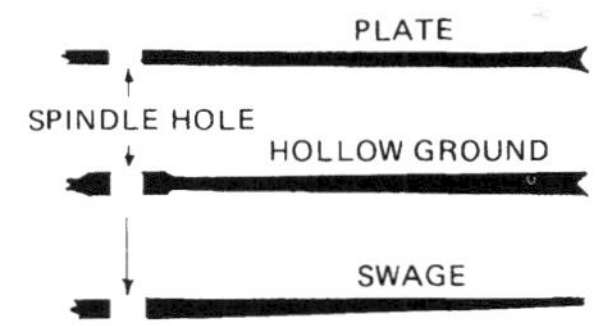

Fig. 11.7. *Sections through saw blades.*

Figure 11.7 shows sections through plate, hollow ground and swage saws, the plate saw being the most commonly used type of blade whilst hollow ground saws give a clean cut without the need of setting. Swage saws are used to cut very thin boards with a minimum of waste, the thin board bending and peeling off easily as it has insufficient thickness to cause binding on the saw blade. With the possible exception of the hollow ground saw and saws with tungsten carbide tipped teeth, circular saw teeth require "setting" as do handsaw teeth to prevent binding. This is done in one of two ways:

(a) spring setting in which the saw teeth are bent alternately in opposite directions as shown in Fig. 11.8*(a)*;

(b) swage setting (not to be confused with swage saw) in which each tooth is spread by means of a special tool to the shape shown in Fig. 11.8*(b)*. As can be seen, swage set teeth have a "square" chisel edge and are thus ideal for ripping.

Fig. 11.8. *Set on saw teeth.*

(a) Spring set. *(b) Swage set.*

Packings
These are inserted into specially formed recesses in the table, one on either side of the saw blade behind the wooden mouthpiece (*see* Fig. 11.9). Packings are made of

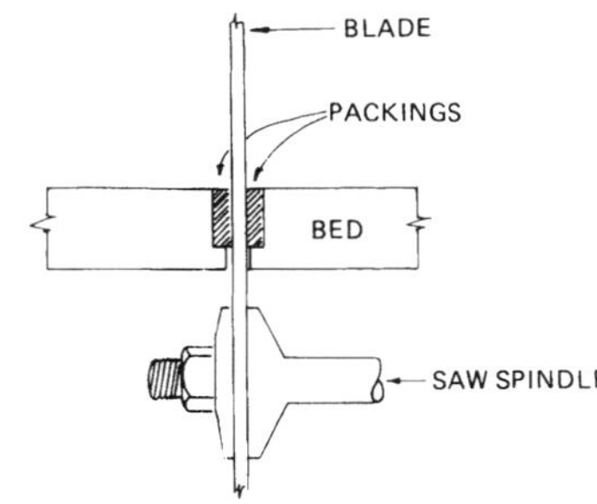

Fig. 11.9. *Saw packings.*

felt, leather or twine wound round a sliver of timber, and serve to steady the saw in its cut and help to preserve its tension when working hard.

Cross cutting and mitring
These operations can be done on the circular saw bench (with a cross cut blade) when the machine is fitted with the cross cut fence as shown in Fig. 11.10.

Peripheral speed
This term refers to the actual linear speed of a tooth, or point of the circumference of a

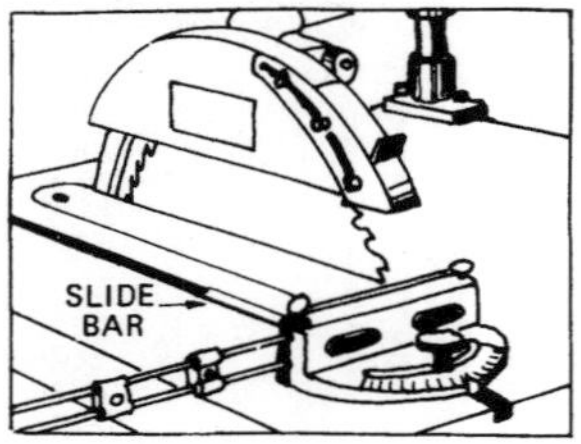

Fig. 11.10. *Cross-cutting attachment for a circular saw.*

saw, when it is revolving under power. Peripheral speed is thus the product of the circumference of the saw blade and the number of revolutions per minute:

$$\text{Peripheral speed of the saw} = \pi \times D \times \text{r.p.m.}$$

Most circular saws are designed to run at a peripheral speed of around 2,200 m/min. and therefore the smaller diameter blades must run at higher r.p.m. than the larger, and vice versa.

Regulations applying to the use of circular saws

Guarding

(a) That part of the saw blade which is below the table shall be guarded (covered in) to the greatest extent that is practicable.

(b) Every circular sawing machine shall have a strong, smooth, rigid and securely fixed riving knife which is easily adjustable. The riving knife shall be so shaped that the edge nearest the saw blade forms an arc of a circle of radius not larger than the radius of the *largest saw blade* with which the saw bench is designed to be fitted. The riving knife shall be adjustable and kept so adjusted

Fig. 11.11. *Regulations applying to circular saws.*

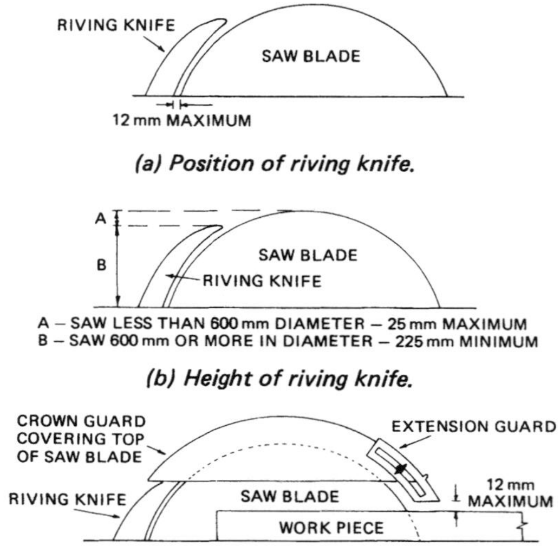

(c) Correct position of crown guard.

that it is as close as practicable to the saw blade, and at the level of the saw table not further than 12 mm away from the saw blade (*see* Fig. 11.11*(a)*). For a saw blade of less than 600 mm diameter, the riving knife shall project above the saw table to a height not less than a level line 25 mm below the highest point of the saw blade, and for a saw blade of 600 mm or more in diameter, the riving knife shall project above the saw table to a height of not less than 225 mm (*see* Fig. 11.11*(b)*). In the case of a parallel plate saw, the riving knife shall be thicker than the plate of the saw blade.

(c) That part of the saw blade which is above the machine table shall be guarded with a strong and easily adjustable guard which shall be kept so adjusted that it extends from the top of the riving knife to a point above the upper surface of the material being cut which is as close as practicable to that surface, or, where squared stock is being fed by hand, to a point which is not more than 12 mm above the surface of the material being cut (*see* Fig. 11.11*(c)*).

Sizes of circular saw blades

In the case of a circular sawing machine the spindle of which is not capable of being operated at more than one working speed, the minimum size of the saw blade which may be used on that machine is six-tenths of the diameter of the largest blade for which the saw bench was designed.

In the case of a circular sawing machine in which the spindle may operate at more than one working speed, the minimum size of saw which may be used is six-tenths the diameter of the largest blade which can be properly used at the *fastest* working speed of the spindle of that saw bench.

There shall be affixed to every circular sawing machine a notice specifying the diameter of the smallest saw blade which may be used with that machine.

Limitations on the use of circular sawing machines

No circular sawing machine shall be used for cutting any rebate, tenon, mould or groove unless that part of the saw blade or cutter which is above the table is effectively guarded.

No circular sawing machine shall be used for a ripping operation (other than cutting a tenon, rebate, mould or groove) unless the teeth of the saw blade project, throughout the operation, through the upper surface of the material being cut.

Provision of push sticks
A suitable push stick shall be provided and kept available for use at every circular sawing machine which is fed by hand.

Removal of timber cut by circular sawing machines
Where any person (other than the operator) is employed at a circular sawing machine in removing, whilst the blade is in motion, material which has been cut, the machine table shall be constructed or extended over its whole width, so that the distance between the delivery end of the table or extension and the up-running part of the saw blade shall not be less than 1,200 mm.

Dimension saw
This machine, illustrated in Fig. 11.12, is basically similar to the hand fed saw bench described previously, but has several refinements.

Generally the machine table is larger (wider) than on the hand fed saw bench to facilitate the cutting of wide panels, whilst part of the table is designed to slide back- and forwards for accurate cross cutting and mitring as shown in Fig. 11.13. In addition, the saw has a rise and fall spindle which can also be tilted up to 45° for splay and compound angle cuts as shown in Fig. 11.14*(b)* and *(e)*. As its name implies, the dimension saw is used mainly for fine accurate work involving ripping, bevelling, cross cutting, mitring, panel cutting, etc., as also shown in

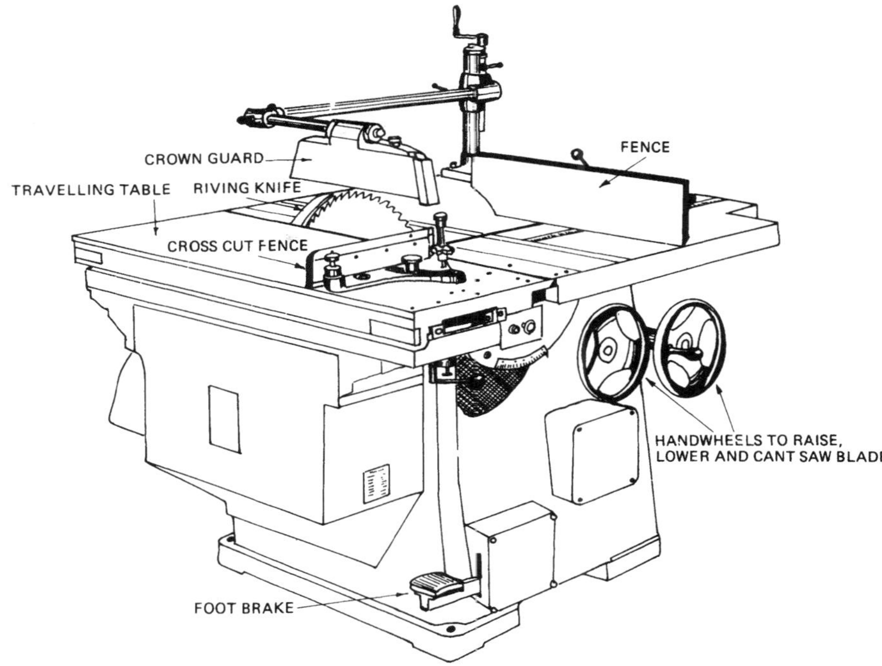

Fig. 11.12. *Dimension saw.*

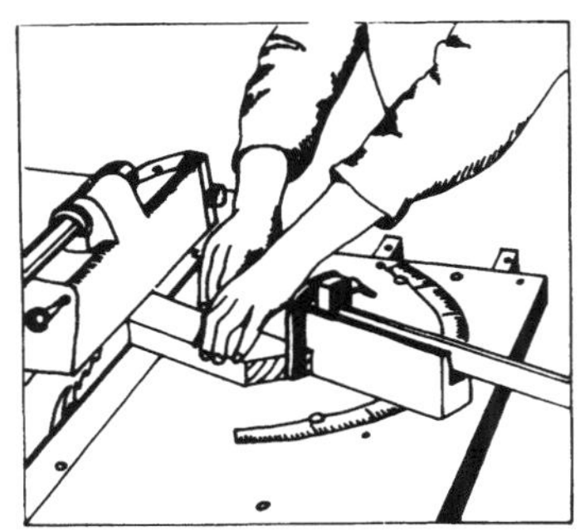
Fig. 11.13. *Cutting mitres on a dimension saw.*

Fig. 11.14. This machine is used with a fairly small diameter saw blade of heavy gauge, sometimes hollow ground, and often with tungsten carbide tipped teeth. Because of the sliding table, packings cannot be used.

As can be seen from the illustration, the fence, cross cut fence and tilting arbor are furnished with finely engraved calibrations to facilitate accurate setting.

Regulations
All the regulations applying to circular saw benches apply to the dimension saw.

Fig. 11.14. *Work done on dimension saw.*

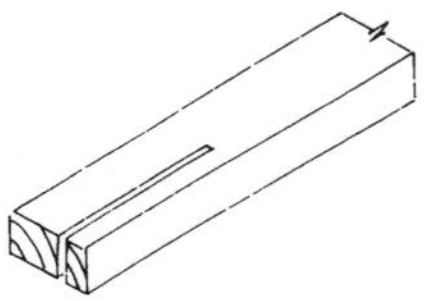

(a) Ripping.

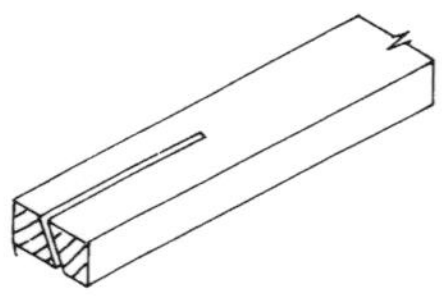

(b) Splay cutting.

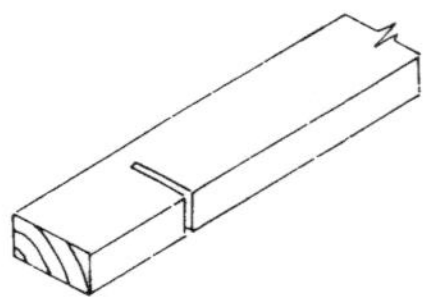

(c) Cross cutting.

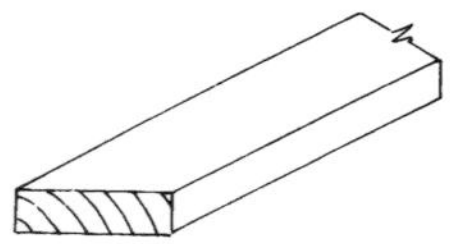

(d) Mitring — angle cutting.

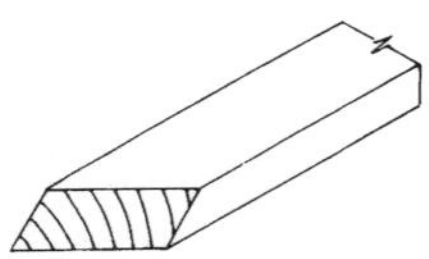

(e) Cutting compound angles.

PLANING MACHINES

Planing machines, of which there are several types, work on the principle of a rotary cutter block above or below which the timber is fed, the rotary action of the cutters removing a series of curved chips from the surface of the timber to plane it smooth and flat.

Figure 11.15 shows the cutting action of a planing machine, and a brief study of this diagram should make it clear that a machine planed surface is not, in fact, really true and flat, but rather a series of small ripples. Generally, these ripples or cutter marks are so small and close together that for most

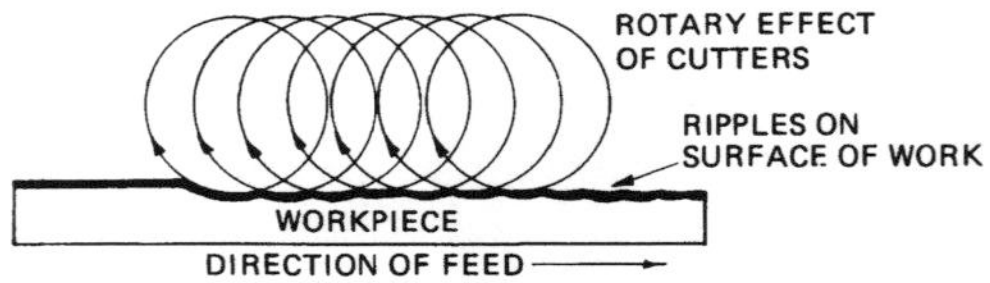

Fig. 11.15. *Cutting action of a planing machine.*

purposes the surface can be regarded as being smooth and flat. However, since on most planing machines the cutter block revolves at a constant fixed speed of about 5,000 r.p.m. it will be appreciated that the size of pitch of the machine ripples so produced is dependent largely upon the speed at which the timber is fed past the cutter, i.e. if the speed of a cutter block is constant, then the faster the work is fed past the cutters, the larger will be the pitch of the ripples and vice versa.

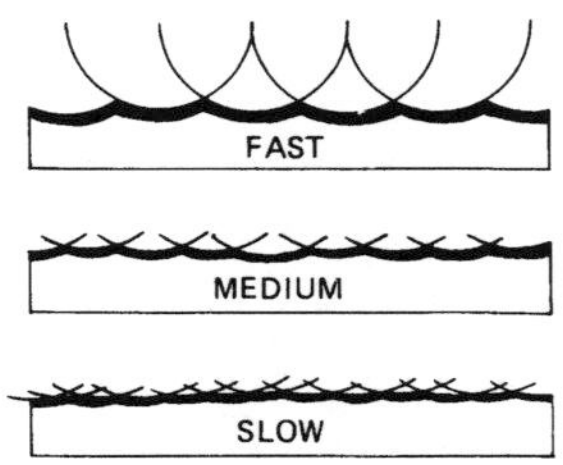

Fig. 11.16. *Effect of varying feed speed on finish of planed surface.*

Figure 11.16 shows the effect on the timber of fast, medium and slow feed speeds respectively. Generally speaking a cutter mark pitch of 1 mm or less can be regarded as a fine finish suitable for joinery and cabinet work, whilst a pitch of more than 3 mm becomes rather less satisfactory and more noticeable, suitable perhaps for work where

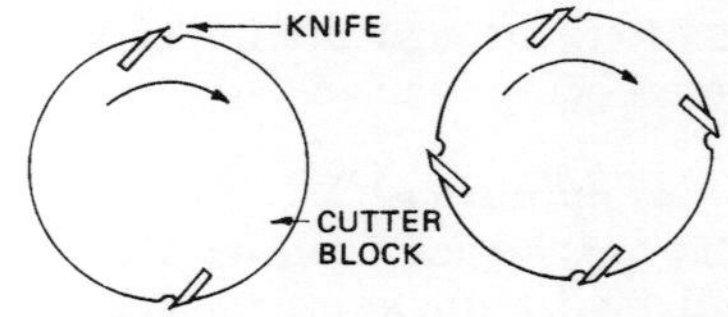

Fig. 11.17. *Two and four knife circular cutter blocks.*

the finish is less important. In any case, machine ripples should be removed with a smoothing plane where a good finish is required, i.e. for polished work.

Figure 11.17 shows diagrammatic sections of cutter blocks which have two and four cutters respectively. Given that the cutter blocks are revolving at the same speed, the four-bladed cutter block will make *twice* as many cuts as the two-bladed cutter block within the same period of time, giving smaller ripples and a better finish. Conversely, it should be possible to obtain the *same* cutter mark pitch and thus an *equally* acceptable finish by feeding past a four-bladed cutter block at twice the speed of the former.

Surface planer

This machine, sometimes also referred to as an "overhand planer" or "jointer", is illustrated in Fig. 11.18. The machine comprises two independently adjustable tables mounted on a rigid steel base frame. The cutter block between the infeed and outfeed tables is driven by an electric motor via an endless vee belt. The fence can be set to any desired position on the machine table and can also be tilted or canted as shown in Fig. 11.19.

Note the conveniently placed on/off switch and the "bridge guard". The position of the bridge guard is most important when using this machine, as is explained later.

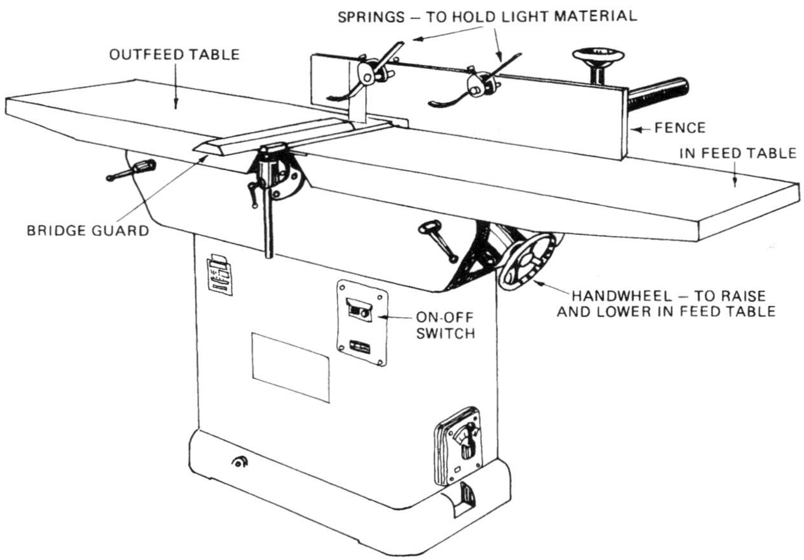

Fig. 11.18. *Surface planer.*

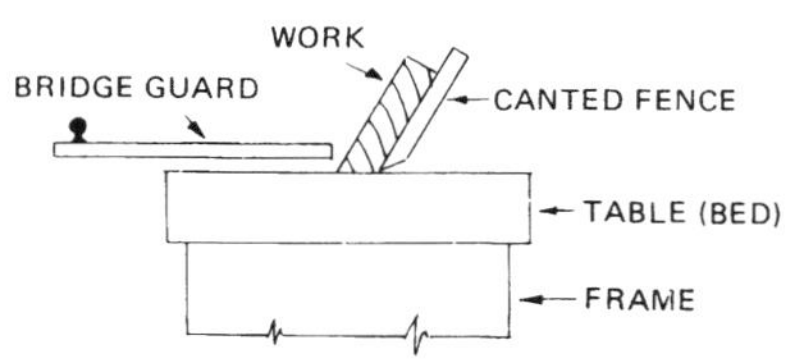

Fig. 11.19. *Bevelling on the surface planer.*

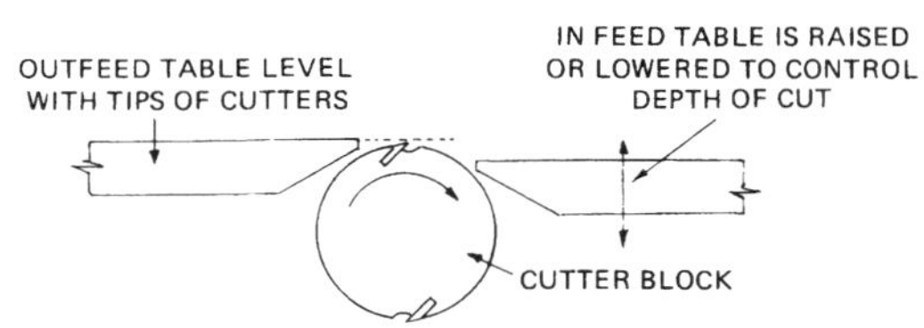

Fig. 11.20. *Adjustment of infeed and outfeed tables.*

Although both tables are independently adjustable, it is the infeed table which is adjusted to give the required depth of cut, the outfeed table being set (for general work) level with the top of the cutting circle, as shown in Fig. 11.20.

Function of a surface planer
The basic functions of the surface planer are:

(*a*) quick, accurate removal of waste material;

(*b*) accurate "shooting" of faces and edges – the length and accuracy of the tables enable long edges and surfaces to be planed perfectly straight;

(*c*) accurate squaring of the face/face edge corner (prior to thicknessing);

(*d*) rebating, which can be done on the surface planer provided the cutters are adequately guarded.

Safe operation of the surface planer
The following safety precautions must be constantly borne in mind when using the surface planer.

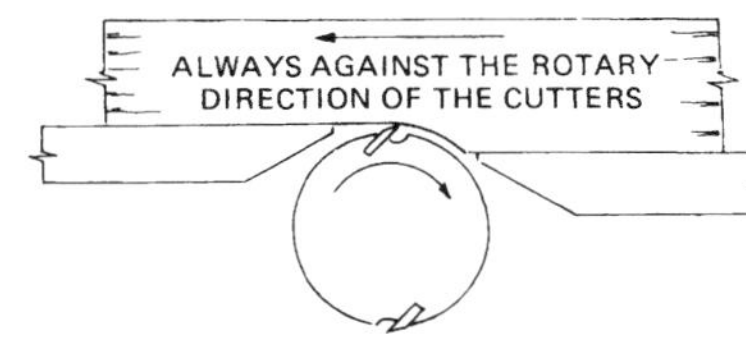

Fig. 11.21. *Direction of feed on a surface planer.*

(*a*) The timber should always be fed "against" the rotary direction of the cutters (*see* Fig. 11.21).

(*b*) The bridge guard must be firmly and properly positioned.

(*c*) The fingers should not be brought close to the cutters – use push blocks or push sticks for short pieces (*see* Fig. 11.22).

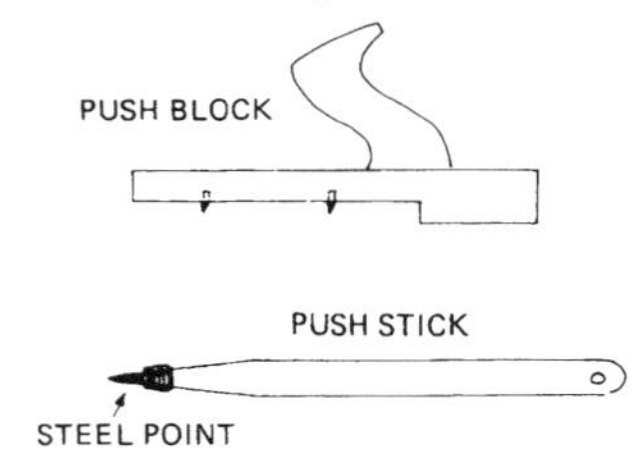

Fig. 11.22. *Planing short pieces.*

(*d*) Assistance is required when shooting long, heavy timbers.

(*e*) Rebating must not be done unless the cutters are effectively guarded, and the timber adequately supported. Figure 11.23(*a*)

Fig. 11.23. *Rebating on the surface planer.*

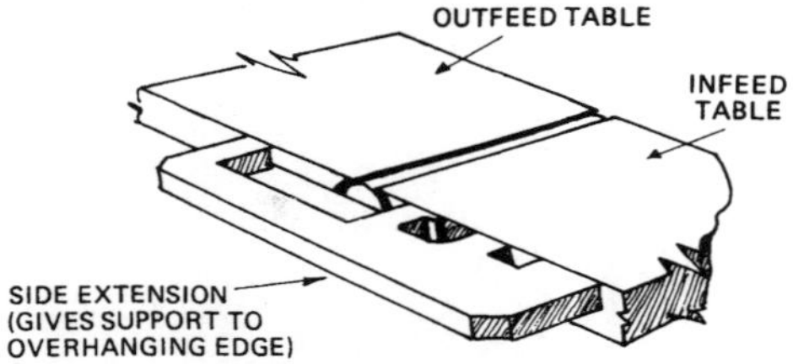

(a) Side extension to table.

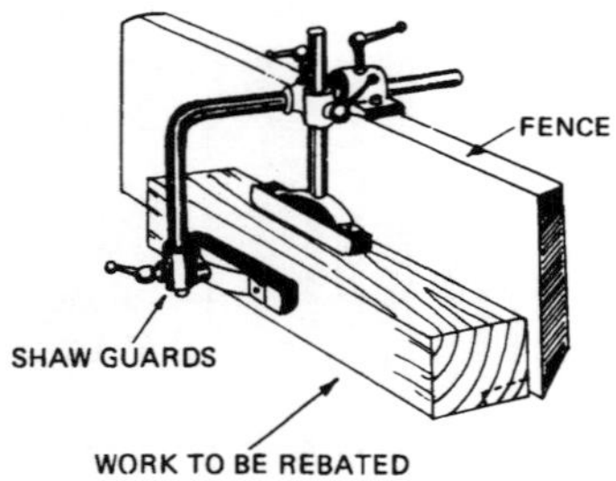

(b) Shaw guards in use on surface planer.

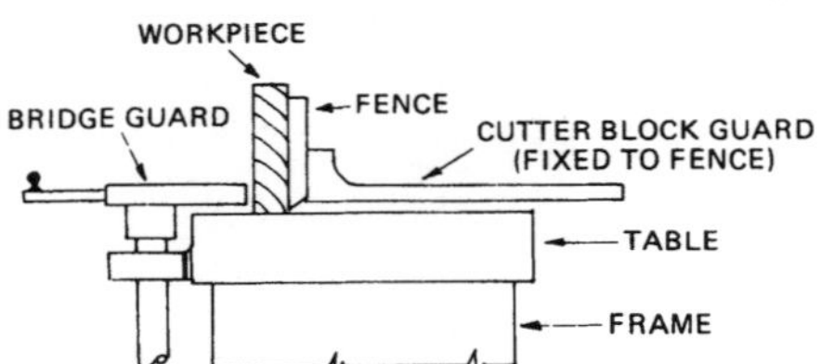

Fig. 11.24. *Guarding of the cutter block.*

and *(b)* shows a side extension table and a pair of Shaw guards designed for this purpose.

(f) The fence must be secure, and should not be adjusted whilst the cutters are rotating.

(g) The cutter block must be guarded on the far side of the fence (*see* Fig. 11.24).

(h) Timber with difficult grain, short grain, large knots or badly twisted surfaces must be machined with due caution, care and attention, otherwise a "kick-back" may result.

Woodworking Machines Regulations (applying to surface planers)

(a) No planing machine shall be used for cutting any rebate, recess, tenon or mould unless the cutter is effectively guarded (*see* Fig. 11.23).

(b) Every planing machine for surfacing which is not mechanically fed shall be fitted with a cylindrical cutter block.

(c) Every planing machine for surfacing which is not mechanically fed shall be so designed and constructed that the clearance between the cutters and the front edge of the delivery table does not exceed 6 mm (measured radially from the centre of the cutter block) and the gap between the feed table and the delivery table is as small as practicable (*see* Fig. 11.25).

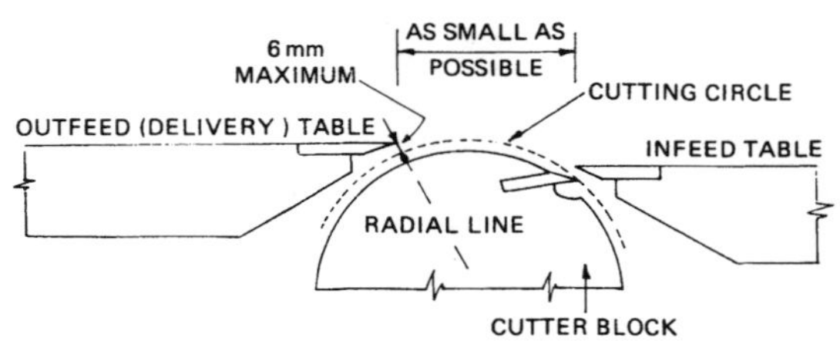

Fig. 11.25. *Diagram showing table gap on a surface planer.*

(d) Every planing machine for surfacing which is not mechanically fed shall be provided with a bridge guard which is strong and rigid, have a length not less than that of

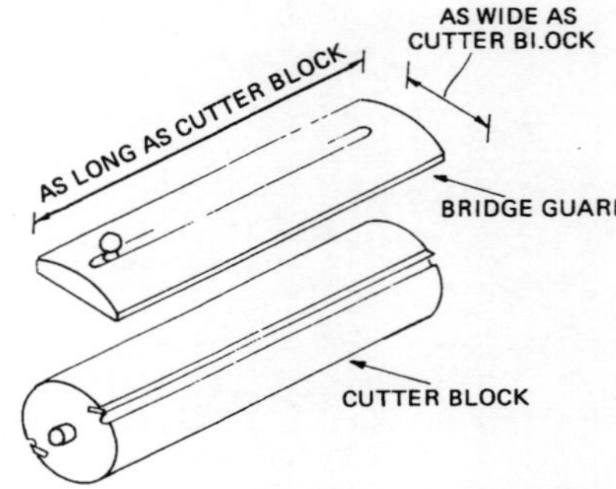

Fig. 11.26. *Dimensions of bridge guard.*

the cutter block and a width not less than the diameter of the cutter block (*see* Fig. 11.26). The bridge guard must be mounted centrally over the axis of the cutter block and must be capable of easy vertical and horizontal adjustment.

(e) Adjustment of a bridge guard: while a planing machine which is not mechanically fed is used for surfacing, the bridge guard shall be positioned as indicated in the following.

(i) Figure 11.27*(a)* shows the correct position of the bridge guard when planing a wide surface.

(ii) Figure 11.27*(b)* shows the correct position of the bridge guard when planing an edge (when shooting edges for jointing etc.).

(iii) Figure 11.27*(c)* shows the correct position of the bridge guard when the planing of the face and the face edge are done consecutively (one operation following the other).

(iv) Figure 11.27*(d)* shows the correct position of the bridge guard when planing timber of square section.

Fig. 11.27. *Adjustment of bridge guard.*

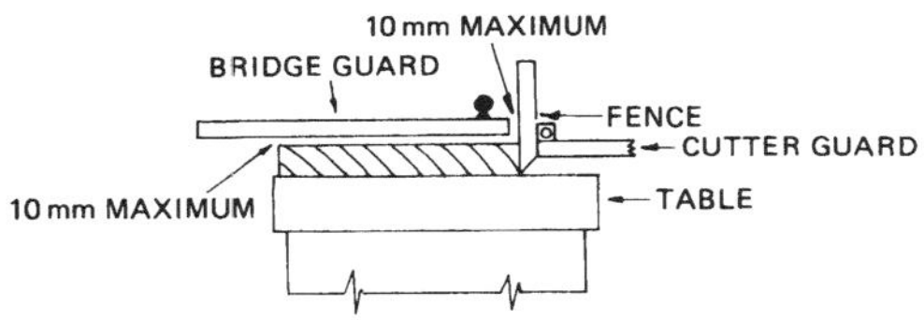

(a) Guard set for facing.

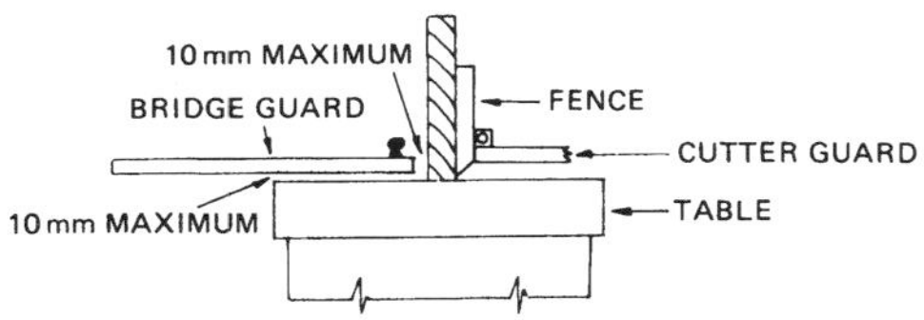

(b) Guard set for edging.

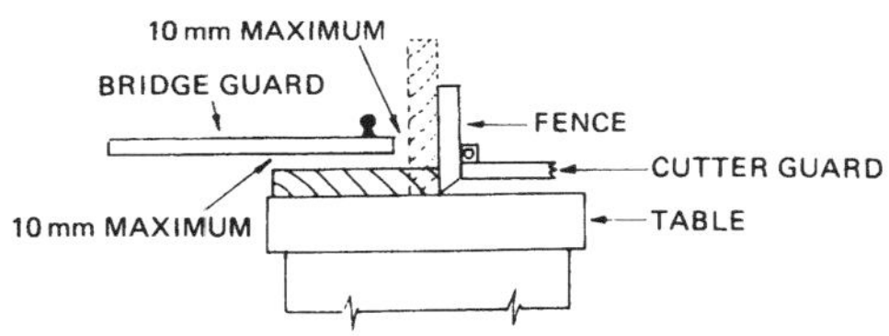

(c) Guard set for facing and edging (consecutive operations).

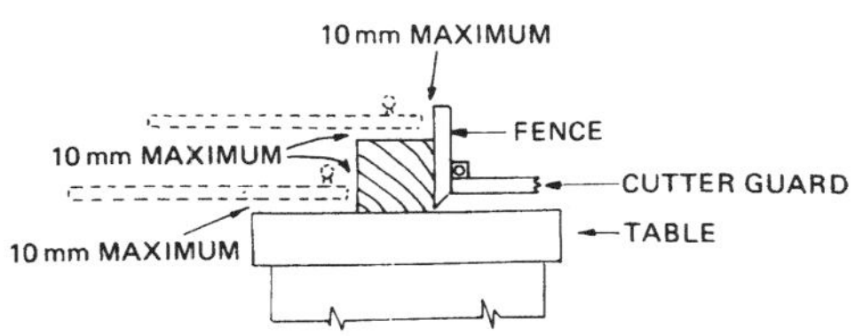

(d) Alternative positions of guard for planing square sections.

(*f*) Cutter block guard: in addition to being provided with a bridge guard, every planing machine for surfacing which is not mechanically fed shall be provided with a strong, effective and easily adjustable guard for that part of the cutter block which is on the side of the fence remote from the bridge guard (*see* Fig. 11.24).

(*g*) Combined machines (over and under): that part of the cutter block of a combined machine which is exposed in the table gap shall, when the machine is used for thicknessing, be effectively guarded.

Thicknesser

The thicknesser, or panel planer as it is sometimes called, is shown in Fig. 11.28 and is used to plane timber accurately to the required thickness. It can also be used to plane

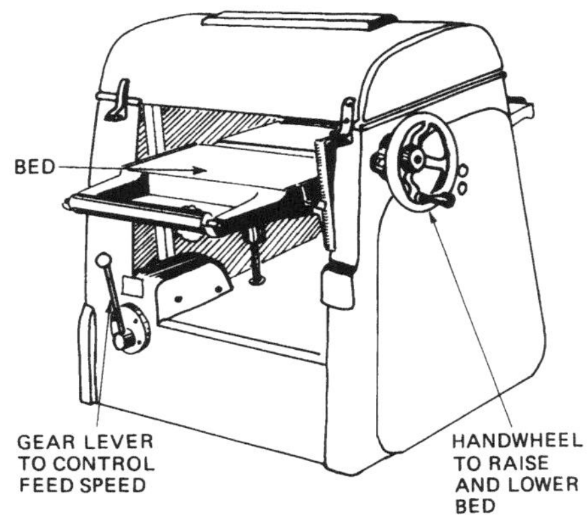

Fig. 11.28. *Thicknesser.*

the edge of a piece of timber to bring the timber to an exact width. This latter purpose becomes less efficient where wide narrow boards are concerned, owing to the tendency of the boards to tip over. As can be seen from the explanatory diagram (Fig. 11.29) the timber is power fed *underneath* the cutter block and the top is thus planed exactly parallel to the lower surface. It should be

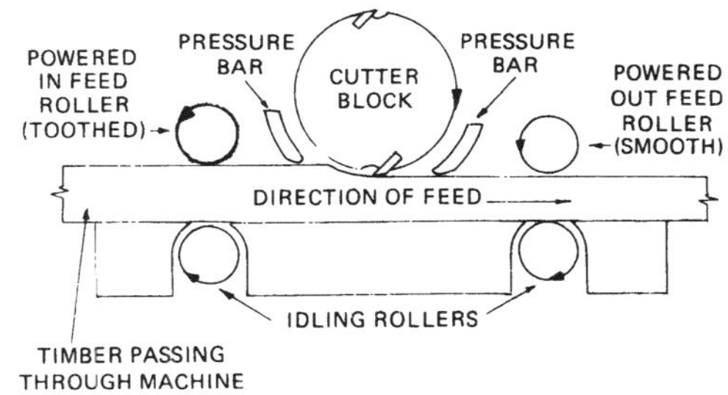

Fig. 11.29. *Section showing working parts of a thicknesser.*

realised that the thicknesser, having no fence and a limited length of bed, cannot be used to straighten or square up the timber, this process being carried out beforehand on the surface planer. Timber must therefore be fed into the machine face or face edge downwards, the machine then producing a similarly accurate surface on the opposite side.

If a piece of twisted or bowed timber is fed into the machine it will emerge planed – but still twisted or bowed; hence, as useful as the machine undoubtedly is, it can only be used effectively in conjunction with the surface planer as previously stated.

Most thicknessers have a facility for varying the speed at which the timber is fed below the cutter block, this being accomplished either by means of a gear box, or electrically. It should be noted that such a variable feed speed merely alters the *speed at which the timber is fed*. It may not affect the speed of the cutter block.

Cutter blocks

The type used in the thicknesser may have two, three or four knives, and since the

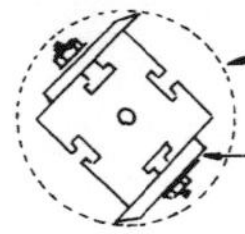

Fig. 11.30. *Two knife square block.*

working parts are totally enclosed, are sometimes "square" as shown in Fig. 11.30. (This would be illegal on the surface planer or on a combination machine.)

Jointing the cutters

Where more than two cutters are employed in the cutter block, it becomes extremely difficult to set them in such a way that they are all working effectively. Obviously, if one of the cutters has a greater projection than the others it will do more of the work, acting in fact, almost like a single bladed

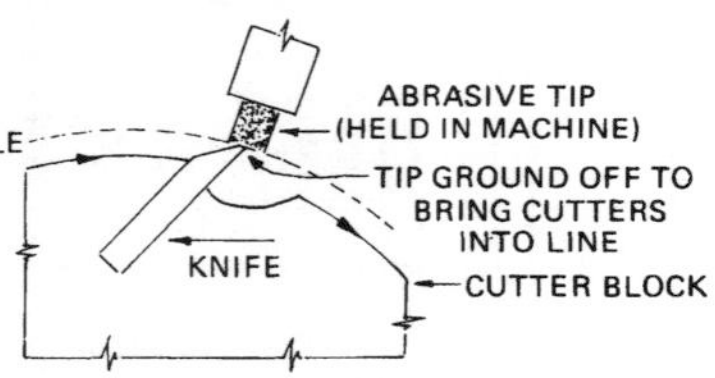

Fig. 11.31. *Diagram showing method of jointing cutters.*

cutter block. To ensure the effectiveness of all the cutters they are often "jointed", a process which involves passing an abrasive stone across the tips of the revolving cutters and thus bringing them exactly into line (*see* Fig. 11.31). Jointing does in fact remove the sharp edge from the tips of the cutters, but this snag is more than offset by the fact that all the cutters will be working equally well.

NOTE: Jointing is carried out by means of a special attachment fitted to the machine and *must not, under any circumstances, be attempted by hand.* Neither must jointing be attempted on a surface planer.

Feed rollers

As can be seen in Fig. 11.29 the thicknesser has two power driven feed rollers — a serrated in-feed roller designed to grip the timber for feeding in, and a smooth out-feed roller for delivering out at the far end without marking the planed surface. The in-feed

Fig. 11.32. *Types of feed roller.*

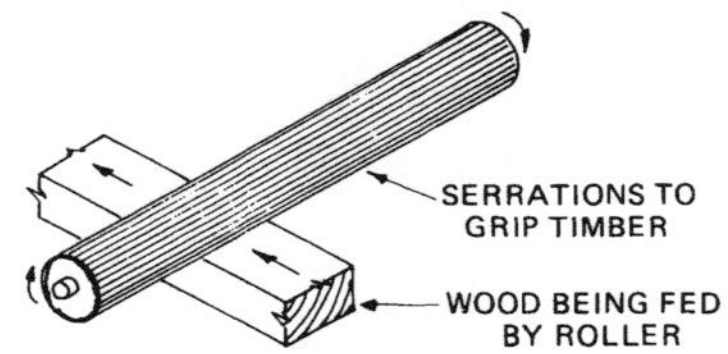

(a) Solid feed roller.

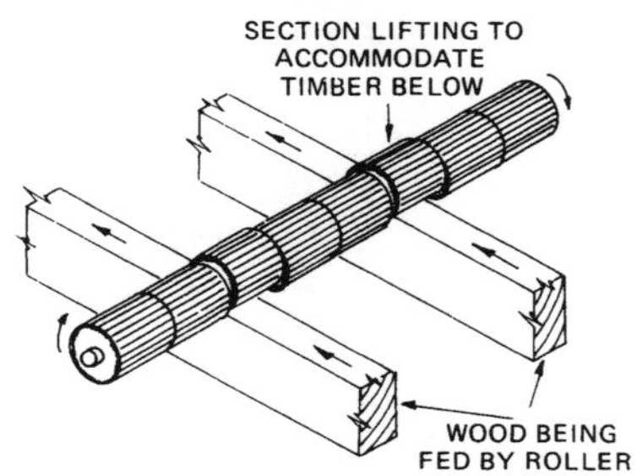

(b) Sectional feed roller.

roller may be either a solid roller, or alternatively a sectional roller as shown in Fig. 11.32*(a)* and *(b)* respectively, the latter type consisting of a series of independently sprung segments. It should be noted that when a piece of timber is fed under a solid feed roller, the roller will lift and therefore be unable to grip effectively any further timber which is fed into the machine whilst the first is still being planed. Thus unless the machine is fitted with an anti-kick-back device, there would be a risk of the second piece of timber being forcibly ejected from the machine by the cutters. The sectional roller, on the other hand, is capable of adjusting itself to accommodate a variation in thickness of two or more pieces of timber being planed simultaneously, and the danger of kick-back is thereby eliminated.

Woodworking Machines Regulations (applying to thicknessers)

The following abridged regulations apply to those aspects of feed roller design previously explained.

(a) Every thicknessing machine shall be provided on the operator's side with sectional feed rollers or other suitable device (anti-kick-back) to restrain any workpiece ejected by the machine.

(b) Paragraph *(a)* of these regulations shall not apply to any machine manufactured before the date of these regulations (November 1974), provided that:

(i) not more than one work piece shall be fed to any such machine; and

(ii) there shall be a notice fixed securely to the machine specifying that only single pieces shall be fed.

MORTICING MACHINES

Morticing machines are designed almost expressly for the purpose of cutting mortices quickly and accurately, and are essential items of equipment in any workshop where

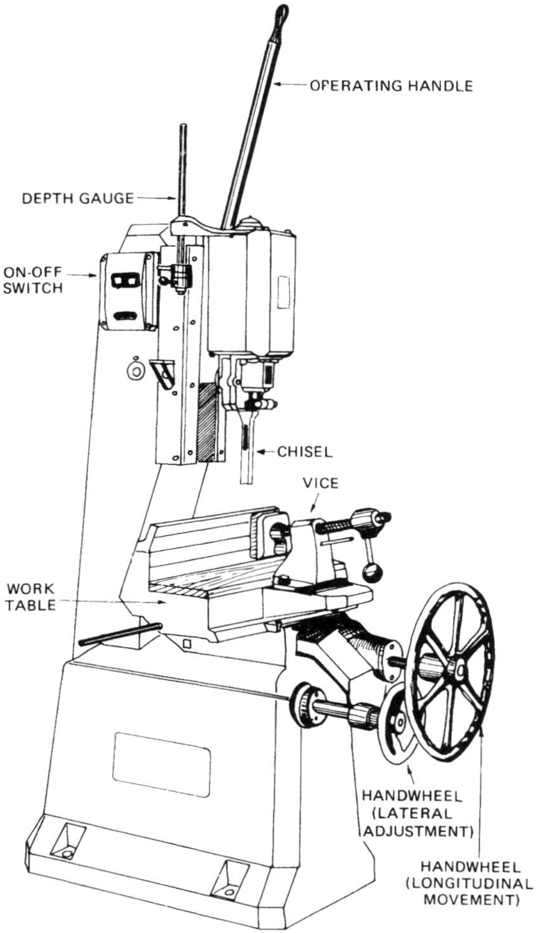

Fig. 11.33. *Hollow square chisel morticing machine.*

production work is undertaken. There are several types of morticing machine, each designed to cut the mortice in a different way and to suit different types of work, but two types have proved their worth over the years and still remain unsurpassed for general work.

Hollow square chisel morticer

This machine, illustrated in Fig. 11.33, is probably the most generally useful of the two main types. The work is held on the work table by means of a clamp, and can be moved laterally – to line up with the chisel – and longitudinally – to traverse the length of the mortice – by means of the controls shown. The mortice is cut by means of a revolving auger encased within a hollow square chisel, the whole assembly being driven into the work piece by the downward movement of the lever handle. The mortice is thus really cut in two stages, the one closely following the other:

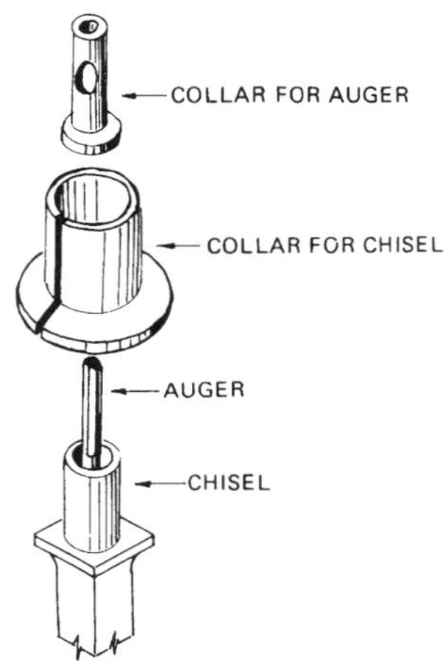

Fig. 11.34. *Parts of a hollow square chisel.*

(*a*) the auger bores the hole;
(*b*) the chisel converts the round hole to a square mortice.

To cater for varying sizes of mortice, the chisels and their respective augers range in size from 6 mm to 25 mm (more in larger machines), each chisel and its auger being fitted with a pair of collars as shown in Fig. 11.34.

Using the morticer

It is standard practice to mark out the mortices in the normal way as for hand morticing, place the work in the machine with the face mark in contact with the vertical face of the machine table, and then proceed to cut the mortice rather more than half way through, reversing the timber to complete the mortice from the opposite side to avoid spelching. Adjustable depth stops are provided to ensure the correct depth of blind mortices and haunches. It is a good policy to insert a 300 mm length of wood between the work and the cramp to reduce the risk of pinching the chisel.

Setting up the machine

It is very important to ensure a 2-3 mm clearance between the tip of the auger and the chisel as shown in Fig. 11.35, otherwise the tips of both chisel and auger are likely to overheat. Some care should be taken to ensure that the chisel is set with two of its sides exactly parallel with the mortice gauge lines, otherwise a "stepped" mortice, as shown in Fig. 11.36(*a*) will result. When properly set up, the hollow square chisel morticing machine cuts a clean, flat bottomed mortice as shown in Fig. 11.36(*b*).

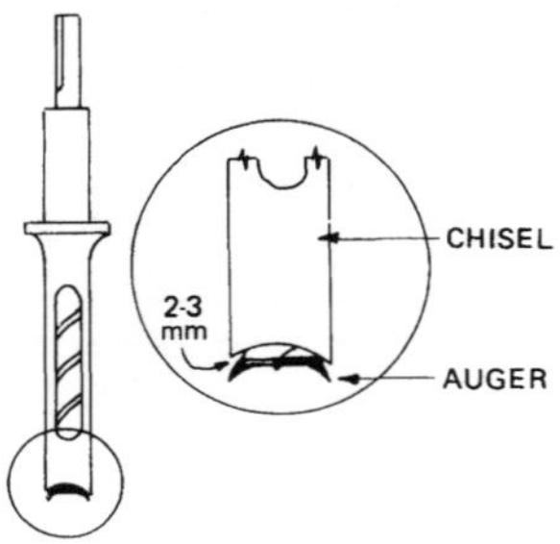

Fig. 11.35. *Auger clearance.*

Fig. 11.36.

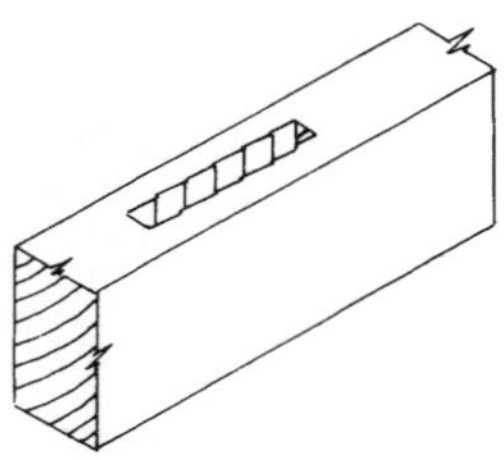

(a) "Stepped" mortice caused by chisel which is not set squarely in machine.

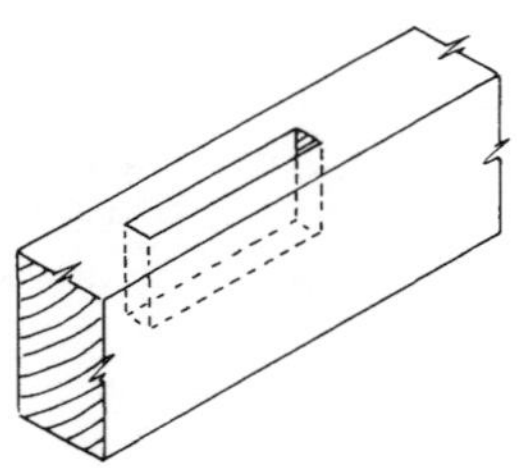

(b) "Blind" mortice cut on hollow square chisel morticer.

Chain morticer

This machine bears a superficial resemblance to the previously described type, having all

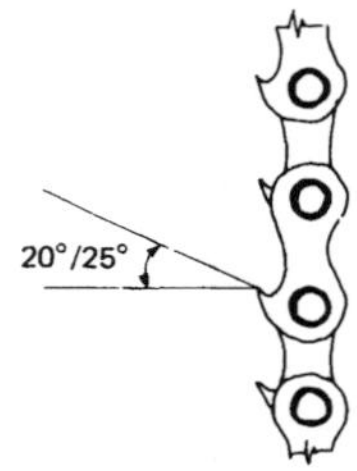

Fig. 11.37. *Sketch of mortice chain links.*

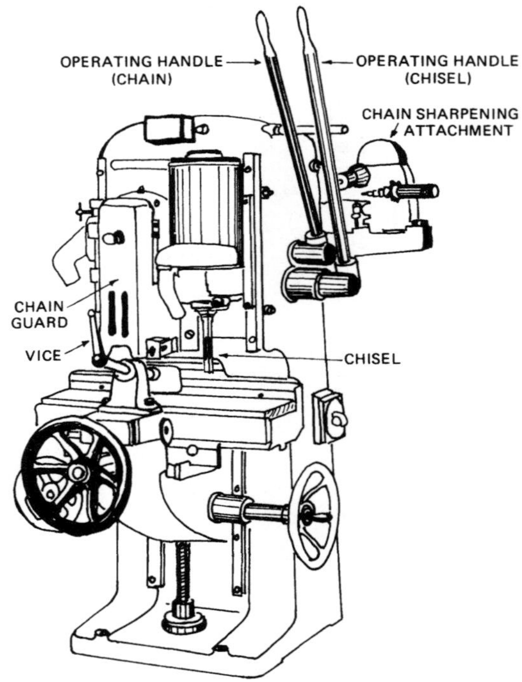

Fig. 11.38. *Dual purpose chain and chisel morticer.*

the controls and adjustments of the former, but it has a very different cutting action. Mortices cut on this machine are produced by means of a chain, similar to a bicycle chain but with sharp hooked teeth on the outer edge as shown in Fig. 11.37. The chain is driven round by means of a toothed sprocket wheel attached to an electric motor and is supported by a finely engineered guide bar around which the chain runs. Often, a morticing machine is equipped with both hollow square chisel and chain cutting heads as illustrated in Fig. 11.38. Chains are available in different sizes, each with its own guide bar and sprocket, to cater for the various widths of mortice which may be required. When not in use, the chain and its guide bar should be kept in an oil bath.

Cutting action of a chain morticer

As can be seen in Fig. 11.39*(a)*, the chain cuts a round bottomed mortice (unless the mortice is cut right through) and, of some importance, the chain teeth cut downwards into the wood on one side, and upwards *out of the wood* on the other. This has a tendency to "spelch" one side of the mortice, a tendency which must be counteracted by the wooden chip breaker shown in Fig. 11.39*(b)*, which is fixed to the sliding guard and bears on the wood on the up-cutting side throughout the cut.

Mortices are cut by the up and down action of the chain (via the operating handle). No attempt should be made to elongate the mortice by traversing the table whilst the chain is in the wood; this would impose an undue strain on the chain and guide bar.

Tension of chain

The amount of tension on the chain is adjustable by means of the guide bar, and whilst the chain should not be run excessively

Fig. 11.39. *Cutting action of a chain morticer.*

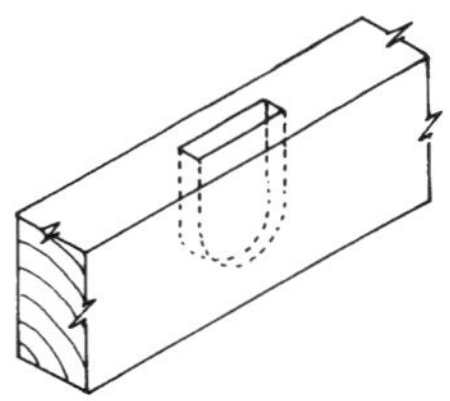

(a) Mortice cut by a chain.

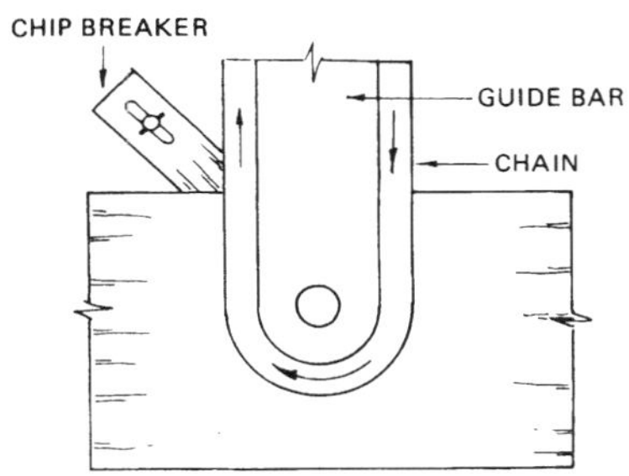

(b) Chip breaker to prevent spelching.

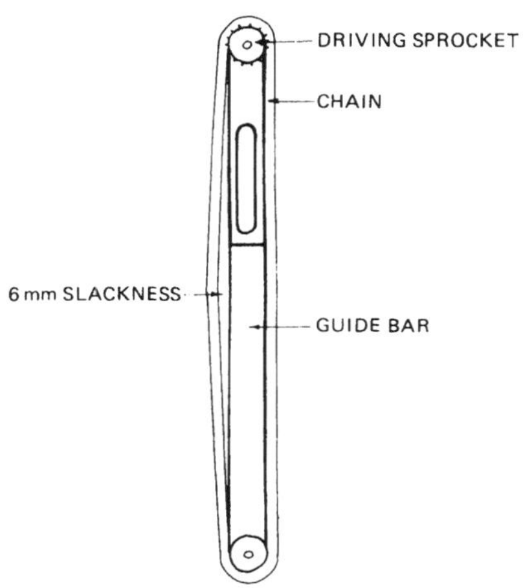

Fig. 11.40. *Sketch showing mortice chain adjustment.*

slack resulting in sloppy, ragged mortices, neither must it be too tight, otherwise it will overheat and wear rapidly. The chain is correctly tensioned when a sideways pull on the centre of the chain (stationary) gives about 6 mm of movement as indicated in Fig. 11.40.

Safety in use

Morticers are possibly the least hazardous of the woodworking machines to operate, but certain common sense precautions need to be observed to ensure safe use. These are listed as follows.

(*a*) The machine should be switched off and *isolated* when changing chisels or chains.

(*b*) Chain morticers must be fitted with an effective sliding guard which covers the chain at all times whilst the machine is working.

(*c*) Keep fingers away from the business end of the machine, whether chain or chisel. Loose chippings obscuring the lines should preferably be blown out of the way (some machines incorporate a blower to do this).

(*d*) Long lengths of timber being morticed should be supported at the overhanging end as otherwise there is a grave danger of the operator's hand being forced upwards into the cutter, when the clamp holding the wood is slackened to move the work along to the next mortice.

(*e*) Chisels, augers and chains should be kept sharp and in good order. Special tools are available for sharpening the hollow square chisel, whilst augers should be kept sharp with a fine saw file. Mortice chains are kept sharp by grinding the hooked teeth to their correct profile (generally 25°) on the specially designed grinding attachment affixed to the machine for that express purpose.

WOOD MACHINING CALCULATIONS

The following examples should prove useful to the student of woodwork.

To calculate the pitch of circular saw teeth

Divide the circumference of the saw by the number of teeth around its periphery:

$$\text{Pitch of tooth} = \frac{\pi \times D}{\text{No. of teeth}}$$

Example

A 600 mm diameter plate saw has 95 teeth around its periphery. Calculate the tooth pitch.

$$\text{Tooth pitch} = \frac{\pi \times D}{\text{No. of teeth}}$$

$$= \frac{3.142 \times 600}{95}$$

$$\therefore \text{Pitch} = \underline{\underline{19.84 \text{ mm}}}$$

To calculate the peripheral speed of a circular saw blade

Multiply the circumference of the saw blade by the number of revolutions it makes per minute:

$$\text{Peripheral speed of saw blade} = \pi \times D \times \text{r.p.m.}$$

Example

Calculate the peripheral speed of a circular saw blade of 600 mm diameter which revolves at 1,200 r.p.m.

Perepheral speed = $\pi \times D \times$ r.p.m.
= $3.142 \times 0.6 \times 1,200$

$\therefore$

Peripheral speed = 2,262.2 metres per min (m/min)

To calculate the pitch of the cutter marks on a piece of machine-planed timber (planed by thicknesser)

Cutter mark pitch =

$$\frac{\text{Feed speed in m/min} \times 1,000}{\text{No. of effective cutters} \times \text{r.p.m.}}$$

Example

Calculate the cutter mark pitch on the timber passed through a thicknesser which has two effective cutters revolving at 4,000 r.p.m. and a feed speed of 5 m/min.

Cutter mark pitch

$$= \frac{5 \times 1,000}{2 \times 4,000}$$

$$= 0.625 \text{ mm}$$

$\therefore$ Pitch of cutter marks
= 0.625 mm

FURTHER READING

Clayton, J.R. *Machine Woodworking.* (Northwood Publications Ltd.)

Health and Safety at Work. Booklet No. 41 "Safety in the use of woodworking machines". (HMSO)

The Woodworking Machines Regulations 1974 (HMSO)

SELF-TESTING QUESTIONS

All the information required to answer the following questions is contained within this chapter. Attempt each section *as fully or as briefly* as the question demands, and then check your answers against the information given in the chapter.

1. List four functions of a woodworking machine.

2. Under the heading of "common sense in the machine shop", list twelve very important aspects of behavioural requirements for persons operating woodworking machines.

3. Explain in the simplest possible terms the "General Requirements" (applicable to all machines) with which it is necessary to comply under the Woodworking Machine Regulations 1974. Your answers should refer to the following: *(a)* guards; *(b)* machine controls; *(c)* working space; *(d)* floors; *(e)* temperature; *(f)* training; *(g)* duties of persons employed; *(h)* lighting; *(j)* noise.

4. *(a)* Sketch the following items related to the use of a hand fed circular saw, indicating dimensions where these are important:

(i) crown guard and front extension guard;

(ii) riving knife;

(iii) push stick.

(b) Explain with sketches the following types of work carried out on a circular saw: *(i)* ripping; *(ii)* deeping; *(iii)* splay cutting.

(c) Draw the shape of circular saw teeth designed for: *(i)* ripping; *(ii)* cross cutting.

(d) List four operations which can be carried out to advantage on a dimension saw.

5. *(a)* Describe the main functions of:

(i) an overhand planer;

(ii) a thicknesser or panel planer.

(b) Illustrate with sketches the correct position of the bridge guard on an overhand planer when set up for the following operations:

(i) facing a piece of 100×25;

(ii) edging a piece of 100×25;

(iii) facing and edging (consecutively) a number of pieces of 100×25.

(c) (i) Draw a diagram to illustrate the working parts of a thicknesser.

(ii) Explain the requirements of the Woodworking Machines Regulations 1974 with respect to feeding timber into a thicknesser.

6. *(a)* List and explain *four* safety factors which should be observed when using morticing machines.

(b) Explain, with diagrammatic sketches, the cutting action of: *(i)* a hollow square chisel morticer; *(ii)* a chain morticer.

(c) State why it is necessary to ensure that there is:

(i) 2-3 mm clearance between the auger tips and the tip of the chisel;

(ii) 6 mm slackness – at the centre of the guide bar – of a mortice chain.

(d) State the purpose of the chip breaker on a chain morticing machine.

7. Calculate:

(a) the pitch of the teeth on a 400 mm diameter plate saw which has 70 teeth;

(b) the peripheral speed of a circular saw 400 mm in diameter which is revolving at 2,300 revolutions per minute;

(c) the pitch of the cutter marks on the surface of timber given the following data:

r.p.m. of cutter block	:	4,500
Number of effective cutters	:	2
Feed speed	:	15 m/min

12. Safety at Work

After studying this chapter the student should be able to:

1. Define an employee's responsibilities under Part 1, s. 7, of the Health and Saftety at Work etc., Act 1974.
2. Fully describe the function and use of safety helmets, eye protectors, ear muffs, face masks and respirators, safety boots, industrial gloves and general protective clothing (overalls).
3. Name four kinds of dangerous or harmful materials used in building construction, outlining the hazards of each and methods of minimising them.
4. Outline the dangers associated with the use of portable electric tools and the reasons for using reduced voltage tools on site.
5. Explain with dimensioned sketches the proper use of ladders, step ladders and trestle scaffolds.
6. Explain with sketches the general forms of putlog and independent tied scaffolds.
7. Outline the main safety requirements for putlog and independent tied scaffolds.
8. Outline the main safety requirements of tower (mobile) scaffolds.

INTRODUCTION

The construction industry has an unenviable record where accidents and personal injury are concerned. Much serious thinking and legislation have been applied to the problem and yet the incidence of serious and crippling accidents still remains very far from being acceptable. One of the root problems, of course, as anyone who has knowledge of the industry is well aware, is the very nature of the work itself. The environment of the building site and the conditions which tend to prevail on it are not yet, and possibly never can be, comparable with those existing in most other major industries. Yet thinking sensibly this is merely good reason why craftsmen should be ever more vigilant and clear thinking where safety is concerned.

Another major reason for the high accident rate is the very insidiousness of the hazards faced. Were these more obvious and more forcibly and constantly brought to the forefront of the mind, were the workers less used and inured to working in such an environment, they would certainly recognise these hazards and take every possible step to minimise them.

A ladder, a scaffold, a high stack of material or a deep excavation are so commonplace and appear so innocent – until something breaks, slips or collapses without warning. Then it is too late; the result – damage to property and materials, time wasted rectifying the trouble, and all too often the pain, suffering, hardship and misery of physical injury. It would be said "This was an accident", and yet most often it would not be. In most instances the accident could have been avoided by following the proper and sensible precautions laid down for this very purpose. Nor is the onus any longer entirely upon the employer to ensure his employees come to no harm. Section 7 of Part 1 of the Health and Safety at Work etc. Act 1974 states quite clearly:

"It shall be the duty of every employee while at work –

(*a*) to take reasonable care for the health and safety of himself and of other persons who may be affected by his acts or omissions at work; and –

(*b*) as regards any duty or requirement imposed on his employer or any other person by or under any of the relevant statutory provisions, to co-operate with him so far as is necessary to enable that duty or requirement to be performed or complied with."

This section of the Act places a legal responsibility upon the employee to work in a safe manner in respect of himself *and* his fellow workmen. Failure to do so could result in legal proceedings being taken against him, terminating in a fine or even imprisonment.

PROTECTIVE CLOTHING

Safety helmet

(*See* Fig. 12.1*(a)*.) This item of equipment should be worn in all construction site situations where there is a danger from falling objects or of striking the head against some part of the structure, scaffolding or equipment. On many sites their use is obligatory — a sensible precaution.

Eye protectors

(*See* Fig. 12.1*(b)*.) These take various forms but are all intended to serve the same purpose: to protect the eyes against pieces of flying debris, dust, grit, sparks, etc. They should be worn whenever there is a natural tendency to squint or turn the head away from the work because of risk to the eyes, i.e. when using a cold chisel and hammer to cut brick or concrete, etc., and when grinding metal or using a mechanical percussion tool. The lens or visor should be kept clean so as not to impair the vision, which, in itself, could present a hazard.

Fig. 12.1. *Protective clothing.*

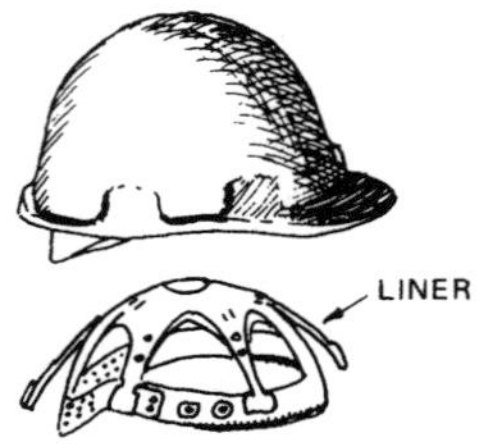

(a) Safety helmet.

(b) Eye shields.

(c) Ear protectors.

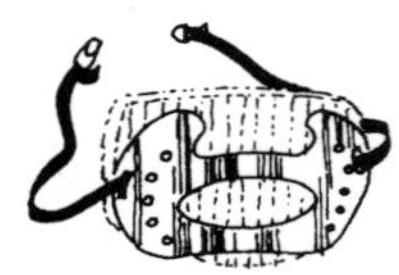

(d) Face mask.

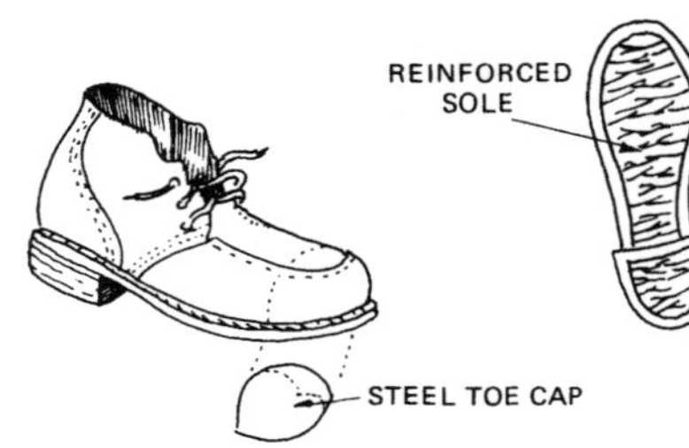

(e) Safety boots.

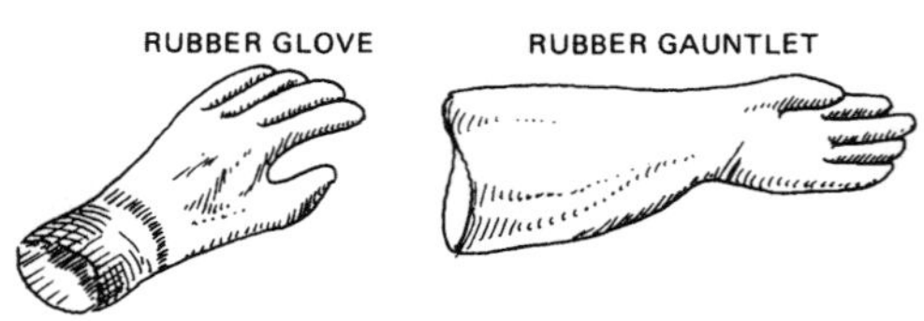

(f) Industrial gloves.

Ear protectors or ear muffs

(*See* Fig. 12.1*(c)*.) These are designed to protect the ears against the effects of prolonged intense noise, particularly high frequency noise. They should be worn whenever the noise reaches or is in excess of 90 decibels (dB). Both eye and ear protectors can, if necessary, be worn under the safety helmet.

Face masks and respirators

(*See* Fig. 12.1*(d)*.) These are designed to prevent the passage of harmful dust and air suspended particles into the lungs. A face mask will not protect the wearer against toxic fumes or gases (for which purpose a respirator with special filters is required), nor is it 100 per cent effective against very minute particles of air suspended dust, but it is at least a means of securing some degree of protection where the atmosphere cannot be cleared effectively by means of a proper dust extraction system.

Safety boots

(*See* Fig. 12.1*(e)*.) These are worn to protect the feet from injury due to walking on rough uneven surfaces (at demolition sites etc.), treading on upstanding nails, and from being crushed or trapped when moving and handling heavy objects. Safety boots have steel reinforced toe caps and intersoles built in for this purpose. Rubber safety boots (knee length) are available for working on wet or muddy ground and are especially necessary when using electrical equipment under these conditions.

Whilst safety boots are not always essential to the average woodworker, good, strong, sensible boots or shoes are! Plimsoles or canvas training shoes are *not* suitable or sensible footwear for woodworkers, whatever the nature of the work being undertaken.

Industrial gloves

(*See* Fig. 12.1(*f*).) These are available in various forms designed either to protect the hand against general rough usage, handling sharp, rough, heavy objects, or to prevent the skin coming into contact with harmful materials, especially liquids such as acids, alkalis, synthetic resins and fungicides. They should be worn whenever the occasion demands.

Overalls, aprons, etc.

These are generally worn as a means of protecting the clothing underneath, and are not, unfortunately, in such common use nowadays. However, where toxic or other injurious materials are handled or used at work they are essential. Such protective clothing should be changed and washed weekly and is best kept in a locker or store at the place of work.

HARMFUL AND DANGEROUS MATERIALS

Quite a few of the materials which are in everyday use in the construction industry are harmful or dangerous to a greater or lesser degree if not properly stored, used and handled.

Highly flammable liquids

These should always be stored safely out of harm's way – either in the open air (protected from snow, ice and direct sunlight) in a fenced-off area under lock and key, or in a specially built inflammable store. The store must be sited at least 6.096 metres away from any other buildings and should bear a bold sign stating the contents are flammable. This sign acts a warning to personnel on the site and also as an aid to the fire brigade should they be called to the site. There must obviously be no smoking or other forms of naked flame in the vicinity of this store, and notices to this effect must be clearly displayed. Inflammable stores should be under the strict supervision of a charge hand who is fully conversant with the dangerous nature of the contents. Small quantities (less than 50 litres) of flammable liquid may be stored in a workroom, provided it is kept in closed containers within a fire resistant cabinet which can be locked.

Generally only small quantities of flammable liquid (i.e. french polish, petroleum based adhesives and volatile solvents) should be available outside of the store or cabinet – enough for the job in hand. Containers for flammable liquids must always indicate the nature of the contents and should be kept closed when not in actual use.

Glass

Many severe injuries are caused by careless handling and storage of glass. Large sheets of glass should be handled with industrial gloves, or with the sharp edges covered with paper or cloth at the gripping point. Glass should be properly stored so that it cannot slip or fall, and broken panes and offcuts should be gathered up and properly disposed of – not left laying about. Glazed windows in a building under construction should be made obvious with whitewash or tape.

Asbestos

Most people are nowadays very well aware of the dangers associated with the inhalation of asbestos fibres, the effects of which are still something of an unknown quantity tending to vary from person to person. It would be prudent, however, to take very great care when using and handling asbestos or any form of material containing it.

Whilst there would appear to be a minimal risk involved in working on asbestos cement sheet in the open air, some form of face mask or respirator is certainly required when using it indoors or when cutting or drilling it with power tools.

Of greater danger are the soft asbestos materials used in asbestos insulation boards and pipe laggings, etc. These are sufficiently dangerous as to warrant always the use of an approved dust respirator (unless an effective exhaust system is in operation).

Blue asbestos (crocidolite)

Usually recognisable by its dark lavender blue colour, blue asbestos should always be treated with the greatest respect and is regarded as sufficiently dangerous to warrant the use of pressurised breathing apparatus when using it. Loose asbestos waste should be kept in airtight containers, and where such waste contains crocidolite, they should be marked *Blue Asbestos – DO NOT inhale dust.* Local Environmental Health Departments have special arrangements for disposing of waste materials containing crocidolite.

Protective clothing should be worn when working on asbestos materials and kept on site. When sent for laundering such clothing should be marked *Asbestos Contaminated Clothing.*

Asbestos dust, swarf and waste resulting from cutting and drilling of the material should be cleared up by a "non-dust method", i.e. by vacuum cleaner or by first wetting down.

Lead paints

The pigments used in lead paints are toxic materials and therefore dangerous if swallowed or inhaled. The effects of lead poisoning are often not apparent for a long time — up to twenty years in some cases — and are sufficiently unpleasant as to require considerable care in their avoidance. No great risk attends the use of lead paint, either in wet or dry film form, provided that sensible precautions and hygiene are observed, such as follows:

(a) No lead paintwork or paintwork which might possibly contain lead should be rubbed down by dry sandpapering methods.

(b) Hands should be carefully and thoroughly washed before taking meals. Particular attention should be paid to cleaning under the finger nails with a good nail brush.

(c) Overalls and protective clothing should not be taken into canteens and mess rooms.

Other materials

There are many other materials such as spirits, solvents, de-greasing agents, synthetic resins, etc., which can cause injury if mishandled. Common sense dictates that such materials should only be handled and used in accordance with the maker's instructions. In the event of a mishap, the Safety Officer or his representative on site should be informed immediately so that any necessary action can be taken.

TOOLS AND EQUIPMENT

Hand tools

A good many accidents occur annually as a result of careless and incorrect use of hand tools or through using ill-maintained or defective tools.

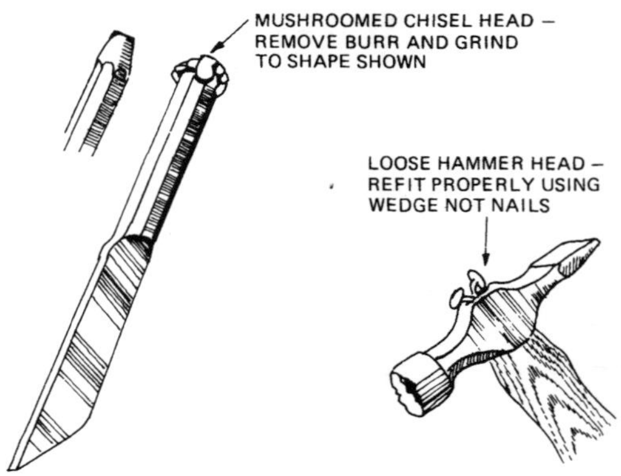

Fig. 12.2. *Defective hand tools.*

Accidents of this nature can easily happen when undue force is being applied to a tool which is not functioning properly because it is not sharp or is incorrectly adjusted. Well maintained tools can usually be relied upon to do what is expected of them; it is the poorly maintained tool that does the unexpected. Figure 12.2 shows two of the more common faults in hand tools — *these faults require rectification.* Tools should not be allowed to fall into this state.

Abrasive wheels

These are used quite frequently by carpenters and joiners to grind a new bevel on a chisel or plane iron when the edge has been damaged or has become "thick" due to constant sharpening on an oilstone. The slowly revolving, "wet" type of grinder often used in schools and colleges, presents no real hazard in use, but the rapidly revolving "dry" type of grinder does!

Accidents which occur as a result of using the latter type of abrasive wheel are generally due to one of three factors —

(a) breakage of the revolving wheel;

(b) hands coming into contact with the revolving wheel;

(c) eye injuries due to flying sparks and particles when using the wheel.

Certain precautions are therefore necessary for the safe use of abrasive wheels and these are fully detailed in the Abrasive Wheels Regulations. The essential safety points outlined in these regulations are as follows.

(a) Operators should be properly trained in the use of a grinding wheel.

(b) The mounting of the abrasive wheel on to the machine shaft must only be carried out by a competent person.

(c) Proper protection must be given to the eyes, either by means of suitable goggles, or a well designed screen — or both.

(d) The wheel must be correctly guarded so as to limit the extent of damage caused by possible breakage during use.

(e) The rest (the support on which the

work bears) should be properly positioned as close to the wheel as possible.

(*f*) Loose clothing and cleaning rags must be kept away from the revolving wheel.

Other important aspects of the safe use of abrasive wheels are of course clearly stated in the Abrasive Wheels Regulations, a copy of which should be displayed prominently near the grinding wheel.

Electric power tools

Electric power tools – saws, drills, routers, sanders, etc. – are very useful and necessary in a modern industry. These generally perform extremely efficiently and safely provided they are properly maintained and correctly used.

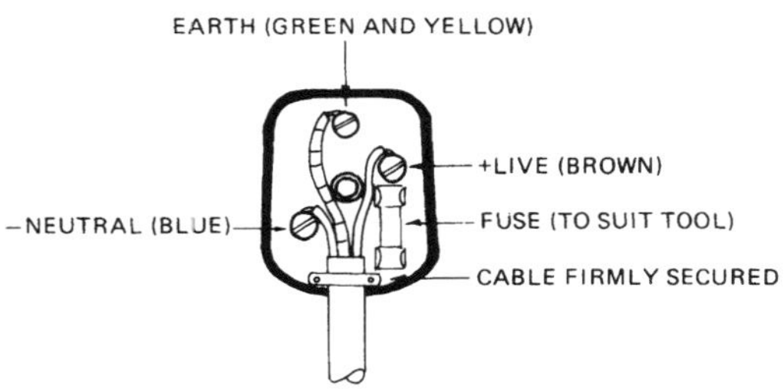

Fig. 12.3. *Wiring to a 240 V 3-pin plug.*

Undoubtedly one of the main dangers associated with the use of electric power tools is the possibility of electric shock, a danger which though seemingly remote, is always present when they are being used. An electric shock, severe or minor, is suffered by a person when the supply of electricity to the tool he is using passes through that person to earth, i.e. the operator becomes the conductor of the electricity. It is essential therefore for all electrical equipment to be efficiently earthed so that the current will by-pass the operator in the event of a short circuit. Figure 12.3 shows the wiring to a 240 V 3-pin plug as is often used on electrical tools. Note that the colour coding of the wires is:

Live – brown
Neutral – blue
Earth – green/yellow

Reduced voltage

Reduced voltage tools work on a 110 V supply, usually through a step-down transformer, and are earthed through the centre tapping of the secondary winding so the nominal voltage to earth is about 55 V. An electric shock from such a tool is not likely to be lethal (though it might well be unpleasant), and therefore reduced voltage tools are strongly recommended for all site work and where wetness or dampness creates a hazardous situation.

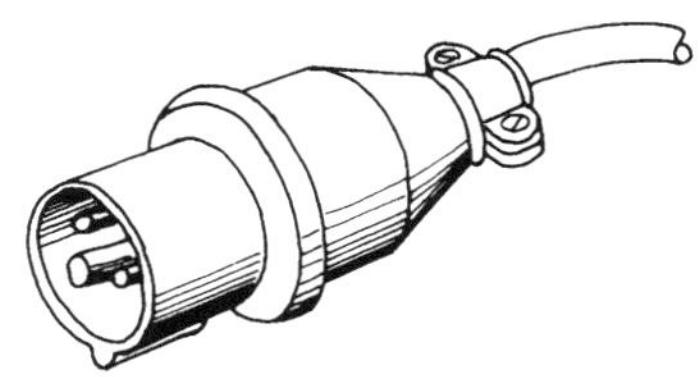

Fig. 12.4. *110 Vx splash proof plug (yellow).*

Supply outlets, 110 V plugs and couplers are normally of the splash proof type (*see* Fig. 12.4) and are so designed as to make connection to an incorrect voltage supply impossible. In addition these connections are coloured *yellow* as a means of ready identification.

Fig. 12.5. *Symbol signifying double insulation.*

Double insulated tools

These are so designed that no part of the tool can become live in the event of a short circuit, and there is therefore no real need for an earth wire. These tools can be obtained for both 240 V and 110 V electricity supplies. Double insulated tools should carry the international symbol shown in Fig. 12.5.

WORKING AT HEIGHTS

A high percentage of accidents occurring each year in the construction industry are due to persons either falling from heights or being struck with objects falling from above, so this is an area where precautions and extra care must be taken to ensure safe working.

Step ladders

These useful items of equipment shown in Fig. 12.6 are used for the many jobs where extra reach is required. Step ladders are made from wood or aluminium and vary in height from 1.5 to 3 metres. The main safety factors applying to the use of step ladders are as follows.

(*a*) They must be properly constructed.

(*b*) They must be frequently inspected to ensure that all parts, including the hinges and cords, are in good condition.

(*c*) It is not good policy to work too near the top of a step ladder where there is no hand hold.

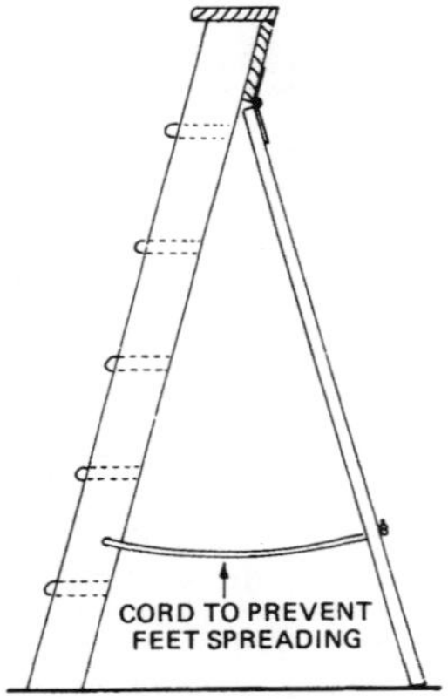

Fig. 12.6. *Builder's step ladder.*

(*d*) Do not attempt to overreach whilst working on a step ladder.

(*e*) Do not attempt to exert undue sideways force when working on a step ladder, e.g. drilling into a wall, unless the step ladder is either fixed or supported by another person.

(*f*) Make sure no one is likely to bump into the step ladder while it is in use, for example, by opening doors or moving wheelbarrows, etc.

(*g*) Never use a step ladder on a scaffold unless it is securely fixed to the platform and has a clear space of at least 450 mm all round.

Trestle scaffolds

(*See* Fig. 12.7.) These are very useful for light work such as shopfitting where the scaffold is required for a limited period only or where it must be moved frequently. Trestle scaffolds must not be used where it

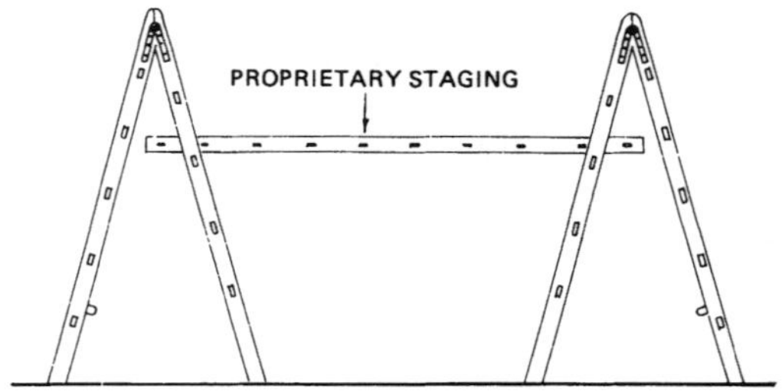

Fig. 12.7. *Trestle scaffold.*

is possible for the user to fall more than 4.570 m to the ground.

As with step ladders, certain precautions *must* be observed when a trestle scaffold is in use. These are as follows.

(*a*) The trestles must stand on firm, level ground.

(*b*) They must be maintained in good condition and inspected frequently for defects.

(*c*) The working platform must be at least two scaffold boards wide (440 m).

(*d*) Scaffold boards for the working platform should not span more than 1.520 m when 38 mm thick boards are used, or more than 2.590 m with 50 mm thick boards.

(*e*) Proprietary type staging should be used whenever possible in preference to scaffold boards, and may safely span a greater distance.

(*f*) A separate access ladder or step ladder should be used to mount the working platform when this is over 1.981 m high.

Ladders

These are essential and very familiar items of equipment on any building site, and may be made of wood or aluminium, builders generally preferring the former. Wooden ladders have steel tie bars at intervals below the rungs and may have a high tensile wire recessed into the underside of the stiles. In use, the tie bars should always be beneath the rungs and the high tensile wire on the underside (the side in tension) of the stiles.

The following safety points *must* be observed when using ladders.

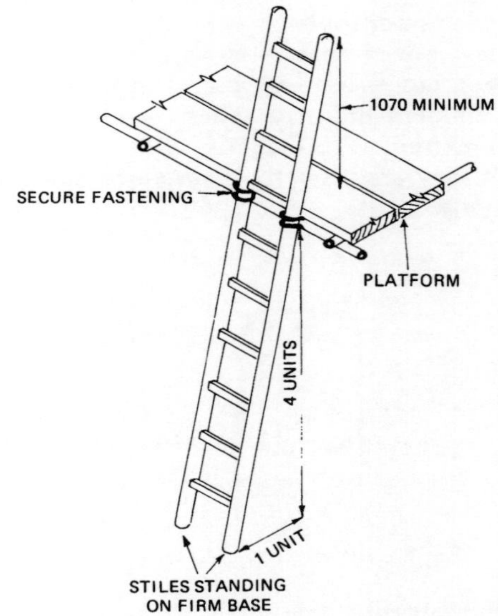

Fig. 12.8. *Proper use of ladder.*

(*a*) The bottom of the ladder stiles must rest on firm ground – no odd bits of packing are permissible.

(*b*) The ladder *stiles* must be lashed near the top to a *firm* anchorage to prevent sideways slipping. If it is impossible to fix the top of the ladder with lashings, then it must be supported during use by another person standing at the bottom.

(*c*) Long ladders should have an intermediate tie rope to prevent undue swaying.

(*d*) The top of the ladder should extend at least 1.070 m above the stepping off place (*see* Fig. 12.8).

(*e*) The ladder should be so positioned that it makes an angle of about 75° with the ground, i.e. 1 unit out to 4 units up (*see* Fig. 12.8).

(*f*) There must be no obstructions behind any of the rungs which could prevent a secure foothold.

(*g*) Ladders must be inspected frequently to check for loose or faulty rungs or tie bars, cracked, bruised or frayed stiles and faulty ropes and fittings (on extending ladders).

(*h*) Ladders should be protected from the elements with a wood preservative and a clear varnish. They *must not* be painted or otherwise coated with anything which could hide a defect.

NOTE: Defective and improperly used ladders account for more accidents than any other single item of equipment.

Scaffolding

The main types of scaffolding are illustrated in Fig. 12.9.

Putlog scaffold

This type of structure is often referred to as a bricklayer's scaffold as it is the one most often used during the construction of brick buildings. Its purpose is simply to provide a working platform for the bricklayer (and other craftsmen) as the building increases in height, each stage in height being known as a "lift".

Fig. 12.9. *Types of scaffold.*

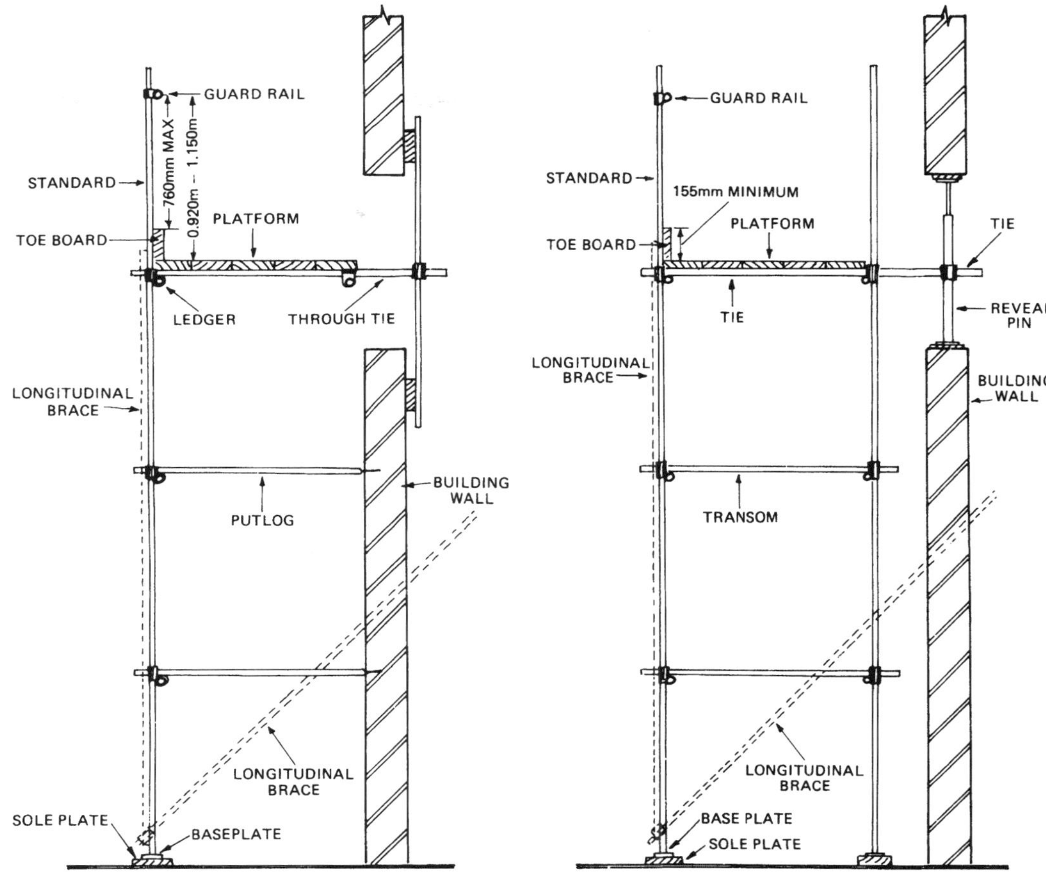

(a) Section through a putlog scaffold showing a through tie.

(b) Section through an independent tied scaffold showing a reveal tie.

The main components of the scaffold are shown in Fig. 12.9(*a*) and comprise the following.

(*a*) *Standards.* These are upright tubular members on the outside of the scaffold.

(*b*) *Ledgers.* These are horizontal longitudinal members which stiffen the standards and support the putlogs.

(*c*) *Putlogs.* These rest on the ledgers on the outside of the scaffold and on the wall

Fig. 12.9 (contd.)

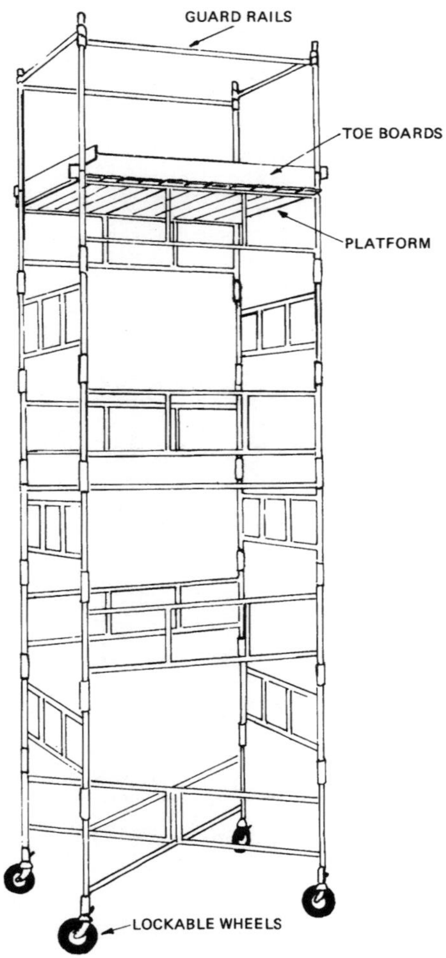

(c) Tower (mobile) scaffold.

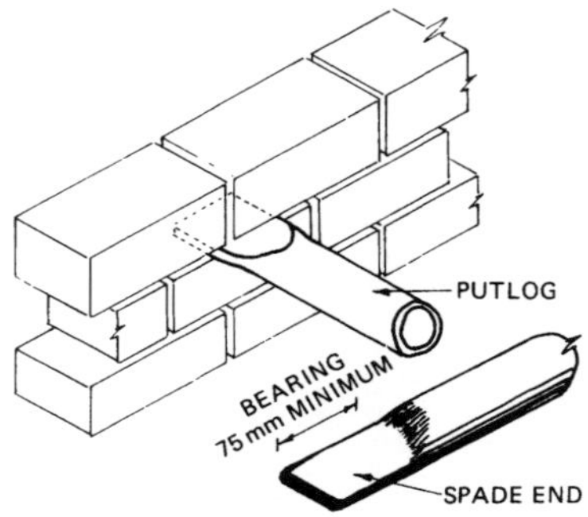

Fig. 12.10. *Putlog bearing in wall.*

on the inside, the spade end giving the putlog a 75 mm bearing on the brickwork (*see* Fig. 12.10). The putlogs support the working platform of scaffold planks.

(d) Braces. These normally slope at about 45° and are used to stiffen (triangulate) the structure.

(e) Ties. These are used to couple the scaffold to the walls since the spade ends of the putlogs do not effectively perform this function. The "through ties" shown in Fig. 12.9*(a)*, are recommended for a putlog scaffold.

(f) Working platforms. These are normally four or five scaffold boards wide (870 mm-1,066 mm) when used for footways and for stacking materials.

(g) Toe boards. As shown in Fig. 12.9*(a)*, these are fitted on the inside of the standards and should not be less than 152 mm high (above the working platform). They are required on all scaffolds more than 1.981 m high. Toe boards are used to prevent materials falling from the platform, and also serve to prevent a person's feet slipping off the edge of the outer plank.

(h) Guard rails. These are used to provide a degree of protection to the workers on the scaffold. The guard rail should be fixed to the inside of the standards at a height of between 0.920 m and 1.150 m above the working platform. The distance between the top of the toe board and the underside of the guard rail must not exceed 760 mm.

Independent tied scaffold

This type of scaffold shown in Fig. 12.9*(b)* is more or less self supporting since it has two rows of standards to carry the weight of men and materials. It is the type of scaffold most often used to provide a working platform to the outside of an existing building. As can be seen from the illustration, the members are similar to those used on a putlog scaffold, but the transverse horizontal members are referred to as "transoms". These do not have spade ends like putlogs since they do not rest in the brickwork joints, but are coupled instead to the inside ledgers and standards.

Independent tied scaffolds require tying to the walls of the building, as do putlog scaffolds. This is accomplished by means of through ties as shown in Fig. 12.9*(a)*, or by reveal ties as shown in Fig. 12.9*(b)* (at least 50 per cent should be through ties). Guard rails, toe boards and bracing are required as before.

Safety requirements on scaffolding

In order that men can work safely upon scaffolding the following precautions *must* be observed (these are simplified extracts from the Construction (Working Places) Regulations).

(a) Scaffolding must be erected, dismantled

or altered only by a competent person, i.e. a person who has thorough knowledge and experience of the work involved.

(b) Scaffolding must be effectively tied to the building at a minimum of 6.096 m intervals on alternate lifts. Ties are required under the working platform.

(c) Scaffolding components – tubes, couplers, planks, etc. – must be sound and of good construction. Defective items must be discarded (or effectively repaired).

(d) All scaffolding should be inspected by a competent person at frequent intervals to ensure the security and good condition of all components.

(e) Guard rails and toe boards must be fitted on all scaffolds more than 1.981 m high.

(f) Partly erected or dismantled scaffolds must be blocked off so as to prevent accidental usage – or a bold warning notice fixed to indicate danger.

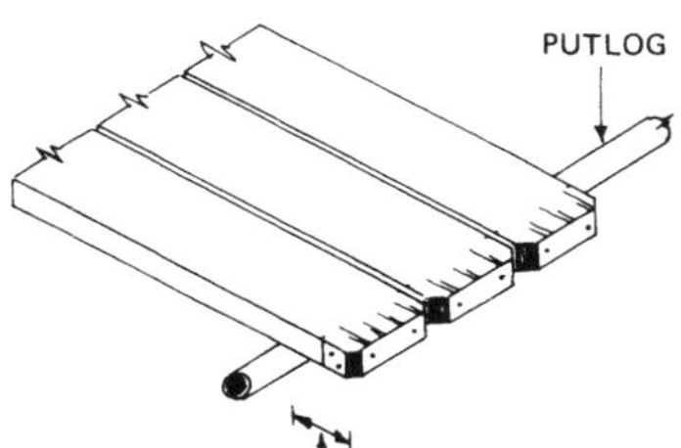

Fig. 12.11. *Clearhang of scaffold board.*

(g) Standards must rest upon a firm base. Baseplates are required under the feet of the standards and wooden sole plates are advisable to prevent the base plates sinking into the ground. Loose rubble is totally unsuitable as a base for scaffolding, and must not be so used.

(h) The ends of scaffold boards must not overhang their support (putlog or transom) by more than 4 times their thickness (*see* Fig. 12.11).

(i) Loads on a scaffold should be evenly distributed and shock-loading avoided. The maximum permissible load on a bricklayer's scaffold is 275 kg/m^2. Loads are best carried near the standards.

Tower (mobile) scaffolds

Shown in Fig. 12.9*(c)* these are very useful and convenient for light duty work, especially where frequent movement of the scaffold is necessary. The main safety aspects of this type of scaffold are as follows.

(a) The height should not exceed 3 times the shortest base dimension when used outside, or 3½ times this dimension when used inside.

(b) Guard rails and toe boards are required as for normal scaffolding, except at the point of access.

(c) Access to the scaffold should be by a vertical ladder lashed to one of the narrower sides.

(d) They must only be used on level, firm ground.

(e) They must not be moved whilst men or materials are still on the working platform.

(f) Wheels must be locked whilst the scaffold is in use and turned to point outwards.

Roof ladders or crawling boards

These should be used whenever work is being carried out on a sloping roof surface or one covered with a fragile material such as asbestos sheeting. Asbestos sheeting is a particularly treacherous material to walk upon (even when stepping from purlin to purlin) as it may give way suddenly and without warning.

Crawling boards, shown in Fig. 12.12 should be well constructed of substantial material and be not less than 380 mm wide, with adequate foot holds.

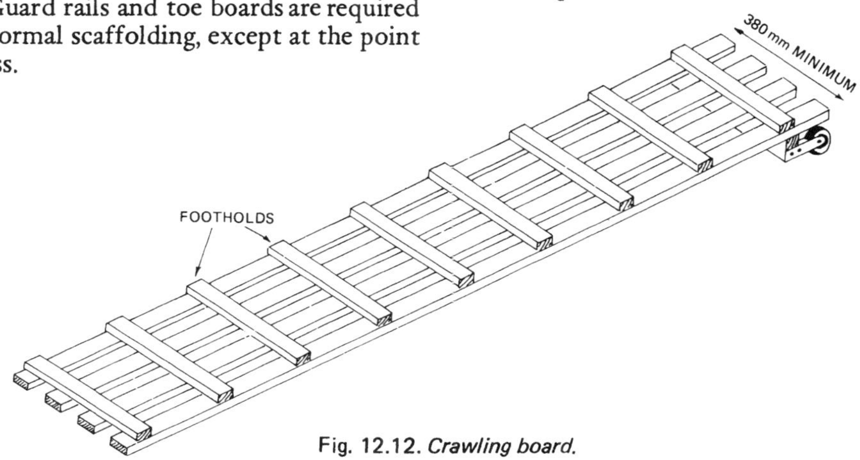

Fig. 12.12. *Crawling board.*

GENERAL SITE SAFETY

A typical building site is full of traps and pitfalls for the unwary, and therefore the newcomer to a site should keep his wits about him and his eyes open. If the site is well organised (as it should be) he will be warned of possible dangers by the site foreman or safety officer. The following are some of the hazards to watch out for — and, where necessary, report to the site foreman so that remedial action can be taken.

Stacking materials

Never stack materials where they may be in the way of other workmen, vehicular traffic or further work. The stack should be safe and firm so that it is unlikely to collapse, either on to a passer-by or someone taking material from the stack itself. Unsafe stacks should be reported to the site foreman.

Fig. 12.13. *Warning of excavation or hole in floor.*

Excavations

Excavations more than 1.981 m deep should be protected by a guard rail or barrier. Never work in an excavation (with steep sides) more than 1.219 m deep unless the sides are adequately timbered and strutted. Holes in the ground or in floors should either have a protective barrier or be marked and covered as shown in Fig. 12.13.

Site tidiness

Every effort should be made to keep the building site clean and tidy. This is not merely a matter of pride in work or providing a more pleasant working environment, but is also a matter of safety. Untidy, littered sites abounding in rubble are awkward for walking, working or moving materials. Thus they make for inefficient working.

Nailed timbers

If left laying about the site as shown in Fig. 12.14 these are particularly dangerous. When dismantling nailed structures, the nails should be withdrawn and the timber stacked neatly ready for re-use. At the very least the nails should be knocked down and made safe.

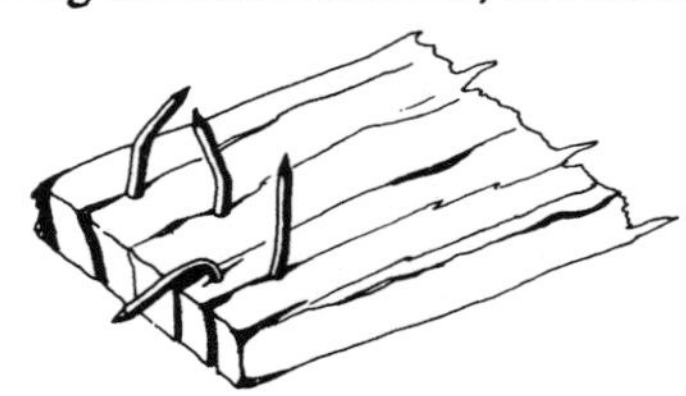

Fig. 12.14. *Nailed timbers. The sketch shows a dreadful example of careless working.*

Site traffic

Keep a wary eye open for dumpers and lorries delivering or moving materials, especially when these are reversing as the driver may not have a clear view to the rear, and even where he is being guided, relayed instructions take time.

Bad weather

Heavy rainfall, snow, frost and mud always add to the dangers inherent on the building site. Extra care should be taken when working in such conditions as they can make a secure foothold on ladders and scaffolds more difficult, in addition to which it is less easy to obtain a secure hand grip on tools and materials. High winds tend to make working at heights more hazardous and, again, extra care must be taken.

In cold weather, clothing should be suitably warm without being too restrictive of physical movement. Remember that both cold and fatigue increase the risk of an accident.

FIRST AID

However carefully a craftsman conducts himself on a building site or in a workshop, there is always an element of risk which manifests itself as an accident from time to time. Most of these accidents are fortunately of a minor category — a splinter in the hand or the hammered thumb — but occasionally a more serious type of accident or occurrence takes place. This could perhaps be a bad fall from a scaffold — or even a heart attack.

In instances such as this, a knowledge of first aid by one or more fellow workers on the site could prove vital to the welfare of the unfortunate person involved. This is not to say that minor injuries should be neglected — these often have a nasty habit of becoming more serious unless properly treated.

First aid may be regarded as the *skilled* treatment of a patient who has suffered an accident or sudden illness by making the best possible use of whatever facilities exist at the time for rendering assistance until the patient can be placed in the hands of a doctor. A knowledge of first aid is therefore a form of skill which is well worth while acquiring. To become a qualified "first aider", it is necessary to undergo a proper course of instruction, and subsequently to pass the examinations set by one of the following Voluntary Aid Societies:

(a) The St. John Ambulance Association and Brigade;

(b) St. Andrew's Ambulance Association;

(c) The British Red Cross Society.

Certain requirements are laid down in the Constructions (Health and Welfare) Regulations in respect to first aid on building sites, and these concern the following areas:

(a) first aid boxes – number required and contents;

(b) ambulance arrangements – regulations applying to any contractor who employs more than 25 persons on a site;

(c) first aid rooms – the provision of these is also dependent on the number of persons employed on the site;

(d) trained first aiders – employer's obligations and the qualifications required by the first aider.

All persons employed in the construction industry should study and *understand* this area of the Construction Regulations since it could vitally affect their personal welfare.

LIFE ON SITE

In spite of the ominous nature of the material in this chapter, life on a building site can be both pleasant and interesting. All that is required is a modicum of sound common sense and a willingness to work as one of a team, having the same consideration for the welfare and safety of fellow workmen as you would hope and expect them to show for you.

FURTHER READING

Building Research Establishment Digest

No. 179 Electricity distribution on site

Other literature

Construction Safety. (NFBTE)

Health and Safety at Work Booklets. (HMSO)

No. 4	Safety in the use of abrasive wheels
No. 6A	Safety in construction work: general site safety practice
No. 6B	Safety in construction work: roofing
No. 6C	Safety in construction work: excavations
No. 6D	Safety in construction work: scaffolding
No. 6E	Safety in construction work: demolitions
No. 6F	Safety in construction work: system building
No. 41	Safety in the use of woodworking machines
No. 44	Asbestos: health precautions in industry

SELF-TESTING QUESTIONS

All the information required to answer the following questions is contained within this chapter. Attempt each section *as fully or as briefly* as the question demands, and then check your answers against the information given in the chapter.

1. Explain an employee's responsibility under Part 1, s. 7, of the Health and Safety at Work etc. Act 1974.

2. Describe and give examples where the following items of safety equipment/clothing should be worn: *(a)* safety helmet; *(b)* eye shields (eye protectors); *(c)* ear protectors (ear muffs); *(d)* face mask/respirator; *(e)* safety boots; *(f)* industrial gloves; *(g)* protective clothing (overalls, etc.).

3. Name four dangerous or harmful materials used in building construction and state for each:

(a) why the material is harmful or dangerous;

(b) what precautions should be taken to reduce or eliminate the risks of using the material.

4. *(a)* Outline the dangers associated with the use of abrasive wheels for grinding steel.

(b) State six important safety aspects concerning the use of grinding machines which are laid down in the Abrasive Wheels Regulations.

5. *(a)* State the *main* danger associated with the use of portable electric tools.

(b) Explain the importance of an effective "earth" when using electric tools.

(c) State the main advantage – or reason for use – of reduced voltage (110 V) power supply.

(d) Explain briefly why certain electrically powered tools are known as "double insulated".

(e) Draw the international symbol which indicates a double insulated tool.

(f) State the colour coding for the various wires supplying electricity to 240 V power tools.

6. Describe with dimensioned sketches where necessary the safe and proper use of: *(a)* step ladders; *(b)* ladders (pole or extending); *(c)* trestle scaffolds.

7. *(a)* Show by means of rule assisted sketches the general arrangement of:

(i) a bricklayer's (putlog) scaffold;

(ii) an independent tied scaffold.

(b) Indicate on your drawings to 7*(a)(i)* and *(ii)* the dimensions which must be observed to meet the requirements of the Construction (Working Places) Regulations.

(c) State the special precautions to be observed when using a tower (mobile) scaffold.

8. Under the heading of "General Site Safety", outline the advice which an experienced worker might give to a young person starting work on a building site. Special reference should be made to:

(a) stacking of materials;
(b) excavations;
(c) site tidiness;
(d) nailed timbers;
(e) site traffic;
(f) personal clothing, including footwear.

Index

INDEX

INDEX

In the same series

PLUMBING: MECHANICAL SERVICES BOOK 1

G J Blower

PLUMBING is intended to provide an introduction for all students of plumbing in Technical Colleges and Government Training Centres and in particular those studying for the City and Guilds Craft Certificate. The Student is led to a sound understanding of plumbing principles and processes using both traditional and modern materials. The author has laid great emphasis on the practical nature of the subject, and the chapters on working processes in particular have a strong practical bias in order to enable the student to tackle the manual skills in which he must become proficient in order to qualify successfully as a plumber.

The book is well illustrated with explanatory diagrams and the learning objectives and self-testing questions associated with each chapter will not only be useful to students, but also to lecturers wishing to integrate the book into their teaching programme.

The book will also be suitable for the amateur craftsman seeking to improve his knowledge of the subject.

The author has many years experience of teaching at this level, and currently lectures in plumbing at Tottenham College of Technology. He is also an assessor and examiner for the City and Guilds examining body.

Illustrated

For Product Safety Concerns and Information please contact our EU representative GPSR@taylorandfrancis.com Taylor & Francis Verlag GmbH, Kaufingerstraße 24, 80331 München, Germany

Printed and bound by CPI Group (UK) Ltd, Croydon, CR0 4YY
14/09/2026
02216255-0001